# Signaling and Communication in Plants

Volume 24

**Series Editor**

František Baluška

Department of Plant Cell Biology, IZMB, University of Bonn, Bonn, Germany

More information about this series at http://www.springer.com/series/8094

Sanjib Kumar Panda • František Baluška
Editors

# Aluminum Stress Adaptation in Plants

Springer

*Editors*
Sanjib Kumar Panda
Dep. of Life Science and Bioinformatics
Plant Molecular Biotechnology Laboratory
Assam University
Silchar, India

František Baluška
Department of Plant Cell Biology, IZMB
University of Bonn
Bonn, Germany

ISSN 1867-9048 ISSN 1867-9056 (electronic)
Signaling and Communication in Plants
ISBN 978-3-319-19967-2 ISBN 978-3-319-19968-9 (eBook)
DOI 10.1007/978-3-319-19968-9

Library of Congress Control Number: 2015948865

Springer Cham Heidelberg New York Dordrecht London


Printed on acid-free paper

Springer International Publishing AG Switzerland is part of Springer Science+Business Media (www.springer.com)

# Preface

Almost 40 % of world arable soil is acidic that limits plant growth and productivity in a complex manner. Al is one of the major constituents of acid soil. It is toxic in the trivalent ($Al^{3+}$) cationic form when the soil pH drops below 5.0 and $Al^{3+}$ is solubilized in the soil. It has been estimated that approximately 50 % of the arable land is negatively impacted by the Al toxicity due to acidic soil. The soluble $Al^{3+}$ inhibits root growth, which subsequently affects a plant's ability to take up water and nutrients. The mechanism of Al response and tolerance is well studied in model plants like Arabidopsis, and the knowledge thus obtained has been also tested in crop plants which are the actual targets of Al stress. Some of the mechanisms are conserved across plant clades showing the early evolution of strategies to tolerate Al stress by land plants. In some crop species, there is genetic variation for $Al^{3+}$ resistance (exclusion of Al) or tolerance (ability to tolerate internal Al). This has been exploited by breeders to develop cultivars that maintain productivity on acid soils. The potential for developing improved cultivars for acid soils has recently been augmented by the isolation of genes responsible for the natural $Al^{3+}$ resistance of some species as well as the over-expression of genes that enhance Al tolerance of plants. Various physiological and molecular approaches have been adopted to understand the Al stress adaptation in plants. However, a lot more needs to be understood in major crops that are under threat of acid soils. Application of functional and translational genomics along with crop physiology can reveal newer facts for crop improvement under acid soils. This book is a sincere and serious attempt in understanding Al stress perception and adaptation in plants that will enhance and augment our understanding of Al stress research.

Silchar, India — Sanjib Kumar Panda
Bonn, Germany — František Baluška

# Contents

# About the Editors

**Dr. Sanjib Kumar Panda** is a Professor in Plant Biology in the Department of Life Science and Bioinformatics, Assam University (a Central University), Silchar, India. He has a research and teaching experience of almost two decades. Understanding Aluminum stress perception and tolerance in crop plants is his research focus. Besides this he also works on molecular biology of abiotic stress tolerance in Plants especially osmotic stress. He uses physiology, molecular biology, and functional and translational genomic approaches in his quest for studying abiotic stress in plants. He is a recipient of prestigious JSPS fellowship (Japan), Boyscast fellowship (USA), Indo-US research fellowship, INSA-DFG fellowship (Germany), and UNESCO-TWAS Associate to name a few. He has numerous research publications published in reputed international journals and is an Associate Editor for the Journal Plant Signaling and Behaviour (Taylor & Francis) and Plant Molecular Biology Reporter (Springer).

**František Baluška** is one of the leading scientists in the fields of root biology, plant actin cytoskeleton, polarity and tip growth, endocytosis, vesicle recycling, and gravitational biology of plant roots. Prof. Baluska has been investigating maize roots since 1985 and Arabidopsis roots since 1998. Besides his expertise on the plant cytoskeleton, endocytosis, polarity, and signaling, he is also expert on the root apex biology of crop maize and the genetic model Arabidopsis. In recent years, he has integrated, together with Prof. Stefano Mancuso and Prof. Liz Van Volkenburg, plant cell biology and plant physiology with the plant sensory ecology and electrophysiology to generate the emerging field of plant neurobiology. In order to foster this new sensory and behavioral view of plants and their roots, he has founded and edits two new scientific journals: Plant Signaling & Behavior and Communicative & Integrative Biology.

# Aluminum Signaling and Potential Links with Safener-Induced Detoxification in Plants

**Hideaki Matsumoto, Dean E. Riechers, Anatoli V. Lygin, František Baluška, and Mayandi Sivaguru**

**Abstract** Aluminum (Al) toxicity limits crop productivity in over 40 % of arable lands on this planet. Understanding the Al signaling and physiological relevance of Al toxicity and tolerance/resistance is fundamental to identify and improve crop productivity which is a better strategy than liming the soils as the latter is labor intensive, ineffective, and expensive. In this chapter, all aspects of Al toxicity and tolerance are discussed in a historic perspective of around a century of research, development, and understanding on this topic; a special section regarding a new function of existing 'safeners' and their potential for protection against Al toxicity is also discussed.

## 1 Introduction

Aluminum (Al) is the most abundant metal in the earth's crust comprising approximately 7 %. The chemical form of Al is dependent on pH. It exists as the octahedral hexahydrate, often abbreviated as $Al^{3+}$, which is the most toxic form at pH < 4.5 in acidic soils. Therefore, the adverse effect of acidic soil on plant growth is strongly related to the toxicity of $Al^{3+}$. It is generally accepted that Al accumulates in root apices including the cap, meristematic, and elongation zone. More recently, Sivaguru and Horst (1998) demonstrated that the distal transition zone (DTZ) locating in 1–2 mm from root tip is the most Al-sensitive zone where various events caused by Al signal are induced. Al-targeted organs or molecules followed

H. Matsumoto
Institute of Plant Science and Resources, Okayama University, Kurashiki 710-0046, Japan

D.E. Riechers • A.V. Lygin
Department of Crop Sciences, University of Illinois, 1102 S. Goodwin, Urbana, IL 61801, USA

F. Baluška
University of Bonn, Kirschallee 1, Bonn, 53115 Germany

M. Sivaguru (✉)
Institute for Genomic Biology, University of Illinois at Urbana-Champaign, 1206 W. Gregory Dr., Urbana, IL 61801, USA
e-mail: sivaguru@illinois.edu



S.K. Panda, F. Baluška (eds.), *Aluminum Stress Adaptation in Plants*, Signaling and Communication in Plants 24, DOI 10.1007/978-3-319-19968-9_1

by Al-induced functional alteration in plant cells are complicated and changeable upon the growth stage or genetic background of the plants. Therefore, the cascade of Al perception, signaling, and its associated signaling networks are important but unresolved research topics. Since the first report by Bennet and Breen (1991), who proposed that the Al signal transduction is perceived in the root cap of *Zea mays*, more than 20 years has passed but a unified understanding of Al signaling is still elusive (Panda et al. 2009; Liu et al. 2014). In this article, an overview of Al signaling related predominantly to the mechanism of Al toxicity is described, and possible links with ROS-mediated signaling and detoxification pathways triggered by chemicals called "safeners" are discussed and compared with Al-induced mechanisms in cereal crops. During the past several decades, our collective understanding of Al-resistance mechanisms, especially from the molecular point of view, has accumulated dramatically. In Chap. 2 of this book, an article regarding "Transcriptional regulation of Al tolerance in plants" is described. Furthermore, an intensive review on "Signaling and sensing in plant Al resistance" was recently published (Liu et al. 2014). Therefore, the impact of Al signaling on Al-resistance mechanisms will not be further addressed.

## 2 Response of Plants to Al Signal

Our knowledge has not accumulated sufficiently to answer the question of precisely how (and how quickly?) plants respond to Al signal in the roots. For example, it is not yet known whether or not specific Al signal receptor(s) exist in plants. It is also not known whether the Al signal is actually $Al^{3+}$ itself, or if other signaling molecules transformed from Al toxicity signal bind to the targeted molecules or organs in the cells exposed to Al, which results in the alteration of the function of targeted molecules. Equilibrium constants for binding to various metal ions had been reported. Log *K* values for ions are these: $Al^{3+}$, 4.30; $La^{3+}$, 3.34; $Cu^{2+}$, 2.60; $Ca^{2+}$ and $Mg^{2+}$, 1.48; $Na^{+}$ and $K^{+}$, 0 $M^{-1}$. These values correlate well with log *K* values for ion binding to many organic and inorganic ligands (Kinraide and Yemiyahu 2007).

Understanding how the Al signal is transformed to other signaling components in the cell, and how the Al signal is transported to the targeted organs is essential for providing a more comprehensive understanding of Al-signaling mechanisms in plant cells. Al signal induces wide ranges of physiological events in plant cells. Common events induced by the Al signal are the inhibition of root/cell elongation at the elongation zone in the root apex, accompanied by typical morphological changes that rupture the cells therein (Kopittke et al. 2008; Yamamoto et al. 2001; Motoda et al. 2010; Matsumoto and Motoda 2012, 2013). Inhibition of cell elongation by Al signal occurs within 30 min in the case of Al-sensitive maize roots (Llugany et al. 1995). Similarly, callose formation triggered by the Al signal is a rapid response of plant cells and has been proposed as an early consequence of Al toxicity. However, the physiological function of callose formation by Al signal is not known entirely. Sivaguru et al. (2000) discovered callose accumulation in the plasmadesmata of root cells in wheat subjected to Al and proposed that root elongation is inhibited by blocking cell-to-cell trafficking of molecules through the plasmadesmata. This might block the transmission of various signals regardless of Al dependent or not between the cells. The regulation of callose level through Al signal is not yet completely understood. One possibility for the regulation is the activity of callose synthetase, which requires free $Ca^{2+}$. Another possibility is the β-1-3 glucanase activity. Recently, Zhang et al. (2015) found that Al-sensitive sweet sorghum had higher callose synthetase activity, lower β-1-3 glucanase activity, and more callose deposition in the root apices during Al treatment compared with Al-tolerant sorghum. They also prepared *SbGlu1* gene, which encodes a β-1-3 glucanase, and expressed it in *Arabidopsis thaliana*. Independent transgenic lines displayed significantly greater Al tolerance than the wild type. This phenotype was associated with greater total β-1-3 glucanase activity, less Al accumulation, and reduced callose deposition in the root. These results indicate that callose production through Al signaling is likely an important component of the overall toxicity pathway leading to root growth inhibition.

## 3 Transport of Al Signal

Although plant responses to Al signal are slow or rapid, depending on growth conditions including different plant age, tissue, and cell, events induced rapidly by Al signal require the rapid uptake of Al into organs or cells. Understanding of uptake and transport mechanism of Al signal including long distance from root to shoot or short distance through plasma membrane in the cell is required. The graded penetration of absorbed Al into maize roots was observed with the Al-specific, fluorescent probe, morin. Al readily accumulates in the epidermal and outer cortical cell layers but does not readily penetrate into the inner cortex. After prolonged exposure, Al had entered all areas of the root apex (Jones et al. 2006). In research with X-ray microanalysis and freeze-dried cryosections, substantial intracellular Al accumulation was detected after a 4-h exposure of soybean root tips to 38 μM $Al^{3+}$ (Lazof et al. 1997). Using secondary ion mass spectrometry (SIMS), Al was found in root tip cells after 30 min exposure of intact soybean roots. Accumulation of Al was greatest within the first 30 μm; i.e., 2–3 cell layers. SIMS analyses clearly show substantial intracellular Al accumulation after Al exposure of only 30 min (Lazof et al. 1997). These results suggest an early effect of Al signaling at the root apex, such as those on cell division, cell expansion, or nutrient transport. In regards to long-distance transport, Al taken up by the 10-mm root apex is rapidly transferred to the xylem, which differentiates in the 10- to 15-mm root zone. Al induces the release of oxalate from root apex but particularly from the sub-apical 6–20 mm root zone, even when Al was applied only to the 5-mm root apex, suggesting a basipetal signal transduction. Oxalic acid was not detected in the xylem sap. Citrate proved to be the most likely ligand for Al in the xylem because Al and citrate transport rates were positively correlated in the buckwheat roots (Klug and Horst 2010). A strategy for Al tolerance in some plants is carried out by the sequestration of Al into vacuole in leaves. Xylem transport of Al signal complex with oxalate from the root apex to the leaves and sequestration into vacuoles associated with an Al-citrate complex was demonstrated in buckwheat (Ma et al. 2001).

As to the transport of Al signal through plasma membrane, several hypotheses including Golgi-associated vesicular pathway, endocytosis, Al-specific pump in plasma membrane, and a specific Al carrier proteins have been proposed. However, distinct evidence for each mechanism has not been obtained. Recently, Nramp Al transporters (*Nrat1*) were found to be associated with Al tolerance in rice (Xia et al. 2010). Nramp proteins are conserved in different species and are involved in divalent ion transport. Nrat1 is a transporter localized in the plasma membrane of rice root apical cells, exhibiting $Al^{3+}$ transport activity, but not for divalent cations or the Al-citrate complex. *Nrat1* expression is induced by Al and is root-specific occurring in all root cells, except for the epidermis. Knockout lines for *Nrat1* showed higher Al sensitivity, higher Al accumulation in the cell wall, and lower Al concentration in root cells in the presence of $Al^{3+}$, suggesting that *Nrat1* controls intracellular $Al^{3+}$ uptake, with subsequent detoxification via transport and Al accumulation into cell vacuoles, possibly mediated by OsALS1 (Xia et al. 2010; Simoes et al. 2012). Al stress signal is sequestered into vacuole by the rice Nramp aluminum transporter (*OsNRAT1*). Tolerant and sensitive NRRT1 alleles were investigated, and lower expression of *NRAT1* in the sensitive allele and reduced Al uptake was exhibited, suggesting that the *NRAT1* transporter plays a major role in rice Al tolerance mechanisms (Li et al. 2014). Similarly, transporters responsible for Al hyperaccumulation were identified: the vacuolar and plasma membrane-localized genes *vacuolar Al transporter (VALT)* and *plasma membrane Al transporter 1 (PALT1)* (Negishi et al. 2012). These genes are members of the aquaporin family. VALT and PALT1 are highly expressed in specific tissues, and their over-expression in *Arabidopsis thaliana* conferred Al tolerance and Al sensitivity, respectively.

# 4 Direct Interaction of Al Signal with Cell Organs or Molecules

## *4.1 Al Signaling and Nuclei*

Several studies reported that absorbed Al accumulated in the nuclei (Matsumoto et al. 1976; Silva et al. 2002; Naidoo et al. 1978). The preferred accumulation of Al in the nuclei is caused by the fact that nuclei are the most negatively charged cell organs due to their high phosphate content (Naora et al. 1961). Thus, Al with a large positive charge in the cytoplasm can be adsorbed by nuclei and move into nuclei through a nuclear hole (Matsumoto 2000). Approximately two-thirds of the total Al incorporated into nuclei in pea root treated with 1 mM Al at pH 5.5 were recovered in the chromatin fraction. 94 % of Al associated with chromatin was recovered in DNA. The Al binding to anionic charged phosphate in DNA strand was confirmed by the in vitro experiment showing that Al binding to DNA was inhibited by 70 % in the presence of equal amount of basic histone to DNA, which masks the negative charge of DNA (Matsumoto et al. 1977). The structural changes of chromatin in the pea root treated with Al in vivo was demonstrated by the shift of absorption spectrum and DNase II digestion of chromatin prepared from Al-treated pea roots (Matsumoto 1988). The results indicate that the Al signaling induced the condensation and/or aggregation of chromatin resulting in its reduced template activity. These events may be related to the inhibition of cell division by Al signaling (Clarkson 1965; Mohanty et al. 2004). It is interesting that similar phenomena were found in animal systems. Chromatin from the cortical area of the brain was much more sensitive to Al than the chromatin from the liver. Moreover, nuclei from the neuron-rich area of the brain were much more sensitive to Al, indicating that chromatin with very short linker region between nucleosomes may be more susceptible to Al that results in an alteration in chromatin structure by Al signaling (Walker et al. 1989).

It is known that the cell cycle is disturbed by Al signal. Sivaguru et al. (1999b) reported that the actively dividing log-phase tobacco cells were characterized by faint and larger phragmoplasts and unusually emerged daughter nuclei after 6 h of Al treatment. Phragmoplasts and spindle microtubules (SMTs) were not observed in cells having metaphase plate chromosomes, suggesting that Al signaling might block cell division directly in metaphase. Damage of DNA was detected by an Al-induced oxidative burst (Achary et al. 2012). Nezames et al. (2012), using *Arabidopsis* mutants, reported that Al might act as a DNA-damaging agent in vivo and affect the cell cycle checkpoint, which halts cell cycle progression and forces the differentiation of the quiescent center. On the other hand, Doncheva et al. (2005), using *Zea mays* differing in Al sensitivity, focused on the behavior of cell cycle and observed Al signaling-induced time-dependent inhibition and stimulation of cell division. They suggested the occurrence of a fast change in cell patterning rather than a general cariotoxic effect after short-term exposure to Al. As a specific response of nuclei in *Allium* roots treated with Al, Fiskesjo (1983) found that nucleolar material was extruded from the nuclei into the cytoplasm, following elongated rod-like bodies that eventually divided into two, one distinct body at each pole of the cell. Recently, Qin et al. (2013) reported the possible association of extrusion of proteins participating in nucleolar organizing regions (NORs) with the extrusion of nucleolar material originally detected (Fiskesjo 1983).

## 4.2 Al Signaling and Plasma Membrane

Plasma membrane (PM) is the first potential target for Al signal (Ishikawa and Wagatsuma 1998) due to high affinity of Al for binding to phosphate groups of phospholipids, which comprise a major component of the PM surrounding the cells. Al-binding sites on the membrane surface are likely to be either carboxylate or phosphate group because Al forms electrostatic bonds preferentially with oxygen donor ligands. $Al^{3+}$ has a greater affinity for phosphate oxyanion than it does for carboxyl oxyanions. These neutral, zwitterionic phospholipids, phosphatidylcholine (PC) and phosphatidyl ethanolamine (PE), together constitute about 80 % of the phospholipids in soybean root PMs. PC is preferentially sequestered in the outer leaflet of the PM. Thus, Al bonding to PC on PM surface plays an important role (Akeson et al. 1989). Upon binding to Al, PM structure and function are affected, resulting in the alteration of ion flux and PM $H^+$-ATPase activity, etc. (Ahn and Matsumoto 2006). $Al^{3+}$ exhibits a 560-fold higher affinity for the phosphatidylcholine surface than that of $Ca^{2+}$, resulting in the reduction of membrane fluidity (Akeson et al. 1989). The decrease of the DBI (double bond index) of total fatty acids may signal a decrease in membrane fluidity (Chaffai et al. 2005). Changes in lipid composition may lead to less ordered PM and compensate for the Al-induced decrease in membrane fluidity. During recovery from Al-induced membrane stiffening, alterations in PM composition, especially of sphingolipids, were reported to be relevant in the Al resistance in yeast and maize (da Silva et al. 2006).

Al signal is reflected in the alteration of electronic signaling of the PM. Electronic signaling caused by Al signal is composed of (1) PM transmembrane potential (Em) and (2) surface membrane potential (zeta potential) of PM (Ahn and Matsumoto 2006; da Silva et al. 2006). Both electronic signaling are caused by the Al-induced changes of the properties of PM differences of the properties and related to the regulation of ionic flux via membrane and signal transduction process. The Al signal-induced PM potential and zeta potential are much more intensive in the cells of the distal than in the proximal portion of the root (Illes et al. 2006; Ahn et al. 2001; Sivaguru et al. 2013).

### 4.2.1 Transmembrane Potential (Em) Regulated by Al Signaling

In early research, Olivetti et al. (1995) reported that Al-tolerant snapbean culture "Dade" rapidly and significantly depolarized PM upon the exposure to increasing Al concentration. In contrast, membrane potential of the Al-sensitive "Romano" was only slightly depolarized. The difference in Al-induced membrane potential indicated that Al reduces $K^+$ efflux channel conductance in Dade but does not affect $K^+$ efflux channel conductance in "Romano." Al-induced depolarization of PM is also caused by malate released from wheat roots (Papernick and Kochian 1997).

### 4.2.2 Zeta Potential and Al Signaling

Cell-surface electrical potential (Zeta potential) is controlled in part by ion binding (especially $H^+$ and metal ions) to PM and cell wall. Cell surface potentials influence ion fluxes across membrane and may control flux rate, saturation, *cis*- and *trans*-inhibition, rectification, voltage gating, shifts in voltage optima (Kinraide and Yemiyahu 2007). The binding of $H^+$ and metal cations including $Al^{3+}$ to PM regulates zeta potential; thereby disturbance of membranous properties is induced. With wheat cultivars, Al-tolerant wheat (ET8) and sensitive wheat (ES8), and squash plant, the relationship between the alteration of zeta potential and PM $H^+$-ATPase is investigated under Al stress (Ahn and Matsumoto 2006; Ahn et al. 2001, 2004). Both zeta potential and $H^+$-ATPase showed the interesting characteristics. Zeta potential was more negative concomitant with higher $H^+$-ATPase activity in the 0–5 mm segments than the other distal portion (5–10, 10–15, and 15–20 mm) in the control squash roots (-Al treatment). A significant increase (from −22.6 to +5 mV) of zeta potential was observed only in the PM vesicles prepared from 0 to 5 mm segments after Al treatment. Interestingly, the $H^+$-ATPase activity of PM vesicles isolated from the 0 to 5 mm root segments decreased significantly under Al stress. The decrease of $H^+$-ATPase activity was accompanied with a decrease in protein level. Al treatment elicited a significant depolarization of the zeta potential of PM vesicles prepared from the root tips of ES8 (from −18.5 to 4.3 mV) but hyperpolarized it in ET8 (from −15.1 to −17.9 mV). $H^+$-ATPase activity in PM vesicles prepared from the root tip of the ET8 increased but decreased in the ES8 under Al stress. The shift of zeta potential toward depolarization correlated with the decline in $H^+$-ATPase activity in PM vesicles prepared from squash root tips under Al stress. The relationship between zeta potential and $H^+$-ATPase can be explained as follows: root cell membrane of the ES8 appears to attract a higher quantity of $Al^{3+}$ than do the root cell membrane of ET8 according to the greater resting negative zeta potential of the native PM of the ES8 when compared to that of ET8. In the presence of $Al^{3+}$, the rate of depolarization of PM zeta potential in the ES8 was always higher than in the ET8 at Al concentrations in excess of 10 μM, which inhibit root elongation in ES8 but not in ET8. Therefore, the PM $H^+$-ATPase-dependent efflux of $H^+$ through plasma membrane might be reduced by the electrical positive surface potential in ES8 under Al stress.

### 4.3 Al Signaling and Cell Wall

Externally added Al rapidly binds to the apoplast, ranging from 30 to 90 % of the total absorbed Al (Horst 1995; Zhang and Yang 2005; Tice et al. 1992; Rengel and Zhang 2003). The amount of Al binding to the cell wall depends on the density of negative charge, which is expressed by cation exchange capacity (CEC). The specific molecule for Al binding in apoplast is thought to be pectin, especially free nonesterified pectin. Al treatment increased pectin content of the root apex and more prominently in the Al-sensitive culture of maize. Pectin and Al contents in 1 mm root section decreased from the apex to the 3–4 mm zone, suggesting the possible binding of Al to pectin. Pectin is methylated by pectin methylesterase, resulting in the decreased binding capacity for Al. Therefore, the decrease in cell wall pectin and its degree of methylation contribute to genotypic difference in Al resistance (Schmohl and Horst 2000; Eticha et al. 2005; Mimmo et al. 2009; Yang et al. 2008). Not only pectin but also hemicelluloses metabolism is more susceptible to Al stress signal (Yang et al. 2011a). Hemicelluloses 1 and 2 in the root apex were significantly higher in Al-sensitive than in Al-tolerant rice in the absence of Al, and Al exposure increased root hemicellulose content more significantly in Al-sensitive rice. In both varieties neither organic acid efflux nor changes in rhizosphere pH appear to be responsible for the Al exclusion. Yang et al. (2008) demonstrated that hemicelluloses are the major pool for Al accumulation in rice, and that xyloglucan endotransglucosylase (XET) activity is inhibited by Al. How are the properties of plant cell walls affected by binding to Al? Al affects the biochemical and biophysical nature of cell wall (Matsumoto and Yamamoto 2013). Al affects the extent of cell wall extensibility resulting in the inhibition of cell elongation (Tabuchi and Mastumoto 2001). The insight into the regulation of extensibility might be related to the alteration of wall-matrix cross-links. Another target molecule of Al binding to cell walls is Ca associated with pectin, which makes the so-called Ca-bridge in pectin. Al may contribute the displacement of Ca from cell wall that results in tight structure of cell wall matrix because the binding force of Al to pectin is much stronger than that of Ca. However, this possibility is unlikely to occur (Ryan et al. 1997; Schofield et al. 1998). Association of Al with other cell wall components such as enzymes, extensin, and xyloglucan may also affect the functional integrity of the cell wall. Al increased the level of covalently bound cell wall proteins in pea root apex.

Extensions are key components responsible for modifying cell wall rigidity. Al-binding experiments in vivo and in vitro suggested that extensin has the highest capacity to bind Al among cell wall-associated proteins (Kenjebaeva et al. 2000). Al induced the cell wall-associated kinase receptors (WAK1), and overexpression of this enzyme renders Al tolerance (Sivaguru et al. 2003a). WAK1 may play an important role in the Al signal transduction since Al signal associated with apoplast can be transferred into symplast via WAK. WAK1 gene in the roots showed typical "on" and "off" pattern. WAK proteins are localized preferentially in the peripheries of cortical cell in the elongation zone of *Arabidopsis*. Furthermore, WAK1-overexpressing *Arabidopsis* enhanced root growth under Al stress. The large amount of absorbed Al localizes in cell wall, and it is known that Al damages the cytoskeleton as described in the following section. Therefore, Al signal seems to involve the extracellular matrix–cell wall–plasma membrane–cytoskeleton continuum as these structures are physically connected. Any effect on the outer cells/cell wall extracellular matrix is sufficient to alter the physiological status of the inner cells (Horst et al. 1999; Baluška et al. 2003).

## 4.4 Al Signaling and Cytoskeleton

Dynamic cytoskeleton-based networks are associated with basic function of plant cells including the cell elongation, cell division, and differentiation. Al toxicity signal is reflected to the disorganization of cytoskeleton proteins followed by their dysfunction (Horst et al. 1999). Al-mediated alterations of microtubules (MTs) and actin microfilament (MF) have been shown to be the most prominent in cells of the distal part of the transition zone (DTZ) of an Al-sensitive maize cultivar (Sivaguru et al. 1999a; Matsumoto and Sivaguru 2008). Inhibition of longitudinal cell expression and cell swelling in the elongation zone of wheat root might be related to the disorder of the cytoskeletal network triggered by Al signaling (Sivaguru et al. 1999a; Matsumoto and Yamamoto 2013). The behavior of actin network is affected markedly in the plant cells under Al toxicity (Frantzios et al. 2005; Ahad and Nick 2007; Amenós et al. 2009). Al-induced significant increase in the tension of actin might be caused by the formation of nonhydrolyzable ($Al^{3+}$-ATP) complexes, and the binding to actin/myosin can modify filament contraction (Grabski and Schindler 1995). Amenós et al. (2009) reported that actin cytoskeleton and vesicle trafficking were primary targets for Al toxicity in root tip of maize. Visualization of boron-cross-linked rhamnogalacturonan II (RGII)-containing brefeldin A (BFA) compartments, which was first evidenced by Baluška et al. (2002), revealed that Al inhibited the formation of these compartments.

Al affects the reorientation of microtubules (MTs) that is clearly related to cell elongation. Longitudinally elongating cells have transversely oriented MTs. MT-disrupting agents promote the lateral expansion but inhibit longitudinal expression. Cortical MTs are known to be involved in the orientation of cellulose microfibers. The disappearance of the cortical MTs in elongating cells of wheat root was observed under Al stress, and this was correlated with the alteration of the ratio of cell width to cell length between control cells and Al-treated cells (Sasaki et al. 1996, 1997). The effect of Al signaling on microtubular cytoskeleton of the suspension cells of tobacco is age dependent, and exponential phase cells are more sensitive compared to stationary cells (Sivaguru et al. 1999b). The increase of α-tubulin and elements of the tubulin-folding chaperone (CCT) was formed by Al signal (Schwarzerová et al. 2002). Furthermore, extensive research examined the effect of Al signaling on alterating microtubules during the cell cycle. Events affected by Al signal are preprophase band of MTs, mitotic spindles, phragmoplast, kinetochore MT bundles, daughter nuclei, and walls that are stage-dependent and tissue-dependent (Frantzios et al. 2005; Sivaguru et al. 1999b; Seju and Lee 1998).

## 4.5 *Case Study of Al-Induced Efflux of Organic Acid*

Al-induced efflux of organic acids is an important Al tolerance mechanism, which is called the Al exclusion mechanism. The kind of organic acid excluded is different among plant species, and different transporters are concerned. It is known that malate exclusion occur without a lag period after plant roots are exposed to Al signal. Osawa and Matsumoto (2001) analyzed the mechanism of malate efflux with excised root tips of wheat. Efflux of malate occurred after 2 min of Al exposure and efflux continued linearly up to 15 min. When the root tips were exposed to Al for only 1 min followed by the removal of Al signal, efflux of malate began almost similarly as the root tips exposed continuously but the efflux gradually decreased after 7 min of removal of Al. These results suggest that root tips memorize Al signal given for only 1 min, which is prerequisite for malate efflux. $Al^{3+}$-resistant cultivars of wheat (*Triticum aestivum L.*) release malate through the $Al^{3+}$-activated anion transport protein *Triticum aestivum* aluminum-activated malate transporter 1 (TaALMT1). ALMT1 gene is constitutively expressed. With the sprit roots of soybean, it was found that the direct contact of Al signal to the roots was essential for the induction of citrate release. This suggests that Al signal is not transported from Al-treated roots to non-treated roots (Yang et al. 2001). The activation of malate efflux through ALMT1 is thought to be the association of Al to transporter. Furuichi et al. (2010) demonstrated that the extracellular C-terminal domain is required for both basal and $Al^{3+}$-dependent TaALMT 1 activity. Furthermore, three acidic amino acids (E274, D275, and E284) within this domain are specifically required for the interaction with Al, which are also specifically required for the activation of transport function by external $Al^{3+}$. Ligaba et al. (2013) further investigated functional, structural, and phylogenetic analysis of domains underlying the Al sensitivity of TaALMT1. They indicate that the N-domain, which is predicted to form the constitutive pathway, mediates ion transport even in the absence of the C-domain. However, segments in both domains are involved in $Al^{3+}$ sensing. They identified two regions, one at the N-terminus and a hydrophobic region at the C-terminus, that jointly contribute to the Al-responsive phenotype and conclude that functional changes observed for TaALMT1 are most likely the result of alterations in the overall structural integrity of ALMT family proteins, rather than modification of specific sites involved in $Al^{3+}$ sensing.

# 5 Al Signal Transduction Pathway

## 5.1 Al Signaling and $Ca^{2+}$

The antagonistic relationship between Al and $Ca^{2+}$ is well known, and many experiments were devoted to investigate the role of $Ca^{2+}$ in Al toxicity (Rengel 1992b; Pineros and Tester 1993). $Ca^{2+}$ associates with plasma membrane and cell wall and play an important role for maintaining their functions. At early research on Al–Ca interaction, inhibitory mechanisms of Al on $Ca^{2+}$ uptake were investigated. The possibility of Al as a Ca channel blocker was proposed as the major mechanism for the inhibition of $Ca^{2+}$ uptake by Al (Pineros and Tester 1993; Jones et al. 1998), although other possibilities are remained. Alterations to the homeostasis of cytosolic free Ca ($[Ca^{2+}]_{cyt}$) as a major event of Al stress signaling has been investigated intensively (Jones et al. 1998; Rengel 1992a; Zhang and Rengel 1999). The source for the increase in $[Ca^{2+}]_{cyt}$ under Al stress is extracellular through the depolarization-activated $Ca^{2+}$ channel and fluxes through $Ca^{2+}$-permeable nonselective cation channel in plasma membrane as well as intracellular sources of $Ca^{2+}$. Increased $[Ca^{2+}]_{cyt}$ from the intracellular stored source might be caused by enhancement of the $Ca^{2+}$ release channels in the tonoplast and the endoplasmic reticulum membrane (Rengel and Zhang 2003). More recently, Al-induced fluctuation of free $[Ca^{2+}]_{cyt}$ was investigated intensively after short period of Al treatment (from only seconds to several minutes) (Rengel 1992a). Evidence obtained suggests that disruption of $[Ca^{2+}]_{cyt}$ homeostasis plays a decisive role in the earliest stage of Al toxicity.

The fine mechanism for the fluctuation of $[Ca^{2+}]_{cyt}$ is still obscure, but Al-induced elevation in $[Ca^{2+}]_{cyt}$ suggests that the phytotoxic Al action in root hair is not through the blockage of $Ca^{2+}$-permeable channels (Jones et al. 1998). The release of $[Ca^{2+}]_{cyt}$ from stored $Ca^{2+}$in endoplasmic reticulum might be affected by inositol-1,4,5-triphophate ($IP_3$) formed from phosphatidylinositol-4,5-bisphosphate ($PIP_2$) through the phosphoinositide-associated signal transduction pathway which is inhibited by Al toxicity. Sivaguru et al. (2003b) reported that $[Ca^{2+}]_{cyt}$ influx might be associated with glutamate receptors, which in animals are ligand-gated cation channel and also known to be present in the genome of *Arabidopsis*. They demonstrated that glutamate depolymerized MTs and depolarized the plasma membrane. These responses, as well as the inhibition of root elongation, occurred within the first few minutes of Al treatment, but were evoked more rapidly by glutamate than Al. They speculated that Al induces the secretion of glutamate, which binds to its receptor and triggers the influx of $Ca^{2+}$ resulting in the observed depolymerization of microtubules and the depolarization of plasma membrane. On the other hand, transient elevation of $[Ca^{2+}]_{cyt}$ was observed in the response to glutamate, ATP, and Al in *Arabidopsis*. Each chemical induced $[Ca^{2+}]_{cyt}$ signature. Glutamate and ATP triggered pattern among the three treatments in regard to the onset duration and shape of the response.

Glutamate and ATP triggered patterns of $[Ca^{2+}]_{cyt}$ increase that were similar among the different root zone, whereas Al evoked $[Ca^{2+}]_{cyt}$ transients had monophasic and biphasic shapes, most probably in the root transition zone. The $Al^{3+}$-induced $[Ca^{2+}]_{cyt}$ increases generally started in the maturation zone and propagated toward the cap, while the earliest $[Ca^{2+}]_{cyt}$ response after glutamate or ATP treatment occurred in an area that encompassed the meristem and elongation zone. The biphasic $[Ca^{2+}]_{cyt}$ signature resulting from $Al^{3+}$ treatment originated mostly from cortical cells, which could be triggered in part through ligand-gated glutamate receptors. With the comparison of other trivalent cations, the trivalent ion-induced $[Ca^{2+}]_{cyt}$ signatures in roots of an Al-tolerant or Al-sensitive mutant were similar to those of wild-type plants, indicating that the early $[Ca^{2+}]_{cyt}$ changes may not be tightly linked to $Al^{3+}$ toxicity but rather to a general response to trivalent cations (Rincón-Zachary et al. 2010). Elevation of $[Ca^{2+}]_{cyt}$ in tobacco suspension cells results from activation of the plasma membrane $Ca^{2+}$-permeable channels by ethylene, but it is inhibited by $La^{3+}$, $Gd^{3+}$, and $Al^{3+}$ (Zhao et al. 2007). The effects of Al on Ca-signaling were assessed in tobacco expressing a $Ca^{2+}$ monitoring luminescent protein, aequorin, as well as in a newly isolated putative plant $Ca^{2+}$ channel protein from *Arabidopsis* AtTPC1 (two-pore channel). Lin et al. (2005)

reported the involvement of an Al-sensitive signaling pathway requiring TPC1-type channel-dependent $Ca^{2+}$ influx in the presence of salicylic acid, a key plant defense-inducing agent. TPC1 channels were demonstrated to be the only Al-sensitive channel which was involved in Ca signaling (Lin et al. 2005). Furthermore, $Ca^{2+}$ is important for the signal transduction with Ca-binding protein-like calmodulin, and μmol level of free $Ca^{2+}$ in cytoplasm $[Ca^{2+}]_{cyt}$ work as second messenger in signal network in both plant and animal. Al signal can be transduced to $Ca^{2+}$ signal through different metabolic pathways that are not completely dissolved. Therefore, maintaining homeostasis of free $[Ca^{2+}]_{cyt}$ is critical for the survival of plant cells under Al stress.

Interactions of Al with the elements of phosphoinositide-associated signal transduction pathway are primary events in Al toxicity (Jones and Kochian 1995; Haug et al. 1994; Pejchar et al. 2010). Al inhibits a key signal transduction enzyme, phospholipase C (PLC) (Jones and Kochian 1995). $I_{50}$ was 15–20 μM Al in wheat roots. Binding of $Al^{3+}$ to microsomes and liposomes was found to be lipid dependent with the signal transduction element. Phosphatidylinostol-4,5-diphosphtate ($PIP_2$)-specific phospholipase Cs are probable interaction sites for inhibitory actions of Al ions. Following interiorization of Al, alteration of inositol-1,4,5-tiphosphate ($IP_3$) which is produced from $PIP_2$ hydrolysis by PLC activity regulates $Ca^{2+}$ release from endoplasmic reticulum concomitant with derangements of intracellular $Ca^{2+}$ homeostasis (Jones and Kochian 1995). This brake of Ca-homeostasis by Al signaling might be related to the mechanism of Al toxicity. Al disrupted production of second messengers such as $IP_3$ and phosphatidic acid (PA) by blocking PLC induces the degradation of $Ca^{2+}$ homeostasis (Chee-Gonzalez et al. 2009). Al has been shown to affect the phospholipid-signaling pathway as well. Phospholipase D (PLD) and diacylglycerol kinase C (DGK) activities are stimulated by Al in *Coffea arabica* (Chee-Gonzalez et al. 2009). Phosphatidic acid (PA), which is a multifunctional stress-signaling molecule in plants (Testerink and Munnik 2006), is generated via Al-inhibited phospahtidylinositol-specific phospholipase C (PI-PLC) and the diacylglycerol (DAG) kinase pathway. Al caused almost 30 % inhibition of PA which is formed by diacylglycerol kinase pathway in *Coffea arabica*. The evidence was obtained that PI-PLC activity is affected by Al results in reduced diaylglycerol formation. Taken together, PC-PLC is a target of Al signaling in plants. Besides PI-PLC, it was found that phosphatidylcholine hydrolyzing phospholipase C (PC-PLC), known as nonspecific phospholipase C (NPC) which generates DAG through glycerophospholipid hydrolysis [mainly phosphatidyl choline (PC)], is affected by Al signal in tobacco cells, which results in reduced DAG formation. Therefore, PC-PLC is also a target of Al signal in plants (Pejchar et al. 2010).

## 5.2 Al Signaling and Oxidative Stress

Al-induced oxidative stress including lipid peroxidation in plants was reported (Yamamoto et al. 2001, 2002; Cakmak and Horst 1991). Unlike Fe, which is a transition metal acting as a catalyst for redox reactions, Al enhanced peroxidation of lipids indirectly. Al has a high affinity for the head group (serine) of phosphatidyl serine, increasing membrane rigidity, fusion, and fatty acid chain packing. Such conformational changes in phosphatidylserine may substantially increase its susceptibility to oxidation in animal system (Matsumoto 2000; Gutteridge et al. 1985; Yin et al. 2010a, b). Similarly in animals, Al and Fe synergistically enhanced the peroxidation of lipids in cultured tobacco cells, which leads to alterations in membrane permeability and eventually causes apoptosis-like cell death (Ono et al. 1995; Yamaguchi et al. 1999) and programmed cell death (Zhan et al. 2013). In whole-plant systems, the enhancement of peroxidation of lipids induced by only Al without Fe has been reported. However, the peroxidation of lipids seems not to be the primary cause leading to inhibition of root elongation in pea roots by Al (Yamamoto et al. 2001). In many studies investigating lipid peroxidation by Al, the detection of malondialdehyde (MDA), a product of lipid peroxidation, was measured as an indicator of lipid peroxidation. However, the production of MDA is not a specific indicator for lipid peroxidation, since other oxidative stresses can also disrupt membrane integrity. Recently, Yin et al. (2010a) reported a possible involvement of a part of the lipid-derived aldehydes such as highly electrophilic alpha, beta-unsaturated aldehyde (2-alkeral), in the Al-induced root elongation inhibition in tobacco. Furthermore, the transgenic tobacco plants over-expressing *Arabidopsis* 2-alkenal reductase showed less accumulation of the aldehydes, less retardation of root elongation by Al, and higher regrowth after Al treatment, suggesting that lipid peroxide-derived aldehydes such as a 4-hydroxyl-(E)-2-nonenal and (E1)-2-hexanol could injure cells directly under Al stress (Yin et al. 2010a, b). Plants as well as other aerobic organisms require oxygen for the effective production of energy. During the reduction of $O_2$ to $H_2O$, reactive oxygen species (ROS) such as superoxide anion radical ($O_2-$), hydrogennperoxide ($H_2O_2$), hydroxyl radical (•OH), and singlet oxygen atoms ($^1O_2$) can be formed. One percent $O_2$ consumed by plants is diverted to produce ROS. ROS have the capacity to oxidize cellular compounds which lead to cell death. So far many researchers reported that Al toxicity mechanism is caused by Al-induced oxidative stress, and correlation of the level of ROS is detected in the process of Al toxicity recovery in pea roots (Matsumoto and Motoda 2012, 2013). The level of ROS increased in the pea roots apex during Al toxicity and decreased during recovery process from Al toxicity (Motoda et al. 2010; Matsumoto and Motoda 2012, 2013). Superoxide anion is a major ROS that is formed by plasma membrane NADPH oxidase under Al stress. NADPH oxidase of *Arabidopsis* is a membrane-associated protein that has six membrane-penetrating domain and EF hand, which is the motif of $Ca^{2+}$ binding. Therefore, the regulation of NADPH oxidase may be induced by the changes of surrounding microenvironments with phospholipids in the plasma membrane and the level of cytoplasmic free $Ca^{2+}$ regulated by Al signal (Matsumoto and Motoda 2012). $H_2O_2$ is a versatile molecule of the ROS network in plant and plays an important role under severe environmental condition including Al stress. $H_2O_2$ is formed in various cell organs including cell wall, cytosol, mitochondria, and chloroplasts. $H_2O_2$ is produced in part by superoxide dismutase (SOD) from superoxide anion under stress conditions. One of the important functions of $H_2O_2$ under Al stress signal may be the possible participation in lignin formation through peroxidase activity. $H_2O_2$ plays an important role as an electron donor for coniferyl alcohol peroxidase participating in lignin formation. The basal level of $H_2O_2$ in the control pea root apex without Al treatment is very low, but increased dramatically during a 24 h Al treatment and decreased during the recovery process after Al treatment (Matsumoto and Motoda 2012). This suggests that $H_2O_2$ associates with the mechanism of Al toxicity and recovery in pea root. Lignin formation causes the decreased cell elongation. Phenolic compounds as antioxidant play a role as a protectant for the increasing production of ROS (Devi et al. 2003). The evolution of all aerobic organisms is dependent on the development of different $H_2O_2$-scavenging mechanism including catalase (CAT), peroxidase (POD), ascorbate peroxidase (APX), and glutathione reductase (GR). Conversely, SOD plays a role in producing $H_2O_2$ in plants, prokaryotes, and eukaryotes. SOD plays an alternative important role in the antioxidant system of the cells because of degradation of toxic superoxide anions concomitant with the formation of $H_2O_2$. Nonenzymatic antioxidants, such as tocophenols, ascorbic acid, and glutathione (GSH), work in concert to detoxify $H_2O_2$ or lipid peroxides. Protein oxidation is defined as a covalent modification of a protein induced by ROS or by-products of oxidative stress. Al-induced ROS are likely to target proteins that contain sulfur-containing amino acids and thiol group. Several researchers reported the Al-induced program cell death is carried out possibly via a ROS-activated signal transduction pathway (Yamaguchi et al. 1999; Pan et al. 2001). Oxidative stress can be induced by several stress signals. So far multiple studies suggest that oxidative stress induced by Al signaling participates in the direct mechanism of Al toxicity. However, Novascués et al. (2012) reported that oxidative stress is a consequence but not a cause of Al toxicity in forage legume *Lotus corniculatus*.

## 5.3 *Reactive Nitrogen Species*

Nitric oxide (NO) is a reactive compound and functions as a signaling molecule in plants (Wilson et al. 2008). NO is an important signal molecule in the response of Al toxicity (He et al. 2012b). NO may enhance Al tolerance by regulating hormonal equilibrium in plants as a regulator of plant hormones signaling (He et al. 2012a). The extent of Al internalization during the recovery from Al stress in living roots of *Arabidopsis thaliana* was investigated. There was no detectable uptake of Al into cells of the proximal part of the transition zone and the whole elongation region. Moreover, cells of the distal portion of the transition zone emitted large amounts of

NO, which was blocked by Al treatment. These data suggest that Al internalization is related to the most sensitive status of the distal portion of the transition zone towards Al. Al in these root cells has an impact on endosomes and NO production (Illes et al. 2006). In roots of *Hibiscus moscheutos* L., exogenous NO ameliorated the Al-induced inhibition of root elongation, and Al inhibited the activity of NO synthase (NOS) and reduced NO concentration (Tian et al. 2007). Other treatments including NO scavengers, inhibitors of NOS, and nitrate reductase caused the inhibition of root elongation, suggesting that the reduction of endogenous NO could lead to inhibition of root elongation under Al stress (He et al. 2012b). He et al. (2012b) also reported that Al toxicity might disrupt NO homeostasis, leading to endogenous NO concentration being lower than required for root elongation in plants. Furthermore, it was confirmed that NO treatment reduced $Al^{3+}$-induced ROS and RNS (reactive nitrogen species) toxicities by increasing the activities and protein expression of antioxidant enzymes as well as *S*-nitrosoglutathione reductase (GSNOR). Suppressing GSNOR enzymatic activity aggravated $Al^{3+}$ damage to rice and increased the accumulation of RNS (Yang et al. 2013).

## 5.4 *Al Signaling and Hormones*

Hormone function is involved in the several events induced rapidly by Al signal in plant cells. This may suggest that Al signal is transduced to signal of hormones. Changes in ethylene evolution and changes in the content and composition of cytokinins (CKs) in the roots of *Phaseolus vulgaris* L. cv *Strike* were determined. The ethylene evolution reached a maximum after 30 min of Al treatment. Levels of CK nucleotides declined after 5 min of Al treatment whereas zeatin increased sixfold (Massot et al. 2002). Similarly, Cizkova (1995) reported that phytohormonal levels induced by Al, decline of cytokinin-like substance, and stimulation of indole-3-acetic acid are undoubtedly causes for changes in spruce roots under Al stress (Čizková 1995). An early study discovered that application of Al to the root caps of Al-sensitive maize strongly promoted acropetal transport of auxin, resulting in the reduction of auxin transport polarity (Hasenstein and Evans 1988). Basipetal transport of exogenously applied [$3H^+$]-indole-3-acetic acid to the meristematic zone significantly alleviated the inhibition of root elongation induced by the application of Al to DTZ. The primary mechanism of genotypical differences in Al resistance is located within the DTZ and suggest a signaling pathway in the root apex mediating the Al signal between the DTZ and elongation zone (EZ) through basipetal auxin transport (Kollmeier et al. 2000). In wheat roots, Al treatment increased the accumulation of endogenous IAA, but decreased the activity of IAA oxidase in a dose-dependent manner. A strong correlation between the data of malic acid efflux rate and endogenous IAA content was obtained (Yang et al. 2011b). With the blocker of auxin transporter, it is speculated that the anion channel might have been activated when IAA was involved in malic acid efflux, suggesting that IAA was involved in Al-induced efflux of malic acid from wheat root. Contrary to auxin, Tian et al. (2014) found that ethylene may behave as a negative regulator on Al-induced malate efflux by targeting *TaALMT1*-mediated malate efflux by an unknown mechanism. Recently, in *Arabidopsis* mutants with an internal Al detoxification mechanism, it was found that the level of endogenous IAA negatively regulates Al tolerance by the alteration of the transport of symplastic $Al^{3+}$ to the vacuole (Zhu et al. 2013). When the wild-type *Arabidopsis* was exposed to $AlCl_3$, marked inhibition of root elongation was observed and elicited a rapid ethylene evolution and enhanced activity of the ethylene reporter EBS:GUS in root apices (Sun et al. 2010). Ethylene synthesis inhibitors, $Co^{2+}$ and aminoethoxyvinylglycine (AVG), and the antagonist of ethylene perception ($Ag^+$) abolished the $Al^{3+}$-induced inhibition of root elongation. There was less inhibition of root elongation by $Al^{3+}$ in mutant of auxin polar transport (aux 1–7 and pin 2) than in the wild type. Results indicate that $Al^{3+}$-induced ethylene production is likely to act as a signal to alter auxin distribution in roots by disrupting AUX1- and PIN2-mediated auxin polar transport, leading to arrest of root elongation (Sun et al. 2010). Massot et al. (2002) reported that $Al^{3+}$-induced inhibition of root growth may be predicted by significant changes in cytokinin content and composition and enhanced ethylene biosynthesis. Ethylene production was closely associated with Al-induced root growth inhibition in lotus (Sun et al. 2007). Abscisic acid (ABA) increased in barley roots treated with $AlCl_3$. Treatment with $AlCl_3$ or ABA increased both ATP-dependent and PPi-dependent $H^+$-pumping activity in tonoplast-enriched membrane vesicle prepared from barley roots (Kasai et al. 1993). These results may contribute towards maintaining the homeostasis of $[H^+]$ in the cytoplasm by which plants defend themselves against hostile environments, since levels of ionized $Al^{3+}$ due to acidification are the key determinants in Al toxicity. Contrary to the effect of ABA on the tonoplast, plasma membranes of *Arabidopsis* were affected differently. ABA induced depolarization and reduced the proton pumping of the plasma membrane, which are $Ca^{2+}$-dependent processes (Brault et al. 2004). Exogenous application of ABA and the ABA synthesis inhibitor, fluoride, respectively, increased and reduced endogenous ABA content in root apices of soybean and results in the corresponding reduction and aggravation of Al toxicity. In a split-root experiment, Al treatment in two parts of roots (Part A, +Al; Part B, +Al) both decreased root elongation and increased ABA accumulation in root apices of soybean. Whereas when only Part A was exposed to Al (Part A, +Al; Part B, −Al), endogenous ABA content in root apices increased in Part A but inversely in Part B, and root elongation inhibition only was found in Part A (Hou et al. 2010). Using $[3H^+]$-ABA, it was determined that $[3H^+]$-ABA is transported at a rate of more than 3.2 cm $min^{-1}$ in whole soybean plants, and $Al^{3+}$ accelerates this rate. In addition, $[3H^+]$-ABA was distributed in the root under Al stress (Hou et al. 2010). ABA might therefore play an important role in regulating Al resistance of soybean, including the possible involvement of ABA in the exudation of citrate from soybean (Shen et al. 2004).

## 5.5 Al Signaling and Phosphorylation

Protein phosphorylation plays an important role in the regulation of various biological functions in plants. Phosphorylation of particular proteins by Al has been reported (Osawa and Matsumoto 2001; Martinez-Estévez et al. 2001). K-252a, a broad range inhibitor of protein kinase blocked the Al-induced malate efflux in Al-resistant wheat (cv Atlas). A transient activation of a 48-kDa protein kinase in wheat root exposed to $Al^{3+}$ was observed preceding the initiation of malate efflux, which was canceled by K252a. The results firstly suggested that protein phosphorylation is involved in the Al-responsive malate efflux and anion-specific channel might be a terminal target that responds to Al signaling mediated by protein phosphorylation (Osawa and Matsumoto 2001). Furthermore, Ligava et al. (2009) reported that S385 of TaALMT1, an Al-activated malate transporter (ALMT1) in wheat root apex, is an essential residue regulating basal transport of malate as well as Al activation of transport activity through TaALMT1 via direct protein phosphorylation. With *Coffea Arabica* suspension cells exposed to $AlCl_3$, in vitro phosphorylation with cell extracts was investigated. Protein phosphorylation patterns dramatically increased for 18, 31, and 53 kDa proteins, but there were no changes detected in these protein levels in cells treated with $AlCl_3$ compared to untreated cells (Martinez-Estévez et al. 2001). Al activated the mitogen-activated kinase (MAP kinases) with an apparent molecular mass of 58 kDa whose substance was 58 kDa myelin basic protein. Since the activity of the 58 kDa MAP kinase was enhanced dramatically after addition of $AlCl_3$ to the medium, it is speculated that Al toxicity in coffee could be perceived through the MAP kinase signal transduction pathway (Arroyo-Serralta et al. 2005). Shen et al. (2004) reported the effect of ABA on citrate efflux from soybean roots. Pretreatment or treatment with K-252a induced a severe inhibition of Al-induced citrate efflux. Al increased endogenous ABA levels, whereas K-252a exerted no detectable inhibitory effects on the Al-induced increase in ABA levels. ABA is probably involved in early response after which a K-252a-sensitive protein kinase plays a key regulatory role in the activity of an anion channel within the plasma membrane, through which citrate is released. Furthermore, citrate efflux from soybean root associated with the function of PM $H^+$-ATPase and upregulation of transcription, translation, and threonine-oriented phosphorylation of PM $H^+$-ATPase was detected (Shen et al. 2005). Chen et al. (2013) reported that citrate exclusion coupled with a concomitant release of protons increased in broad bean under Al stress. They suggested that phosphorylation and interaction with the vf14-3-3 protein of the VHA2 were enhanced in Al-resistant broad bean but not in Al-sensitive variety. Al enhanced the expression and interaction of the PM $H^+$-ATPase and 14-3-3 proteins that lead to higher activity of the PM $H^+$-ATPase and more citrate exudation in Al-resistant broad bean (Chen et al. 2013). Protein kinase and phosphatase inhibitor studies showed that reversible phosphorylation was important for the transcriptional and posttranslational regulation of *AtALMT1* encoding malate transporter in *Arabidopsis* (Kobayashi et al. 2007).

# 6 Herbicide Safeners, Detoxification, and Potential Links with Al Signaling and Resistance Mechanisms in Plants

## *6.1 Background on Herbicide Safeners*

Plants are frequently exposed to synthetic toxins and abiotic stress compounds that elicit the rapid production of reactive oxygen species (ROS) and activation of plant defense mechanisms for adaptation and survival. There are multiple scenarios in which Al also elicits ROS production at the root apex, specifically at the distal part of the transition zone (Sivaguru et al. 2013 and references therein). One such mechanism of ROS adaptation is the metabolism of ROS-generating organic compounds from abiotic origin (*xenobiotics*). Herbicide safeners are nontoxic compounds that confer protection from herbicide injury in cereal crops by inducing detoxification systems for endogenous toxins, xenobiotics, and ROS (Riechers et al. 2010). Safeners tap into preexisting signaling pathways to promote the expression of detoxification genes such as glutathione *S*-transferases (GSTs) and cytochrome P450 monooxygenases (P450s). Safeners are thus unique and valuable tools for studying early signaling and stress-response genes, the regulation of GST and P450 gene expression, and for inducing the expression of other essential components of the well-documented three-phase detoxification pathway (Kreuz et al. 1996; Zhang et al. 2007; Riechers et al. 2010) in agricultural plants in the absence of phytotoxicity. Recent studies in my laboratory have indicated that safeners induce the expression of GSTs that detoxify xenobiotics mainly in the outer three cell layers (i.e., epidermal and subepidermal) of cereal crop seedling coleoptiles (Riechers et al. 2003). These findings have led to new hypotheses that (a) safeners are tapping into an unidentified, preexisting signaling pathway for detoxification of endogenous toxins, xenobiotics, and ROS in a *tissue-specific* manner, and (b) safeners may be utilizing an oxidized lipid ("oxylipin" or cyclo-pentenone)-mediated signaling pathway (Mueller 2004; Dave and Graham 2012) in the coleoptile, which subsequently leads to dramatic but specific induction of plant defense genes involved in detoxification in epidermal and subepidermal cell layers.

Safener treatment of cereal crops results in an increase in gene expression and subsequent enzymatic activity of herbicide detoxification enzymes, such as GSTs, P450s, and UDP-glycosyl-transferases (uGTs). Much information is known about herbicide safener uptake and translocation in the plant, their effects on herbicide metabolism rates, and the increases in GST and P450 enzymatic activity with various xenobiotic substrates (Davies and Caseley 1999; Riechers et al. 2010). However, detailed information is lacking regarding the biochemical and molecular events that occur between the initial safener application and the end result (i.e., increased GST activity and enhanced herbicide metabolism) in safener-responsive cereal crops.

## 6.2 *Safeners Enhance Herbicide Selectivity and Protect Cereal Crops*

Safeners are chemical compounds that increase the tolerance of certain grass crops (e.g., maize, grain sorghum, wheat, and rice) to herbicides frequently used in the Midwest for selective control of annual grass weeds (Davies and Caseley 1999). Safeners protect cereal crop plants by increasing rates of herbicide metabolism through induction of detoxification pathways (Kreuz et al. 1996; Davies and Caseley 1999). The increase in metabolism results from an increase in the activity of several key detoxification enzymes, such as GSTs, P450s, and uGTs (Riechers et al. 2010). GST expression is induced following exposure to many stresses (reviewed by Dixon et al. 2010; Cummins et al. 2011). GST enzymatic activity may involve direct glutathione (GSH) conjugation to toxic electrophilic molecules, or glutathione-dependent peroxidase activity, using GSH as reductant for the detoxification of reactive oxygen (ROS) and lipid peroxides formed during or after plant stress (Edwards and Dixon 2005; Dixon et al. 2010; Cummins et al. 2011). In addition to increasing the expression of GST and P450-encoding genes, safeners increase the enzymatic activity of vacuolar transporters of xenobiotic-GSH conjugates (reviewed by Kreuz et al. 1996; Riechers et al. 2010). As a result, it has become apparent that the ultimate function of GSH-mediated detoxification in plants is to remove GSH-xenobiotic conjugates from the cytosol by transporting them into the vacuole for further metabolic processing (Cummins et al. 2011).

## 6.3 *Mechanisms of Safener-Regulated Gene Expression in Plants*

Several potential models for abiotic stress signaling in response to xenobiotics have been proposed, but none have been clearly tested using safeners (Ramel et al. 2012). It remains difficult to distinguish between contrasting hypotheses of direct xenobiotic sensing and indirect sensing of xenobiotic-related modifications (Ramel et al. 2012). In one proposed model, xenobiotic-induced gene expression may result from oxidative stress in the cell. Possible triggers include an alteration in the reduced glutathione to oxidized glutathione ratio (GSH: GSSG), indicating perturbed glutathione homeostasis (Ramel et al. 2012), or the production of ROS (Baxter et al. 2014). Safener-GSH conjugates have been identified in maize and *Arabidopsis* (reviewed by Riechers et al. 2010), although it is not known if this is nonenzymatic or GST-catalyzed GSH conjugation of safeners. Information about the molecular components of the putative signal transduction pathway induced by safeners is extremely limited. The biochemical and molecular events that occur between initial safener treatment and the end result (e.g., increased GST/P450/uGT activity and enhanced herbicide metabolism) are unknown. Possible mechanisms for safener-regulated signaling involving oxidized lipids ("oxylipins") have been proposed (Riechers et al. 2010; Skipsey et al. 2011), as has the involvement of TGA transcription factors (Behringer et al. 2011). The potential roles of oxylipins in safener-regulated signaling mechanisms will be described in more detail below.

## *6.4 Structures, Synthesis, and Roles of Oxylipins in Signaling*

Oxylipins are structurally diverse metabolites derived from fatty acid oxidation and can be formed through either nonenzymatic or enzymatic reactions. Nonenzymatically generated oxylipins are formed via free radical-catalyzed reactions in or near cell membranes, where polyunsaturated fatty acids (such as α-linolenic acid; Christeller and Galis 2014) serve as precursors for their synthesis, and include different types of phytoprostanes ($A_1$ and $B_1$), malondialdehyde (MDA), and 4-hydroxy-2*E*-nonenal (Mueller 2004; Loeffler et al. 2005; Mosblech et al. 2009). Enzymatically produced oxylipins include jasmonic acid (JA) and 12-oxo-phytodienoic acid (OPDA); this pathway has been well studied due to the hormonal activity of JA and defense gene activation (Mosblech et al. 2009; Schaller and Stintzi 2009). JA is synthesized via a series of steps starting with the oxygenation of α-linolenic acid from membrane lipids via lipoxygenase activity (Vicente et al. 2012; Grebner et al. 2013) and subsequent conversion to OPDA (Schaller and Stintzi 2009). It has become clear that biological activity differs among the various JA metabolites, as well as its biosynthetic precursors (reviewed by Schaller and Stintzi 2009; Dave and Graham 2012; Vicente et al. 2012; Zhou et al. 2014).

Lipase-induced release of specific lipid substrates (such as α-linolenic acid) in response to abiotic and biotic stresses enables their rapid conversion into various classes of oxylipins, which may perceive and respond to a wide range of environmental stimuli (Bonaventure et al. 2011; Schuck et al. 2014). Membranes in the epidermal and subepidermal cell layers of plant tissues and organs are likely the initial sites of perception for a wide range of environmental stressors (Javelle et al. 2011; Okazaki and Saito 2014). Polyunsaturated fatty acids are major structural constituents of cell membranes that also function as modulators of diverse signal transduction pathways triggered by abiotic stresses (Okazaki and Saito 2014; Savchenko et al. 2014). Different stresses induce the production of different classes of oxylipins with distinct, yet partially overlapping, transcriptional responses (Taki et al. 2005; Mueller et al. 2008; Schuck et al. 2014). Recent results suggest that enzymatically formed oxylipins (e.g., OPDA and jasmonates) and nonenzymatically formed oxylipins (e.g., phytoprostanes) perform important but distinct functions in plant defense responses (Mueller and Berger 2009; Dave and Graham 2012; Savchenko et al. 2014). However, the specific roles of different classes of oxylipins in plant defense and detoxification mechanisms, most notably their roles in xenobiotic sensing and signaling in cereal crops, remain unclear.

Oxylipins differ not only in their origin and structures but also in their electrophilicity. For example, strong reactive electrophilic species (RES) include the $A_1$-type phytoprostanes ($PPA_1$) and OPDA, and weak RES include the $B_1$-type phytoprostanes ($PPB_1$) and JA (Almeras et al. 2003; Farmer and Davoine 2007). RES oxylipins share several common properties, including a lipophilic nature that aids in binding to hydrophobic pockets or active sites of proteins, thiol reactivity due to an electrophilic site, and the ability to modify proteins and strongly induce genes and enzymes involved in detoxification (Mueller and Berger 2009). Common genes and proteins induced by both safeners and various classes of oxylipins include several P450s, uGTs, GSTs, and glutathione-conjugate ABC transporters (Taki et al. 2005; Dueckershoff et al. 2008; Mueller et al. 2008; Riechers et al. 2010).

## *6.5 Critical Knowledge Gaps in Molecular Aspects of Safener-Mediated Signaling in Cereal Crops*

In spite of the wealth of physiological, biochemical, and phenotypic information regarding use of safeners to protect cereal crops from herbicides (Davies and Caseley 1999), there is comparatively little known regarding safener-mediated signaling pathways in safener-responsive cereal crops (Riechers et al. 2010). Most information on safener-regulated signaling mechanisms is derived from *Arabidopsis* (Behringer et al. 2011) and *Populus* (Rishi et al. 2004), particularly with regard to the roles of TGA transcription factors (Behringer et al. 2011) and oxylipins (Skipsey et al. 2011). Root cultures from wild-type *Arabidopsis* plants and mutants defective in fatty acid desaturation (*fad3-2*/*fad7-2*/*fad8*), which are defective in forming the oxylipin precursor linolenic acid, demonstrated an attenuated ability to respond to safener treatment when measuring *AtGSTU24* expression (Skipsey et al. 2011). Since these *fad* mutants accumulate linoleic acid instead of α-linolenic acid, they are unable to synthesize OPDA or phytoprostanes from α-linolenic acid substrates released via lipase activities (Christeller and Galis 2014). The decreased ability of these mutant lines to respond to safener treatment via induction of GST expression is consistent with a link between safener-regulated responses and endogenous oxylipin signaling. Therefore, attenuated GST induction by safener in mutants displaying a reduction in polyunsaturated fatty acids suggests that safeners must act either in parallel or upstream of oxylipin signaling, potentially through regulating the availability of these endogenous molecules via hydrolytic lipase activities (Skipsey et al. 2011; Okazaki and Saito 2014). As a result, we propose a unifying model (Fig. 1) that proposes and integrates the roles of lipase activities, α-linolenic acid, and oxylipins in safener-regulated signaling of plant defense gene expression and detoxification reactions in cereal crops. This diagram will assist researchers in forming testable hypotheses to drive future mechanistic

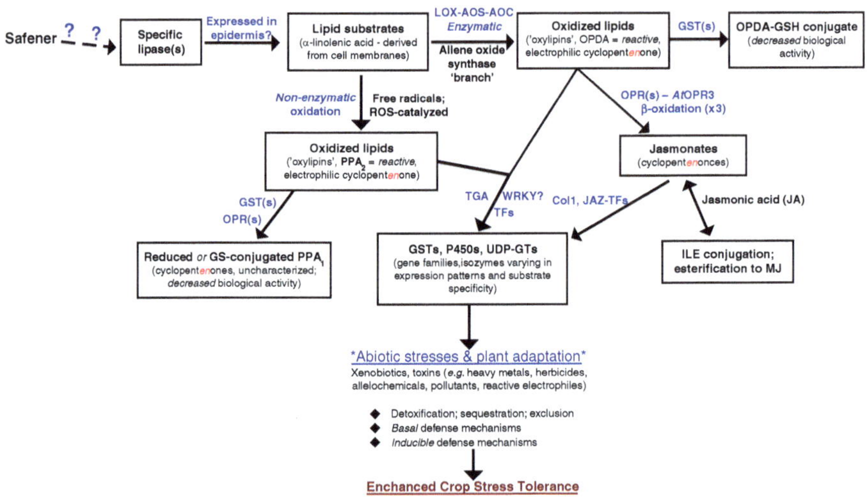


**Fig. 1** Proposed events leading to safener-enhanced abiotic stress tolerance in cereal crops

research regarding safeners, herbicide detoxification, and tolerance to other forms of abiotic stress (described below).

# 7 Similarities Between Al and Herbicide Tolerance Mechanisms in Grain Sorghum and Maize

Specific biochemical and molecular-genetic mechanisms to overcome aluminum toxicity at low soil pH (<5) in sorghum and maize roots involve aluminum-activated citrate extrusion into the rhizosphere (Magalhaes et al. 2007, and references cited therein; Maron et al. 2013). Recent cytochemical research has shown that expression of *SbMATE* (multidrug and toxic compound extrusion) genes in root tips, specifically in the root distal transition zone (DTZ) (Sivaguru et al. 2013), is associated with aluminum tolerance in sorghum. Within the root DTZ, aluminum-induced ROS production and *SbMATE* expression were localized primarily to the epidermal and outer cortex cell layers of an aluminum-resistant sorghum line (Sivaguru et al. 2013). Recently, aluminum tolerance in maize was shown to associate with copy number variation (CNV) in the *ZmMATE1* gene resulting from a tandem triplication in aluminum-tolerant maize lines (Maron et al. 2013). The cell- and tissue-specific expression pattern of *SbMATE* genes in aluminum-treated sorghum roots (Sivaguru et al. 2013) is very similar to our previous findings in *Aegilops tauschii* (synonymous with *Triticum tauschii*) coleoptiles in which *TtGST* transcripts, proteins, and enzymatic activities are localized to the outer cell layers of safener-treated *Ae. tauschii* shoots (Xu et al. 2002; Riechers et al. 2003). Moreover, recent research (Sivaguru, Riechers, and Lygin; unpublished results) has

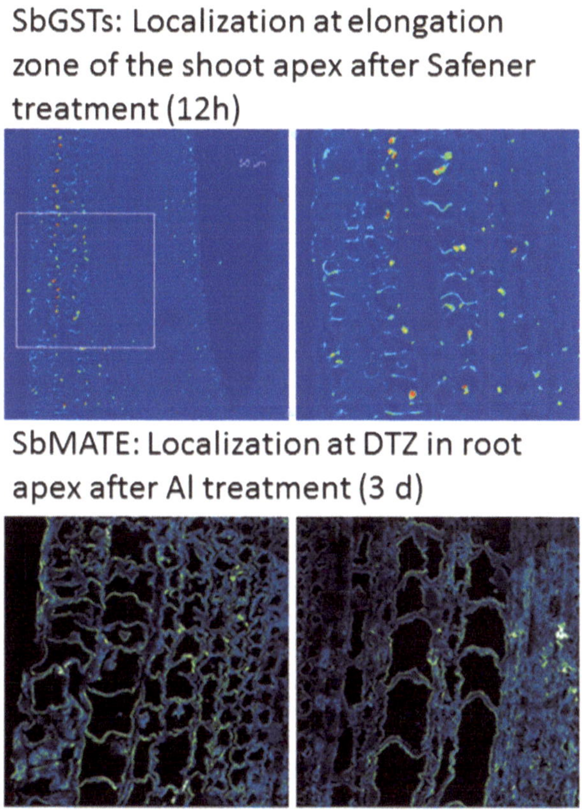


**Fig. 2** Localization of *Sb*GSTs at elongation zone of the shoot apex, 12 h after safener treatment (*top panel*), and *Sb*MATE at DTZ in the root apex, 3 days after Al treatment (*bottom panel*)

shown a very similar labeling pattern of safener-induced GST proteins in grain sorghum shoot sections [Fig. 2; using the same wheat GST antiserum as used previously in *Ae. tauschii* shoots (Riechers et al. 2003)]. By comparison, results of MATE localization patterns in root apices from aluminum-tolerant maize and sorghum lines, combined with GST localization in safener-treated wheat and sorghum shoot apices, indicate that highly conserved signaling and detoxification pathways are present in cereal crop seedling tissues. Future research using protein localization methods and RNAseq expression profiling in safener-treated sorghum coleoptile tissues will investigate the hypothesis that safeners may protect against abiotic stresses other than herbicides (Fig. 1), including aluminum, heavy metals, or other ROS-generating compounds, utilizing similar or parallel signaling pathways for aluminum tolerance and plant defense (Kong et al. 2014).

## 8 Potential Biochemical and Molecular Mechanisms That Confer Herbicide Tolerance in Cereals

We hypothesize that duplication of xenobiotic-detoxifying GST genes within the sorghum genome may confer natural herbicide tolerance in tolerant genotypes due to increased GST copy number and basal expression levels, as well as higher tolerance levels following safener treatment due to the combination of increased GST copy number and transcriptional activation of these genes. Our hypothesis is based on an analogous molecular-genetic mechanism involving *ZmMATE* copy number variation that confers natural tolerance to aluminum toxicity in maize varieties originating from tropical areas with acidic soils (Maron et al. 2013). Although the hypothesis of GST copy number variation (CNV) has not been tested previously as a possible mechanism for herbicide tolerance in cereal crops, sorghum herbicide tolerance (either natural or safener-enhanced) demonstrates many similarities with mechanisms for aluminum tolerance in sorghum roots (Sivaguru et al. 2013; Maron et al. 2013). Of particular interest in our current research are the tolerant (T) and sensitive (S) phenotypic classes that possess either increased or decreased relative GST activity with a chloroacetamide herbicide substrate. Since some safener-induced *SbGSTs* are derived from gene clusters located within the sorghum reference genome (BTx623—a safener-responsive sorghum genotype), both T and S phenotypes might be explained by GST CNVs.

Another possible mechanism that could account for the massive induction of GST transcripts, protein, and enzymatic activity following safener treatment (Riechers et al. 1997, 2003) is endo-reduplication, or DNA replication without cell division (Larkins et al. 2001). This genetic mechanism leads to multiple genomes per cell and is typically found in cells with terminal growth and development, such as the shoot epidermis or seed endosperm (Larkins et al. 2001; Lee et al. 2009). Endo-reduplication thus confers the genetic equivalent of polyploidy to certain cells and tissues, potentially in response to various physiological stresses (Lee et al. 2009), and might help explain the dramatic transcriptional responses to safener treatment observed with initial RNAseq analyses in dissected sorghum coleoptiles (Riechers, Lygin, Brown, and Moose, unpublished).

## 9 Plant GSTs Play Important Roles in Diverse Stress Responses, Adaptation, and Tolerance

GST expression in plants is induced following exposure to many environmental stresses (Wagner et al. 2002; Frova 2003), including heat shock, cold, high salt, and UV light exposure; biotic stresses such as pathogen attack and fungal elicitors; abiotic stresses such as heavy metals, herbicides, and safeners; and phytohormone treatments such as ethylene, auxins, abscisic acid, methyl jasmonate, and salicylic acid (reviewed by Riechers et al. 2010; Cummins et al. 2011). Induction of GST expression by so many diverse stimuli indicates that plant GSTs are critical for plant responses and tolerance to stresses, either by directly participating in the signal transduction process and/or via detoxification of harmful compounds produced in response to or as a result of a given stress (Cummins et al. 2011; Diao et al. 2010). Alternatively, plant GSTs may function indirectly through playing a central role related to antioxidant function. GST enzymatic activities can involve direct GSH conjugation to toxic electrophilic molecules (Farmer and Davoine 2007) via nucleophilic substitution reactions (such as with numerous herbicide substrates; Riechers et al. 2010; Cummins et al. 2011), or via GSH-dependent peroxidase activities, using GSH as reductant for the detoxification of ROS, radicals, and lipid peroxides formed during or after exposure to various stress agents (Edwards and Dixon 2005).

Certain GST isozymes confer herbicide (Davies and Caseley 1999) or Al tolerance (Ezaki et al. 2000) in plants, but possibly via different detoxification mechanisms. As mentioned above, GSTs are well known for their enzymatic activities in rapidly detoxifying herbicides via direct conjugation with GSH (Kreuz et al. 1996; Riechers et al. 2010; Cummins et al. 2011). However, GST involvement in ameliorating Al stress may involve a role associated with antioxidant function following exposure to Al under acidic conditions. For example, several Al-inducible genes have been identified in *Arabidopsis* plants and cultured tobacco cells, including GSTs and other genes associated with oxidative stress responses (Ezaki et al. 2000). In the root tips of transgenic *Arabidopsis* overexpressing Al-inducible genes, Al-induced callose deposition decreased markedly. Moreover, expression of a tobacco GST (*parB*) and tobacco anionic peroxidase in transgenic *Arabidopsis* lines decreased the oxidative damage caused by Al treatment (Ezaki et al. 2000, 2001), including a significant decrease in levels of Al-induced lipid peroxidation (Ezaki et al. 2001). Additionally, a GST (named *GST27.2*) was isolated from the root tips of Al-treated maize plants (Concado et al. 2005) via differential display of mRNAs. This transcript accumulated during a time course following Al and cadmium treatment in roots of both Al-sensitive and Al-tolerant maize lines (Concado et al. 2005). Interestingly, this same maize GST gene (originally termed *GST-27* or *GST IV*) was biochemically characterized and cloned from safener-treated maize shoots (Irzyk and Fuerst 1993, 1997; Jepson et al. 1994) and determined to utilize the chloroacetamide herbicide metolachlor as substrate via GSH conjugation (Irzyk and Fuerst 1997). Subsequent research with transgenic tobacco plants that constitutively over-expressed the maize *GST IV* (or maize *GST-27*) gene displayed tolerance to soil-applied rates of metolachlor that would typically injure sensitive tobacco plants (Jepson et al. 1997). Comparison of the maize *GST27.2* and maize *GST-27* gene coding regions showed 99 % nucleotide and amino acid sequence identities, with differences in only two amino acids, suggesting that these proteins may be encoded by a recent gene duplication in the maize genome or that they are allelic at a single *GST27* locus (Concado et al. 2005). Whether or not the amino acid substitutions identified between GST27.2 and GST-27 affect enzymatic activities, specificity with various substrates

(Concado et al. 2005), subcellular localization, gene expression patterns, or tolerance to abiotic stresses (Fig. 1) in maize seedling tissues remains to be determined.

## 10 Summary, Conclusions, and Future Research Directions

Al-resistance mechanisms are still unfolding at the molecular-genetic and biochemical levels, and mechanisms to manipulate plant tolerance or resistance at these levels offer promising areas for further crop improvement under stressful growing conditions. The safener-enhanced expression of GSTs in grain sorghum shoots, which shows a remarkably similar localization pattern as SbMATE in Al-treated roots (Fig. 2), offers an exciting new area for Al signaling research. For example, if a safener pretreatment (applied as a seed coating; Davies and Caseley 1999) could provide Al resistance or tolerance in cereal crops growing in acidic soils, then several new and exciting research areas for investigating xenobiotic and Al signaling and detoxification mechanisms would be uncovered. Moreover, it would be of great interest for plant stress physiology researchers to utilize a safener to upregulate various molecular aspects associated with Al signaling and resistance. The authors hope that the field of chemically enhanced abiotic stress resistance via safener application is explored further to bring new and novel mechanisms of enhancing Al and abiotic stress resistance to crop plants.

## References

Achary VMM, Parinandi NL, Panda BB (2012) Aluminum induces oxidative burst, cell wall NADH-peroxidase activity, and DNA damage in root cell of *Allium cepa* L. Environ Molec Mutagen 53:550–560

Ahad A, Nick P (2007) Actin is bundled in activation-tagged tobacco mutants that tolerate aluminum. Planta 225:451–468

Ahn SJ, Matsumoto H (2006) The role of the plasma membrane in the response of plant roots to aluminum toxicity. Plant Signal Behav 1–2:37–45

Ahn SJ, Rengel Z, Matsumoto H (2004) Aluminum-induced plasma membrane surface potential and $H^+$-ATPase activity in near-isogenic wheat lines differing in tolerance to aluminum. New Phytol 162:71–79

Ahn SJ, Sivaguru M, Osawa M, Cheng GC, Matsumoto H (2001) Aluminum inhibits the $H^+$-ATPase activity by permanently altering the plasma membrane surface potential in squash roots. Plant Physiol 126:1381–1390

Akeson MA, Munns DN, Burau RG (1989) Adsorption of $Al^{3+}$ to phosphatidylcholine vesicles. Biochem Biophys Acta 986:33–40

Almeras E, Stolz S, Vollenweider S, Reymond P, Mene-Saffrane L, Farmer EE (2003) Reactive electrophile species activate defense gene expression in *Arabidopsis*. Plant J 34:205–216

Amenós M, Corrales I, Poschenrieder C, Illeš P, Baluška F, Barceló J (2009) Different effects of aluminum on the actin cytoskeleton and brefeldin A-sensitive vesicle recycling in root apex cells of two maize varieties differing in root elongation rate and aluminum tolerance. Plant Cell Physiol 50:528–540

Arroyo-Serralta GA, Ku-González A, Hernández-Sotomayor SMT, Aguilar JJZ (2005) Exposure to toxic concentrations of aluminum activates a MAPK-like protein in cell suspension cultures of *Coffea arabica*. Plant Physiol Biochem 43:27–35

Baluška F, Hlavacka A, Saniaj J, Palme K, Robinson G, Matoh T, McCurdy DW, Menzel D, Volkmann D (2002) F-actin-dependent endocytosis of cell wall pectins in meristematic root cells. Insights from brefeldin A-induced compartments. Plant Physiol 130:422–431

Baluška F, Samaj J, Wojtaszek P, Volkann O, Menzel D (2003) Cytoskeleton-plasma membrane-cell wall continuum in plants. Plant Physiol 133:482–491

Baxter A, Mittler R, Suzuki N (2014) ROS as key players in plant stress signaling. J Exp Bot 65:1229–1240

Behringer C, Bartsch K, Schaller A (2011) Safeners recruit multiple signalling pathways for the orchestrated induction of the cellular xenobiotic detoxification machinery in *Arabidopsis*. Plant Cell Environ 34:1970–1985

Bennet RJ, Breen CM (1991) The aluminium signal: new dimensions to mechanisms of aluminium tolerance. Plant Soil 134:153–166

Bonaventure G, Schuck S, Baldwin IT (2011) Revealing complexity and specificity in the activation of lipase-mediated oxylipin biosynthesis: a specific role of the *Nicotiana attenuata* GLA1 lipase in the activation of jasmonic acid biosynthesis in leaves and roots. Plant Cell Environ 34:1507–1520

Brault M, Amiar Z, Pennarun A-M, Monetiez M, Zhang Z, Cornel D, Dellis O, Knight H, Bouteau F, Rona J-P (2004) Plasma membrane depolarization induced by abscisic acid in *Arabidopsis* suspension cells involves reduction of proton pumping in addition to anion channel activation, which are both $Ca^{2+}$ dependent. Plant Physiol 135:231–243

Cakmak I, Horst WJ (1991) Effect of aluminium on lipid peroxidation, superoxide dismutase, catalase, and peroxidase activities in root tips of soybean (*Glycine max*). Physiol Plant 83: 463–468

Chaffai R, Marzouk B, Ferjani E (2005) Aluminum mediates compositional alterations of polar lipid classes in maize seedlings. Phytochem 66:1903–1912

Chee-Gonzalez I, Munoz-Sanchez JA, Racagri-Di Palma G (2009) Effect of phosphate on aluminum-inhibited growth and signal transduction pathway in *Coffea arabica* suspension cells. J Inorg Biochem 103:1497–1503

Chen Q, Guo C-L, Wang P, Chen X-Q, Wu K-H, Li K-Z, Y-X Y, Chen L-M (2013) Up-regulation and interaction of the plasma membrane H+-ATPaseand the 14-3-3 protein are involved in the regulation of citrate exudation from the broad bean (*Vicia faba* L.) under Al stress. Plant Physiol Biochem 10:504–511

Christeller JT, Galis I (2014) alpha-Linolenic acid concentration and not wounding *per se* is the key regulator of octadecanoid (oxylipin) pathway activity in rice (*Oryza sativa* L.) leaves. Plant Physiol Biochem 83:117–125

Čizková R (1995) Phytohormonal levels in spruce roots under aluminum stress. In: Baluška F et al (eds) Structure and function of roots. Kluwer, The Netherlands, p 335–339

Clarkson DT (1965) The effect of aluminium and some other trivalent metal cations on cell division in the root apices of *Allium cepa*. Ann Bot 29:309–315

Concado GMA, De Rosa Jr VE, Fernandez JH, Maron LG, Jorge RA, Menossi M (2005) Glutathione *S*-transferase and aluminum toxicity. Functional Plant Biol 32:1045–1055

Cummins I, Dixon DP, Freitag-Pohl S, Skipsey M, Edwards R (2011) Multiple roles for plant glutathione transferases in xenobiotic detoxification. Drug Metab Rev 43:266–280

da Silva ALS, Sperling P, Horst W, Franke S, Ott C, Becker D, Stass A, Lorz H, Heinz W (2006) A possible role of sphingolipids in the aluminium resistance of yeast and maize. J Plant Physiol 163:26–38
Dave A, Graham IA (2012) Oxylipin signaling: a distinct role for the jasmonic acid precursor cis-(+)-12-oxo-phytodienoic acid (cis-OPDA). Front Plant Sci 3:42
Devi SR, Yamamoto Y, Matsumoto H (2003) An intracellular mechanism of aluminum tolerance associated with antioxidant status in cultured tobacco cells. Inorg Biochem 97:59–68
Davies J, Caseley JC (1999) Herbicide safeners: a review. Pestic Sci 55:1043–1058
Doncheva S, Amenos M, Poschenrieder C, Barcelo J (2005) Root cell patterning: a primary target for aluminium toxicity in maize. J Exp Bot 56:1213–1220
Diao G, Wang Y, Yang C (2010) Functional characterization of a glutathione *S*-transferase gene from *Limonium bicolor* in response to several abiotic stresses. Afric J Biotechnol 32: 5060–5065
Dixon DP, Skipsey M, Edwards R (2010) Roles for glutathione transferases in plant secondary metabolism. Phytochemistry 71:338–350
Dueckershoff K, Mueller S, Mueller MJ, Reinders J (2008) Impact of cyclopentenone-oxylipins on the proteome of *Arabidopsis thaliana*. Biochim Biophys Acta 1784:1975–1985
Eticha D, Stass A, Horst WJ (2005) Cell-wall pectin and its degree of methylation in the maize root apex: significance for genotypic differences in aluminum resistance. Plant Cell Envir 28: 1410–1420
Edwards R, Dixon DP (2005) Plant glutathione transferases. Methods Enzymol 401:169–186
Ezaki B, Gardner RC, Ezaki Y, Matsumoto H (2000) Expression of aluminum-induced genes in transgenic *Arabidopsis* plants can ameliorate aluminum stress and/or oxidative stress. Plant Physiol 122:657–665
Ezaki B, Katsuhara M, Kawamura M, Matsumoto H (2001) Different mechanisms of four aluminum (Al)-resistant transgenes for Al toxicity in *Arabidopsis*. Plant Physiol 127:918–927
Farmer EE, Davoine C (2007) Reactive electrophile species. Curr Opin Plant Biol 10:380–386
Fiskesjo G (1983) Nucleolar dissolution induced by aluminum in root cells of *Allium*. Physiol Plant 59:508–511
Frantzios G, Galati B, Apostolakos P (2005) Aluminum causes variable responses in actin filament cytoskeleton of the root tip cell of *Triticum furigidum*. Protoplasma 225:129–140
Frova C (2003) The plant glutathione transferase gene family: genomic structure, functions, expression, and evolution. Physiol Plant 119:469–479
Furuichi J, Sasaki T, Tsuchiya Y, Ryan PR, Delhaize E, Yamamoto Y (2010) An extracellular hydrophilic carboxyl-terminal domain regulates the activity of TaALMT1, the aluminum-actiavated malate transport protein of wheat. Plant J 64:47–55
Grabski S, Schindler M (1995) Aluminum induces rigor within the actin network of soybean cells. Plant Physiol 108:897–901
Grebner W, Stingl NE, Oenel A, Mueller MJ, Berger S (2013) Lipoxygenase6-dependent oxylipin synthesis in roots is required for abiotic and biotic stress resistance of *Arabidopsis*. Plant Physiol 161:2159–2170
Gutteridge JMC, Quinlan GJ, Clark I, Halliwell B (1985) Aluminium salts accerelate peroxidation of membrane lipids stimulated by iron salts. Biochem Biophys Acta 835:441–447
Hasenstein KH, Evans ML (1988) Effects of cations on hormone transport in primary roots of *Zea mays*. Plant Physiol 86:890–894
Haug A, Shi B, Vitorello V (1994) Aluminum interaction with phosphoinositide-associated signal transduction. Arch Toxicol 68:1–7
He H, He L, Gu M (2012a) Interactions between nitric oxide and plant hormones in aluminum tolerance. Plant Signal Behav 7:469–471
He H, Zhan J, He L, Gu M (2012b) Nitric oxide signaling in aluminum stress in plants. Protoplasma 249:483–492
Horst WJ (1995) The role of the apoplast in aluminum toxicity and resistance of higher plants. Z Pflanzennernahr Bodenkd 158:419–428

Horst WJ, Schmohl N, Kollemeier M, Baluška F, Sivaguru M (1999) Does aluminum affect root growth of maize through interaction with the cell wall-plasma membrane-cytoskeleton continuum? Plant Soil 215:163–174

Hou N, Yan J, Pang J, Xu M, Chen G, Yang Z (2010) The accumulation and transport of abscisic acid in soybean (*Glycine max* L.) under aluminum stress. Plant Soil 330:127–137

Illes P, Schlicht M, Pavlovkin J, Lichtschidl I, Baluška F, Ovecka M (2006) Aluminium toxicity in plants: internalization of aluminium into cells of the transition zone in Arabidopsis root apices related to changes in plasma membrane potential, endosomal behavior, and nitric oxide production. J Exp Bot 57:4201–4213

Irzyk GP, Fuerst EP (1993) Purification and characterization of a glutathione *S*-transferase from benoxacor-treated maize (*Zea mays*). Plant Physiol 102:803–810

Irzyk GP, Fuerst EP (1997) Characterization and induction of maize glutathione *S*-transferases involved in herbicide detoxification. In: Hatzios KK (ed) Regulation of enzymatic systems detoxifying xenobiotics in plants. Kluwer, Dordrecht, pp 155–170

Ishikawa S, Wagatsuma T (1998) Plasma membrane permeability of root-tip cells following temporary exposure to Al ions is a rapid measure of Al tolerance among plant species. Plant Cell Physiol 39:516–525

Javelle M, Vernoud V, Rogowsky PM, Ingram GC (2011) Epidermis: the formation and functions of a fundamental plant tissue. New Phytol 189:17–39

Jepson I, Lay VJ, Holt DC, Bright SWJ, Greenland AJ (1994) Cloning and characterization of maize herbicide safener-induced cDNAs encoding subunits of glutathione *S*-transferase isoforms I, II and IV. Plant Mol Biol 26:1855–1866

Jepson I, Holt DC, Roussel V, Wright SY, Greenland AJ (1997) Transgenic plant analysis as a tool for the study of maize glutathione *S*-transferases. In: Hatzios KK (ed) Regulation of enzymatic systems detoxifying xenobiotics in plants. Kluwer, Dordrecht, pp 313–323

Jones DL, Blancaflor EB, Kochian LV, Gilroy S (2006) Spatial coordination of aluminium uptake, production of reactive oxygen species, callose production and wall rigidification in maize roots. Plant Cell Envirom 29:1309–1318

Jones DL, Gilroy S, Larsen PB, Howell SH, Kochian LV (1998) Effect of aluminum on cytoplasmic $Ca^{2+}$ homeostasis in root hairs of *Arabidopsis thaliana* (L.). Planta 206:378–387

Jones DL, Kochian LV (1995) Aluminum inhibition of the inositol 1,4,5-triphosphate signal transduction pathway in wheat roots: a role in aluminum toxicity? Plant Cell 7:1913–1921

Kasai M, Sasaki M, Tanakamaru S, Yamamoto Y, Matsumoto H (1993) Possible involvement of abscisic acid in activities of two vacuular $H^+$-pumps in barley roots under aluminum stress. Plant Cell Physiol 34:1335–1338

Kenjebaeva S, Yamamoto Y, Matsumoto H (2000) The impact of aluminum on the distribution of cell wall glycoproteins of pea root tip and their Al-binding capacity. Soil Sci Plant Nutr 47: 629–636

Kinraide TB, Yemiyahu U (2007) A scale of metal ion binding strengths correlating with ionic charge, Pauling electronegativity, toxicity, and other physiological effects. Inorg Biochem 101:1201–1213

Klug B, Horst WJ (2010) Spatial characteristics of aluminum uptake and translocation in roots of buckwheat (*Fagopyrum esculentum*). Physiol Plant 139:181–191

Kobayashi Y, Hoekenga OA, Itoh H, Nakashima M, Saito S, Shoff JE, Maron LG, Pineros MA, Kochian LV, Koyama H (2007) Characterization of *AtALMT1* expression in aluminum-inducible malate release and its role for rhizotoxic stress tolerance in *Arabidopsis*. Plant Physiol 145:843–852

Kollmeier M, Felle HH, Horst WJ (2000) Genotypical differences in aluminum resistance of maize are expressed in the distal part of the transition zone. Is reduced basipetal auxin flow involved in inhibition of root elongation by aluminum? Plant Physiol 122:945–956

Kong X, Zhang M, Xu X, Li X, Li C, Ding Z (2014) System analysis of microRNAs in the development and aluminum stress responses of the maize root system. Plant Biotechnol J 12: 1108–1121

Kopittke PM, Blamey FPC, Menzies NW (2008) Toxicities of soluble Al, Cu, and La include ruptures to rhizodermal and root critical cells of cowpea. Plant Soil 303:217–227
Kreuz K, Tommasini R, Martinoia E (1996) Old enzymes for a new job. Herbicide detoxification in plants. Plant Physiol 111:349–353
Larkins BA, Dilkes DP, Dante RA, Coelho CM, Woo Y-M, Liu Y (2001) Investigating the hows and whys of DNA endoreduplication. J Exp Bot 52:183–192
Lazof DB, Goldsmith JG, Linton RW (1997) The *in situ* analysis of intracellular aluminum in plants. Prog Bot 58:112–149
Lee HO, Davidson JM, Duronio RJ (2009) Endoreplication: polyploidy with purpose. Genes Dev 23:2461–2477
Li J-Y, Liu J, Dong D, Jia X, MxCouch SR, Kochian LV (2014) Natural variation underlies alterations in Nramp aluminum transporter (*NRAT1*) expression and function that plays a key role in rice aluminum tolerance. Proc Natl Acad Sci U S A 111:6503–6508
Ligaba A, Dreyer I, Margaryan A, Schneider DJ, Kochian L, Pineros M (2013) Functional, structural and phylogenetic analysis of domains underlying the Al sensitivity of the aluminum-activated malate/anion transporter TaALMT1. Plant J 76:766–780
Ligava A, Kochian LV, Pineros M (2009) Phosphorylation at S384 regulates the activity of the TaALMT1 malate transporter that underlies aluminum resistance in wheat. Plant J 60:411–423
Lin C, Yu Y, Kadono T, Iwata M, Umemura K, Furuichi T, Kase M, Isobe M, Yamamoto Y, Matasumoto H, Yoshizuka K, Kawano T (2005) Action of aluminum, novel TPC1-type channel inhibitor, against silicylate-inducd and cold-shock-induced -calcium influx in tobacco - BY-2 cells. Biochem Biophys Res Commun 332:823–832
Liu J, Pineros MA, Kochian LV (2014) The role of aluminum sensing and signaling in plant aluminum resistance. J Integrative Plant Biol 56:221–230
Llugany M, Poschenrieder L, Barcelo J (1995) Monitoring of aluminium-induced inhibition of root elongation in four maize cultivars differing in tolerance to aluminium and proton toxicity. Physiol Plant 93:265–271
Loeffler C, Berger S, Guy A, Durand T, Bringmann G, Dreyer M, von Rad U, Durner J, Mueller MJ (2005) $B_1$-phytoprostanes trigger plant defense and detoxification responses. Plant Physiol 137:328–340
Ma JF, Ryan PR, Delhaize E (2001) Aluminum tolerance in plants and the complexity of organic acids. Trends Plant Sci 6:273–278
Magalhaes JV, Liu J, Guimaraes CT, Lana UGP, Alves VMC, Wang Y-H, Schaffert RE, Hoekenga OH, Pineros MA, Shaff JE, Klein PE, Carneiro NP, Coelho CM, Trick HN, Kochian LV (2007) A gene in the multidrug and toxic compound extrusion (*MATE*) family confers aluminum tolerance in sorghum. Nat Genet 39:1156–1161
Maron LG, Guimaraes CT, Kirst M, Albert PS, Birchler JA, Bradburya PJ, Buckler ES, Coluccio AE, Danilova TV, Kudrna D, Magalhaes JV, Pineros MA, Schatz MC, Wing RA, Kochian LV (2013) Aluminum tolerance in maize is associated with higher *MATE1* gene copy number. Proc Natl Acad Sci U S A 110:5241–5246
Martinez-Estévez M, Loyola-Vargas VM, Hervández-Sotomayor SMT (2001) Aluminum increase phosphorylation of particular proteins in cellular suspension cultures of coffee (*Coffea arabica*). J Plant Physiol 158:1375–1379
Massot N, Nicander B, Barcelo J, Poschenrieder C, Tillberg E (2002) A rapid increase in cytokinin level, and enhanced ethylene evolution precede $Al^{3+}$-induced inhibition of root growth in bean seedlings (*Phaseolus vulgaris* L.). Plant Growth Regul 37:105–112
Matsumoto H (1988) Changes of the structure of pea chromatin by aluminum. Plant Cell Physiol 29:281–287
Matsumoto H (2000) Cell biology of aluminum toxicity and tolerance in higher plants. Int Rev Cytol 200:1–46
Matsumoto H, Hirasawa E, Torikai H, Takahashi E (1976) Localization of absorbed aluminum in pea root and its binding to nucleic acids. Plant Cell Physiol 17:127–137

Matsumoto H, Morimura S, Takahashi E (1977) Binding of aluminum to DNP in pea root nuclei. Plant Cell Physiol 18:987–993

Matsumoto H, Motoda H (2012) Aluminum toxicity recovery process in root apices. Possible association with oxidative stress. Plant Sci 185–186:1–8

Matsumoto H, Motoda H (2013) Oxidative stress is associated with aluminum toxicity recovery in apex of pea root. Plant Soil 363:399–410

Matsumoto H, Sivaguru M (2008) Advances in the aluminum toxicity and tolerance of plants for increased productivity in acid soil. In: Debios AN (ed) Soil contamination: new research. Nova Science, New York, NY, pp 1–42

Matsumoto H, Yamamoto Y (2013) Plant roots under aluminum stress: toxicity and tolerance. In: Eshel A, Beckman T (eds) Plant roots: the hidden half, 4th edn. CRC Press/Taylor & Francis Group, Boca Raton, FL, 33-1

Mimmo T, Marzadori C, Gessa CE (2009) Does the degree of pectin esterification influence aluminum sorption by the root apoplast? Plant Soil 314:159–168

Mohanty S, Das AB, Da P, Mohanty P (2004) Effect of a low dose of aluminum on mitotic and meiotic activity, 4C DNA content, and pollen sterility in rice, *Oryza sativa* L. cv. Lalat. Ecotoxicol Environ Saf 59:70–75

Mosblech A, Feussner I, Heilmann I (2009) Oxylipins: structurally diverse metabolites from fatty acid oxidation. Plant Physiol Biochem 47:511–517

Motoda H, Kano Y, Hiragami F, Kawamura K, Matsumoto H (2010) Changes in the apex of pea roots during and after recovery from aluminum treatment. Plant Soil 333:49–58

Mueller MJ (2004) Archetype signals in plants: the phytoprostanes. Curr Opin Plant Biol 7: 441–448

Mueller MJ, Berger S (2009) Reactive electrophilic oxylipins: pattern recognition and signalling. Phytochemistry 70:1511–1521

Mueller S, Hilbert B, Dueckershoff K, Roitsch T, Krischke M, Mueller MJ, Berger S (2008) General detoxification and stress responses are mediated by oxidized lipids through TGA transcription factors in Arabidopsis. Plant Cell 20:768–785

Naidoo G, Stewart J, Med LRJ (1978) Accumulation sites of Al in snapbean and cotton roots. Agron J 70:489–492

Naora H, Naora H, Mirsky AE, Allfrey VG (1961) Magnesium and calcium in isolated cell nuclei. J Gen Physiol 44:713–741

Negishi T, Oshima K, Hattori M, Kanai M, Mano S, Nishimura H, Yoshida R (2012) Tonoplast- and plasma membrane-localized aquaporin-family transporters in blue hydrangea sepals of aluminum hyperaccumulating plant. PLoS One 7:e4389

Nezames CD, Sjogren CA, Barajas JF, Larsen PB (2012) The *Arabidopsis* cell cycle checkpoint regulators TANMEI/ALT2 and ATR mediates the active process of aluminum-dependent root growth inhibition. Plant Cell 24:608–621

Novascués J, Pérez-Rontomé C, Sánchez DH, Staudinger C, Wienkoop S, Rellan-Alvarez R, Becana M (2012) Oxidative stress is a consequence, not a cause, of aluminum toxicity in the forage legume *Lotus csrniculatus*. New Phytol 193:625–630

Olivetti GP, Gumming JR, Etherton B (1995) Membrane potential depolarization of root cap cells precedes aluminum tolerance in snap bean. Plant Physiol 109:123–129

Okazaki Y, Saito K (2014) Roles of lipids as signaling molecules and mitigators during stress response in plants. Plant J 79:584–596

Ono K, Yamamoto Y, Haxchiya A, Matsumoto H (1995) Synergistic inhibition of growth by aluminum and iron of tobacco (*Nicotiana tobacum* L.) cells in suspension culture. Plant Cell Physiol 36:115–125

Osawa H, Matsumoto H (2001) Possible involvement of protein phosphorylation in aluminum-responsive malate efflux from wheat root apex. Plant Physiol 126:411–420

Pan J-W, Zhu M-Y, Chen H (2001) Aluminum-induced cell death in root tip cells of barley. Environ Exp Bot 46:71–79
Panda SK, Baluška F, Matsumoto H (2009) Aluminum stress signaling in plants. Plant Signaling Behav 4:592–597
Papernick LA, Kochian LV (1997) Possible involvement of Al-induced electrical signal in Al tolerance in wheat. Plant Physiol 115:657–667
Pejchar P, Potocky M, Novotná Z, Veselková S, Kocourková D, Valentová O, Schwarzerova K, Martinec J (2010) Aluminium ions inhibit the formation of diacylglycerol generated by phosphatidylcholin-hydrolysing phospholipase C in tobacco cells. New Phytol 188:150–160
Pineros M, Tester M (1993) Plasma membrane $Ca^{2+}$ channels in roots of higher plants and their role in aluminium toxicity. Plant Soil 155(156):119–122
Qin R, Jiang WS, Liu DH (2013) Aluminum can induce alteration in the cellular localization and expression of three major nucleolar proteins in root tip cells of *Allium cepa* var. *agrogarum* L. Chemosphere 90:827–834
Ramel F, Sulmon C, Serra A-A, Gouesbet G, Couee I (2012) Xenobiotic sensing and signaling in higher plants. J Exp Bot 63:3999–4014
Rengel Z (1992a) Distribution of cell $Ca^{2+}$ homeostasis as a primary target of Al toxicity syndrome. Plant Cell Environ 15:931–938
Rengel Z (1992b) Role of calcium in aluminium toxicity. New Phytol 121:499–513
Rengel Z, Zhang WG (2003) Role of dynamics of intracellular calcium in aluminium-toxicity syndrome. New Phytol 59:295–314
Riechers DE, Irzyk GP, Jones SS, Fuerst EP (1997) Partial characterization of glutathione *S*-transferases from wheat (*Triticum* spp.) and purification of a safener-induced glutathione *S*-transferase from *Triticum tauschii*. Plant Physiol 114:1461–1470
Riechers DE, Zhang Q, Xu FX, Vaughn KC (2003) Tissue-specific expression and localization of safener-induced glutathione *S*-transferase proteins in *Triticum tauschii*. Planta 217:831–840
Riechers DE, Kreuz K, Zhang Q (2010) Detoxification without intoxication: herbicide safeners activate plant defense gene expression. Plant Physiol 153:3–13
Rincón-Zachary M, Teaster ND, Sparks JA, Valster AH, Motes CM, Blancaflor EB (2010) Fluorescence resonance energy transfer-sensitized emission of yellow cameleon 3.60 reveals root zone specific calcium signatures in *Arabidopsis* in response to aluminum and other trivalent cations. Plant Physiol 152:1442–1458
Rishi A, Muni S, Kapur V, Nelson ND, Goyal A (2004) Identification and analysis of safener-inducible expressed sequence tags in *Populus* using a cDNA microarray. Planta 220:296–306
Ryan PR, Reid RJ, Smitth FA (1997) Direct evaluation of the $Ca^{2+}$-displacement hypothesis for Al toxicity. Plant Physiol 113:1351–1357
Sasaki M, Yamamoto Y, Matsumoto H (1996) Lignin deposition induced by aluminum in wheat (*Triticum aestivum*) roots. Physiol Plant 96:193–198
Sasaki M, Yamamoto Y, Matsumoto H (1997) Aluminum inhibits growth and stability of cortical microtubules in wheat (*Triticum aestivum*) roots. Soil Sci Plant Nutr 43:469–472
Savchenko T, Kolla VA, Wang C-Q, Nasafi Z, Hicks DR, Phadungchob B, Chehab WE, Brandizzi F, Froehlich J, Dehesh K (2014) Functional convergence of oxylipin and abscisic acid pathways controls stomatal closure in response to drought. Plant Physiol 164:1151–1160
Schaller A, Stintzi A (2009) Enzymes in jasmonate biosynthesis: structure, function, regulation. Phytochemistry 70:1532–1538
Schmohl N, Horst WJ (2000) Cell wall pectin content modulates aluminum sensitivity of *Zea mays* (L.) cells grown in suspension culture. Plant Cell Envir 23:735–742
Schofield RMS, Pallon J, Fiskesjo G, Karlsson G, Malmqvist KG (1998) Aluminum and calcium distribution patterns in aluminum-intoxicated roots of *Allium cepa* do not support the calcium-displacement hypothesis and indicate signal-mediated inhibition of root growth. Planta 205:175–180

Schuck S, Kallenbach M, Baldwin IT, Bonaventure G (2014) The *Nicotiana attenuata* GLA1 lipase controls the accumulation of *Phytophthora parasitica*-induced oxylipins and defensive secondary metabolites. Plant Cell Environ 37:1703–1715

Schwarzerová K, Zelenková S, Nick P, Opatrny Z (2002) Aluminum-induced rapid changes in the microtubular cytoskeleton of tobacco cell lines. Plant Cell Physiol 43:207–216

Seju K, Lee Y (1998) Aluminum induces changes in the orientation of microtubules and the division plane in root meristem of *Zea mays*. J Plant Biol 41:269–276

Shen H, He LF, Sasaki T, Yamamoto Y, Zheng SJ, Ligaba A, Yan XL, Ahn SJ, Yamaguchi M, Sasakawa H, Matsumoto H (2005) Citrate secretion coupled with the modulation of soybean root tip under aluminum stress: up-regulation of transcription, translation, and threonine-oriented phosphorylation of plasma membrane H+-ATPase. Plant Physiol 138:287–296

Shen H, Ligaba A, Yamaguchi M, Osawa H, Shibata H, Yan X, Matsumoto H (2004) Effect of K-252a and abscisic acid on the efflux of citrate from soybean roots. J Exp Bot 55:663–671

Simoes CC, Melo JO, Magalhaes JV, Guimaraes CT (2012) Genetic and molecular mechanisms of aluminum tolerance in plants. Genet Mol Res 11:1949–1957

Silva IR, Smyth TJ, Moxley DF, Carter TE, Allen NS, Rufty TW (2002) Aluminum accumulation at nuclei of cells in the root tip. Fluorescence detection using lumogallion and confocal laser scanning microscopy. Plant Physiol 123:543–552

Sivaguru M, Baluska F, Volkmann D, Felle HH, Horst WJ (1999a) Impacts of aluminum on the cytoskeleton of the maize root apex. Short-term effects on the distal part of the transition zone. Plant Physiol 119:1073–1082

Sivaguru M, Ezaki B, He ZH, Tong H, Osawa H, Baluška F, Volkmann D, Matsumoto H (2003a) Aluminum-induced gene expression and protein localization of a cell wall-associated receptor kinase in *Arabidopsis*. Plant Physiol 132:1–11

Sivaguru M, Fujiwara T, Samaj J, Baluška F, Yang Z, Osawa H, Maeda T, Mori T, Volkmann D, Matsumoto H (2000) Aluminum-induced 1→3-β-D-glucan inhibits cell-to-cell trafficking of molecules through plasmadesmata: a new mechanism of Al toxicity in plants. Plant Physiol 124:991–1005

Sivaguru M, Horst WJ (1998) The distal part of the transition zone is the most aluminum-sensitive apical zone of maize. Plant Physiol 116:155–163

Sivaguru M, Liu J, Kochian LV (2013) Targeted expression of *SbMATE* in the root distal transition zone is responsible for sorghum aluminum resistance. Plant J 76:297–307

Sivaguru M, Pike S, Gassmann W, Baskin TI (2003b) Aluminum rapidly depolymerizes cortical microtubules and depolarizes the plasma membrane: evidence that these responses are mediated by a glutamate receptor. Plant Cell Physiol 44:667–6745

Sivaguru M, Yamamoto Y, Matsumoto H (1999b) Diefferential impacts of aluminium on microtubules organization depends on growth phase in suspension-cultured tobacco cells. Physiol Plant 109:110–119

Skipsey M, Knight KM, Brazier-Hicks M, Dixon DP, Steel PG, Edwards R (2011) Xenobiotic responsiveness of *Arabidopsis thaliana* to a chemical series derived from a herbicide safener. J Biol Chem 286:32268–32276

Sun P, Tian QY, Chen J, Zhang WH (2010) Aluminum induced inhibition of root elongation in *Arabidopsis* is mediated by ethylene and auxin. J Exp Bot 65:347–356

Sun P, Tian QY, Zhao MG, Dai XY, Huang JH, Li LH, Zhang WH (2007) Aluminum-induced ethylene production is associated with inhibition of root elongation in *Lotus japonicus* L. Plant Cell Physiol 48:1229–1335

Tabuchi A, Mastumoto H (2001) Changes in cell-wall properties of wheat (*Triticum aestivum*) roots during aluminum-induced growth inhibition. Physiol Plant 112:353–358

Taki N, Sasaki-Sekimoto Y, Obayashi T, Kikuta A, Kobayashi K, Ainai T, Yagi K, Sakurai N, Suzuki H, Masuda T, Takamiya K-I, Shibata D, Kobayashi Y, Ohta H (2005) 12-Oxophytodienoic acid triggers expression of a distinct set of genes and plays a role in wound-induced gene expression in *Arabidopsis*. Plant Physiol 139:1268–1283

Testerink C, Munnik T (2006) Phosphatidic acid: a multifunctional stress-signaling lipid in plants. Trends Plant Sci 10:368–375

Tian Q, Zhang X, Ramesh S, Gilliham M, Tyerman SD, Zhang W-H (2014) Ethylene negatively regulates aluminum-induced malate efflux from wheat roots and tobacco cells transformed with *TaLALMT1*. J Exp Bot 65:2415–2426

Tian Q-Y, Sun D-H, Zhao M-G, Zhang W-H (2007) Inhibition of nitric oxide synthase (NOS) underlines aluminum-induced inhibition of root elongation in *Hibiscus moscheutos*. New Phytol 174:322–331

Tice KP, Parker DR, DeMason DA (1992) Operationally defined apoplastic and symplastic aluminum fractions in root tips of aluminum-intoxicated wheat. Plant Physiol 100:309–318

Vicente J, Cascon T, Vicedo B, Garcia-Agustin P, Hamberg M, Castresana C (2012) Role of 9-lipoxygenase and α-dioxygenase oxylipin pathways as modulators of local and systemic defense. Mol Plant 5:914–928

Wagner U, Edwards R, Dixon DP, Mauch F (2002) Probing the diversity of the *Arabidopsis* glutathione *S*-transferase gene family. Plant Mol Biol 49:515–532

Walker PR, LeBlanc J, Sikorska M (1989) Effects of aluminum and other cations on the structure of brain and liver chromatin. Biochem 28:3911–39152

Wilson ID, Neill SJ, Hancock JT (2008) Nitric oxide synthesis and signalling in plants. Plant Cell Environ 31:622–631

Xia J, Yamaji N, Kasai T, Ma JF (2010) Plasma membrane-localized transporter for aluminum in rice. Proc Natl Acad Sci U S A 107:18381–18385

Xu FX, Lagudah ES, Moose SP, Riechers DE (2002) Tandemly-duplicated safener-induced glutathione *S*-transferase genes from *Triticum tauschii* contribute to genome- and organ-specific expression in hexaploid wheat. Plant Physiol 130:362–373

Yamaguchi Y, Yamamoto Y, Matsumoto H (1999) Cell death process initiates by a combination of aluminum and iron in suspension-cultured tobacco cells (*Nicotiana tobacum*): apoptosis-like cell death mediated by calcium and proteinase. Soil Sci Plant Nutr 45:647–657

Yamamoto Y, Kobayashi Y, Matsumoto H (2001) Lipid peroxidation is an early symptom triggered by aluminum, but not the primary cause of elongation inhibition in pea roots. Plant Physiol 125:199–208

Yamamoto Y, Kobayashi Y, Rama DS, Rikiishi S, Matasumoto H (2002) Aluminum toxicity is associated with mitochondrial dysfunction and the production of reactive oxygen species in plant cells. Plant Physiol 128:63–72

Yang JL, Li YY, Zhang YJ, Zhang SS, Wu YR, Wu P, Zheng SJ (2008) Cell wall polysaccharides are specifically involved in the exclusion of aluminum from the rice root apex. Plant Physiol 146:602–611

Yang JL, Zhu XF, Peng YX, Zheng CJ, Li GX, Liu Y, Shi YZ, Zheng SJ (2011a) Cell wall hemicelluloses contribute significantly to aluminum adsorption and root growth in *Arabidopsis*. Plant Physiol 155:1885–1892

Yang L, Tian D, Todd CD, Luo Y, Hu X (2013) Comparative proteome analyses reveal that nitric oxide is an important signal molecule in the response of rice to aluminum toxicity. J Proteome Res 1:1316–1330

Yang ZM, Nian H, Sivaguru M, Tanakamaru S, Matsumoto H (2001) Characterization of aluminium-induced citrate secretion in aluminium-tolerant soybean (*Glycine max*) plants. Physiol Plant 113:64–71

Yang Y, Wang QL, Geng MJ, Guo ZH, Zhao Z (2011b) Effect of indole-3-acetic acid on aluminum-induced efflux of malic acid from wheat (*Triticum aestivum* L.). Plant Soil 346: 215–230

Yin L, Mano J, Wang S, Tsuji W, Tanaka K (2010a) The involvement of lipid peroxide-derived aldehydes in aluminum toxicity of tobacco roots. Plant Physiol 152:1406–1417

Yin L, Wang S, Eltayeb AE, Uddin Md I, Yamamoto Y, Tsuji W, Takeuchi Y, Tanaka K (2010b) Over-expression of dehydroascorbate reductase, but not monodehydroascorbate reductase, confers tolerance to aluminum stress in transgenic tobacco. Planta 231:609–621

Zhan J, He H-Y, Wang T-J, Wang A-Q, Li C-Z, He L-F (2013) Aluminum-induced programmed cell death promoted by *AhSAG*, a senescence-associated gene in *Arachis hypoganea* L. Plant Sci 210:108–117

Zhang H, Shi W, You JF, Bian MD, Qin XM, Yu H, Liu Q, Ryan PR, Yang ZM (2015) Transgenic *Arabidopsis thaliana* plants expressing a β-1,3-glucanase from sweet sorghum (*Sorghum bicolor* L.) show reduced callose deposition and increase tolerance to aluminum toxicity. Plant Cell Environ 38:1178–1188

Zhang Q, Xu FX, Lambert KN, Riechers DE (2007) Safeners coordinately induce the expression of multiple proteins and MRP transcripts involved in herbicide metabolism and detoxification in *Triticum tauschii* seedling tissues. Proteomics 7:1261–1278

Zhang SJ, Yang JL (2005) Target sites of aluminum phytotoxicity. Biol Plant 49:321–331

Zhang WH, Rengel Z (1999) Aluminum induces an increase in cytoplasmic calcium in intact wheat root apical cells. Aust J Plant Physiol 26:401–409

Zhao M-G, Tian G-Y, Zhang W-H (2007) Ethylene activates a plasma membrane $Ca^{2+}$ permeable channel in tobacco suspension cells. New Phytol 174:507–515

Zhou G, Ren N, Qi J, Lu J, Xiang C, Ju H, Cheng J, Lou Y (2014) The 9-lipoxygenase *Osr9-LOX1* interacts with the 13-lipoxygenase-mediated pathway to regulate resistance to chewing and piercing-sucking herbivores in rice. Physiol Plant 152:59–69

Zhu XF, Lei GJ, Wang ZW, Shi YZ, Braam J, Li GX, Zheng SJ (2013) Coordination between apoplastic and symplastic detoxification confers plant aluminum resistance. Plant Physiol 162:1947–1955

# Transcriptional Regulation of Al Tolerance in Plants

**Kengo Yokosho and Jian Feng Ma**

**Abstract** Great progresses have been made in understating of molecular mechanisms of Al tolerance in plants during last decade. A number of Al-tolerance genes have been functionally characterized, especially in rice and Arabidopsis. Several transcription factors including a C2H2 zinc finger-type ART1/STOP1 have been identified, but only a few downstream genes regulated by ART1 and STOP1 are similar, indicating different regulation mechanism of Al tolerance in rice and Arabidopsis. Transcriptional regulation of the Al-tolerance genes also differs with plant species and genes. Four different patterns have been reported including increase of gene copy number in the genome, insertion of transposon-like sequences, tandem repeat sequences, and increase of cis-acting element of transcription factor in the promoter region. All these alternations in the genome enhance the expression of Al-tolerance genes.

## 1 Al-Tolerance Genes

Plants have developed strategies to cope with Al toxicity. During last decade, great progresses have been made in understanding molecular mechanisms of aluminum (Al) tolerance in plants, especially in Arabidopsis and rice (Delhaize et al. 2012; Ma et al. 2014; Kochian et al. 2015). Since the first Al-tolerance gene, *ALMT1* (Aluminum-activated malate transporter), which is responsible for the Al-induced malate secretion, was identified in wheat (Sasaki et al. 2004), a number of Al-tolerance genes have been identified in different plant species (Figs. 1 and 2). For example, homologs of *ALMT1* have been identified in Arabidopsis (Hoekenga et al. 2006), oilseed rape (Ligaba et al. 2006), rye (Collins et al. 2008), soybean (Liang et al. 2013), yorkshire fog (Chen et al. 2013) in addition to wheat. Genes responsible for Al-induced secretion of citrate, *AACT1/MATE/FRDL* (Aluminum-activated citrate transporter 1/Multidrug and toxic compound extrusion/Ferric

K. Yokosho • J.F. Ma (✉)
Institute of Plant Science and Resources, Okayama University, Chuo 2-20-1, Kurashiki 710-0046, Japan
e-mail: maj@rib.okayama-u.ac.jp

 
S.K. Panda, F. Baluška (eds.), *Aluminum Stress Adaptation in Plants*, Signaling and Communication in Plants 24, DOI 10.1007/978-3-319-19968-9_2

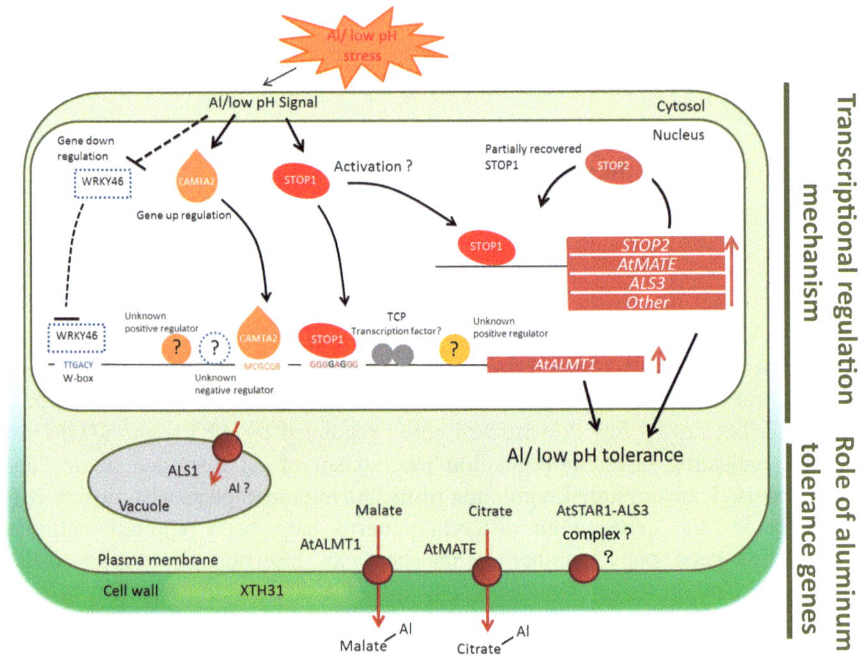


**Fig. 1** Aluminum tolerance genes and their transcriptional regulation mechanisms in Arabidopsis. *AtALMT1*, *ATMATE*, and *ALS3* are involved in Al-induced secretion of malate, citrate, and redistribution of Al from sensitive region, respectively. They are induced by low pH and Al and regulated by the transcription factor, STOP1. By contrast, *AtSTAR1* and *ALS1* do not respond to low pH and Al, and their exact role is unknown. On the other hand, XTH31 is involved in cell wall modification. In addition to STOP1, other transcription factors for Al tolerance have also been identified including STOP2, WRKY46 as a negative regulator, and CAMTA2 as an activator. For details, refer to the text

reductase defective3 like), have also been identified in barley (*HvAACT1*) (Furukawa et al. 2007), sorghum (*SbMATE*) (Magalhaes et al. 2007), rice (*OsFRDL4*) (Yokosho et al. 2011), *Arabidopsis* (Liu et al. 2009), rye (Yokosho et al. 2010), wheat (Ryan et al. 2009), maize (Maron et al. 2010), river red gum (Sawaki et al. 2013), and cabbage (Wu et al. 2014).

In addition to these genes related to organic acid anion secretion, several important genes involved in the external and internal detoxification of Al have also been identified. In rice, a highly Al-tolerant species, seven genes have been functionally demonstrated to be required for high Al tolerance in rice (Fig. 2). *STAR1* and *STAR2* (Sensitive to Al rhizotoxicity 1 and 2) encode an ATP-binding domain and a transmembrane domain, respectively, of a bacterial-type ATP binding cassette (ABC) transporter. The complex of STAR1 and STAR2 transports UDP-glucose (Huang et al. 2009), which is used for cell wall modification, resulting in Al fixation in the root cell wall. *Nrat1* (Nramp aluminum transporter 1) encodes an Al transporter localized at the plasma membrane, which transports trivalent Al

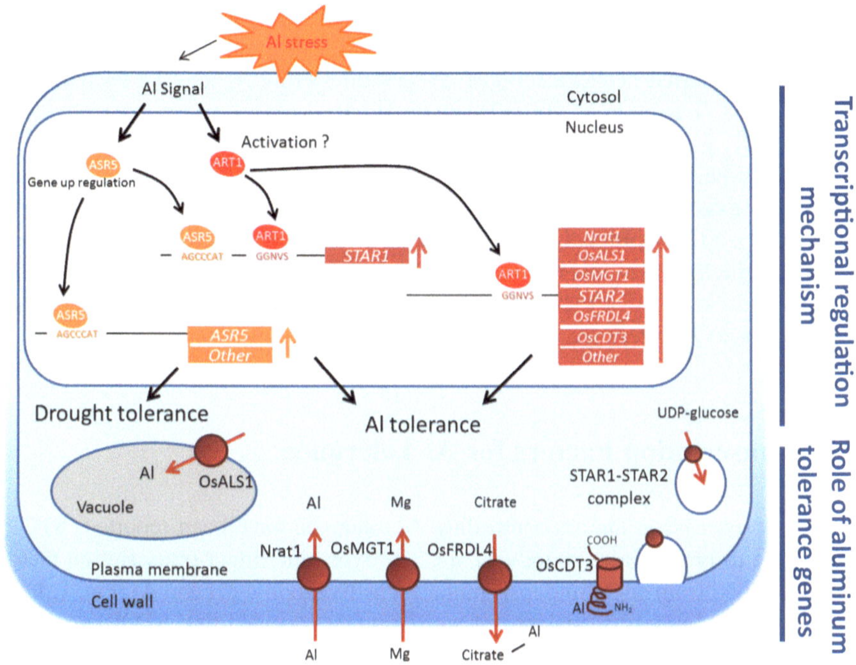


**Fig. 2** Aluminum tolerance genes and their transcriptional regulation mechanisms in rice. Al-tolerance genes including *STAR1*, *STAR2*, *Nrat1*, *OsALS1*, *OsFRDL4*, *OsMGT1*, and *OsCDT3* function in Al detoxification at different cellular levels. All of them are regulated by the transcription factor, ART1. ASR5 is another transcription factor, which regulates STAR1 expression. For details, refer to the text

into the cells (Xia et al. 2010) for subsequent sequestration of Al into the vacuoles by OsALS1 (Aluminum sensitive 1). OsALS1 is a half-size ABC transporter and localized to tonoplast of root cells (Huang et al. 2012). On the other hand, a plasma membrane-localized Mg transporter, OsMGT1 (Magnesium transporter 1), also plays an important role in Al tolerance by increasing Mg uptake (Chen et al. 2012). Recently, *OsCDT3*, which encodes a small peptide rich in cysteine, was also demonstrated to be involved in Al tolerance in rice (Xia et al. 2013). It is anchored to the plasma membrane and confers Al probably by binding Al, therefore stopping entry of Al into the root cells. All these genes are rapidly and specifically induced by Al (Huang et al. 2009, 2012; Xia et al. 2010, 2013; Yokosho et al. 2011; Chen et al. 2012). The proteins encoded by these genes are similarly localized at all root cells. Knockout of either gene results in decreased Al tolerance although the contribution to Al tolerance differs with genes.

In Arabidopsis, in addition to *AtALMT1* and *AtMATE*, two other genes (*ALS1* and *ALS3*) confer Al tolerance (Fig. 1). *ALS3* in Arabidopsis is a homolog of rice *STAR2*, encoding a half-size ATP-binding cassette (ABC) transporter (Larsen et al. 2005). It may form a complex with AtSTAR1 to confer Al tolerance (Huang et al. 2010), which is probably involved in the redistribution process of

Al from sensitive region (Larsen et al. 2005). ALS1, a homolog of rice OsALS1, is a tonoplast-localized ABC transporter in Arabidopsis (Larsen et al. 2007). It may be involved in internal detoxification of Al by sequestering Al into the vacuoles like rice OsALS1 (Larsen et al. 2007; Huang et al. 2012).

Recently, genes involved in xyloglucan metabolism in the cell wall are also reported to be involved in Al tolerance in Arabidopsis (Fig. 1; Zhu et al. 2012, 2014). For example, knockout of *XTH31* resulted in an increased Al tolerance (Zhu et al. 2012). *XTH31* probably encodes xyloglucan endotransglucosylase/hydrolases (XTHs), which cleaves and rejoins hemicellulosic xyloglucan polymers during cell expansion. It seems that reduction of production of cleaved xyloglucans in the cell wall increases Al tolerance.

## 2 Transcription Factors for Al Tolerance

Several transcription factors controlling Al tolerance have been reported. STOP1 (Sensitive to proton rhizotoxicity 1), a C2H2-type zinc finger transcription factor, was identified from an Arabidopsis mutant sensitive to low pH and Al (Fig. 1; Iuchi et al. 2007). Microarray analyses of *stop1* mutant revealed that STOP1 regulates 43 genes including *AtALMT1*, *AtMATE*, and *ALS3* (Sawaki et al. 2009). The expression of *STOP1* is not induced by low pH and Al, although regulated downstream genes rapidly respond to low pH and Al (Sawaki et al. 2009; Liu et al. 2009), suggesting that posttranslational process is required.

Recently, a STOP1 homolog in Arabidopsis, STOP2, was identified (Kobayashi et al. 2014). STOP1 and STOP2 shared 40 % identity at amino acids level (Kobayashi et al. 2014). The expression level of *STOP2* is much lower than that of *STOP1*, but the expression of *STOP2* is regulated by STOP1 (Fig. 1; Sawaki et al. 2009; Kobayashi et al. 2014). Over-expression of *STOP2* in *stop1* mutant resulted in enhanced tolerance to low pH (Kobayashi et al. 2014). Introduction of *STOP2* in *stop1* mutant under the control of *STOP1* promoter also complemented the expression level of *ALS3* and *AtMATE* (Kobayashi et al. 2014). It seems that STOP2 is a minor isoform, but it can activate transcription of some genes regulated by STOP1 (Fig. 1; Kobayashi et al. 2014).

ART1 (Al resistance transcription factor 1) is a transcription factor for Al tolerance identified in rice (Fig. 2; Yamaji et al. 2009). Similar to STOP1 in Arabidopsis, ART1 is also a C2H2-type zinc finger transcription factor in rice. However, different from STOP1, ART1 does not confer low pH (Yamaji et al. 2009). ART1 regulates the expression of at least 31 genes (Yamaji et al. 2009). Among these genes, seven genes, including *STAR1*, *STAR2*, *Nrat1*, *OsALS1*, *OsFRDL4*, *OsMGT1*, *OsCDT3*, have been functionally characterized as described above (Fig. 2). Comparison of downstream genes showed that only two genes (*AtMATE*/*OsFRDL4*, *ALS3*/*STAR2*) are common between STOP1- and ART1-regulated genes, indicating different Al tolerance regulation mechanisms between Arabidopsis and rice. The core *cis*-acting element of ART1 is [GGN(T/g/a/C)V(C/A/g)S(C/G)], which can be found in the promoter region of 29 genes among 31 ART1-regulated genes (Fig. 2; Tsutsui et al. 2011). Similar to STOP1, the expression of *ART1* is also not induced by Al, but the mechanism underlying activation of ART1 is unknown.

The homologs of *STOP1/ART1* homologs were also found in other plant species including monocots, dicots, woody plants, and a bryophyte (Ohyama et al. 2013; Garcia-Oliveira et al. 2013; Chen et al. 2013). Knockout or knockdown of *STOP1/ART1* homolog showed Al sensitive phenotype in tobacco and *Moss* (Ohyama et al. 2013). Interestingly, TaSTOP1 in wheat is involved in Al tolerance, but not through regulating *TaALMT1* (Garcia-Oliveira et al. 2013), since *TaALMT1* expression is not induced by Al.

Recently, WRKY46, a member of the WRKY domain-containing family of transcription factors, was identified as a negative regulator of expression of *AtALMT1*, a key Al-tolerance gene in Arabidopsis (Ding et al. 2013). It could be bound to the *AtALMT1* promoter region as a repressor (Fig. 1; Ding et al. 2013), because there are several putative W-box domains in the *AtALMT1* promoter region. Knockout of *AtWRKY46* resulted in increased expression of *AtALMT1*, root malate secretion, and Al tolerance (Fig. 1; Ding et al. 2013). More recently, a study by Tokizawa et al. (2015) showed that CAMTA2 (Calmodulin binding transcription activator 2) may function as an activator of *AtALMT1* expression (Fig. 1).

ASR5 (Abscisic acid, stress and ripening 5) is another transcription factor for Al tolerance identified in rice (Fig. 2; Arenhart et al. 2013, 2014). ASR5 is localized in the chloroplast, cytoplasm, and nucleus (Arenhart et al. 2013). Its transcript levels increase in response to Al in the roots and shoots, and ASR5-silenced plants are extremely sensitive to Al (Arenhart et al. 2013). ASR5 is the Al-activated factor that binds to the STAR1 promoter to enhance its expression (Fig. 2). Among ASR5-regulated genes, there are three genes (*STAR1*, *Nrat1*, *OsFRDL4*), which are regulated by ART1 (Arenhart et al. 2014). The requirement of both ASR5 and ART1 for Al-induced STAR1 expression suggests that ASR5 and ART1 may interact with each other directly and function cooperatively although further work is required.

Different from STOP1 and ART1, WRKY46 and ASR5 are also involved in tolerance to other abiotic stresses.

## 3 Transcriptional Regulation of Al-Tolerance Genes

Recent several studies have shown that some Al-tolerance genes undergo transcription regulation although unknown translational and posttranslational regulation may also be involved. There are four different patterns for the transcriptional regulation depending on plant species (Fig. 3). In pattern I, the expression of Al-tolerance genes is enhanced through increasing gene copy number in the genome. This is seen in *ScALMT1* in rye and *ZmMATE1* in maize. Rye (*Secale cereale* L.), one of the most

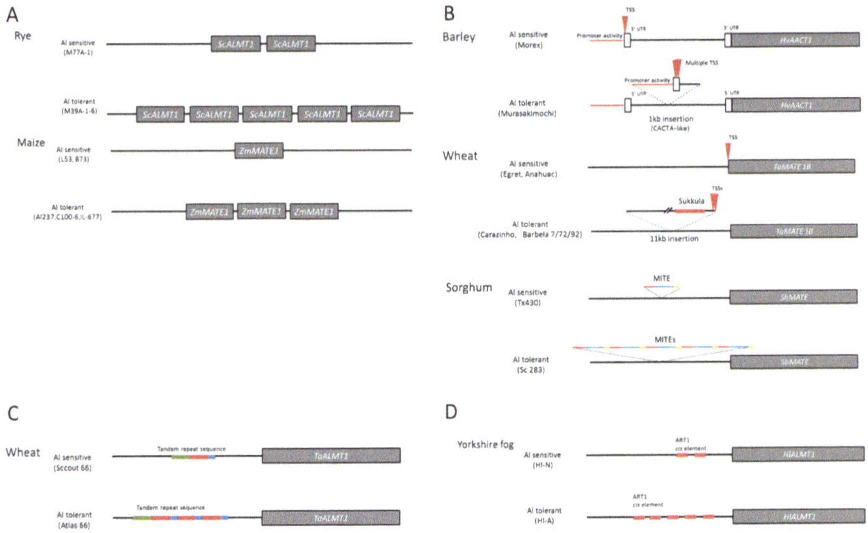


**Fig. 3** Transcriptional regulation of Al-tolerance genes involved in Al-induced organic acid anion secretion in different plant species. (**A**) Pattern I: Increase of gene copy number in the genome. The copy number of *ScALMT1* in rye and *ZmMATE1* in maize are more in the Al-tolerant cultivars than in the Al-sensitive cultivars. (**B**) Pattern II: Transposon-like insertion in the genome. A transposon-like sequence is inserted upstream of *HvAACT1* in barley, *TaMATE1B* in wheat, and *SbMATE* in sorghum, which enhance the expression of these genes. (**C**) Pattern III: Tandem repeat sequences in the promoter region. Tandem repeat sequences are located at upstream of *TaALMT1* in wheat and increase its expression. (**D**) Pattern IV: Increase of ART1 *cis*-acting elements in the promoter region. The number of *cis*-acting element of ART1 is increased in the promoter region of *HlALMT1* in Al-tolerant accession of Yorkshire Fog (*H. lanatus*), resulting in increased expression of *HlALMT1*. Modified according to Delhaize et al. (2012), Ma et al. (2014)

Al-tolerant cereal crops, secretes both malate and citrate from the roots in response to Al (Li et al. 2000). The expression level of *ScALMT1* in the Al-tolerance cultivar is higher than that in the Al-sensitive cultivar, which is derived from the difference in the genomic copy number of *ScALMT1* (Fig. 3A; Collins et al. 2008). Five *ScALMT1* genes are clustered together on chromosome 7R in the tolerance cultivar, of which two are highly expressed in the root tip. On the other hand, only two copies are found in the sensitive cultivar, of which only one copy is highly expressed in the root tip (Collins et al. 2008). In maize, Al-tolerant cultivars have three functional copies of *ZmMATE1* in the genome, which are identical and part of a tandem triplication (Maron et al. 2013). This copy number variation is associated with both gene expression of *ZmMATE1* and Al tolerance (Fig. 3A). Interestingly, maize cultivars carrying the three-copy allele share the same geographical origin in acid-soil regions of the South American tropics, suggesting that copy number increase is an adaptation to acid soil.

By contrast, in pattern II, the expression level of Al-tolerance genes is enhanced through an insertion of transposon-like elements in the upstream region of ORF (Fig. 3B), which can be found in barley, wheat, and sorghum. In barley, the higher expression of *HvAACT1* in the root tips of Al-tolerant cultivars is associated with a 1-kb transposon insertion (CACTA-like transposon) in the upstream of the ORF region (Fujii et al. 2012). This insertion acts as a promoter to enhance the expression level of *HvAACT1* (Fig. 3B). Furthermore, this insertion also alters the expression location of *HvAACT1* from mature root region to the root tips, the site of Al toxicity (Fujii et al. 2012). Interestingly, *HvAACT1* at the mature root region is originally involved in the translocation of Fe from the roots to the shoots, which have similar role as OsFRDL1 in rice (Yokosho et al. 2009). The 1-kb insertion was only found in barley accessions cultivated in Japan, China, and Korea, where acid soils distribute.

In wheat, a Sukkula-like transposable element (11 kb) was found to be inserted at the promoter region of *TaMATE* gene in several Brazilian wheat cultivars, which secrete citrate constitutively in response to Al (Tovkach et al. 2013). This insertion also enhances the expression of the *TaMATE1* gene (Fig. 3B; Tovkach et al. 2013). Furthermore, this insertion was also found in other Al-tolerance bread wheat cultivars (Garcia-Oliveira et al. 2014).

In sorghum, tourist-like miniature inverted repeat transposable elements (MITEs) occur at upstream of the *SbMATE* gene, and the number of these repeats is broadly correlated with the level of *SbMATE* expression (Fig. 3B; Magalhaes et al. 2007). Recent study showed that introduction of *SbMATE* into different background resulted in different expression level, suggesting that *SbMATE* expression is regulated at multiple levels (Melo et al. 2013). It seems that both *cis*- and *trans*-acting elements are involved in regulating *SbMATE* expression.

In pattern III as seen in wheat, the expression level of *TaALMT1* is controlled by tandem repeated elements in the promoter region (Fig. 3C; Sasaki et al. 2006; Ryan et al. 2010). The expression level of *TaALMT1* is not induced by Al, but the expression level is higher in Al-tolerant cultivars than Al-sensitive cultivars (Sasaki et al. 2006). The constitutively greater expression of *TaALMT1* in Al-tolerant genotypes has a series of *cis* mutations in the promoter (Ryan et al. 2010). Among these, alleles with duplications and triplications are associated with enhanced expression of *TaALMT1*, increased malate efflux, and greater Al tolerance compared with alleles that lack these repeats (Ryan et al. 2010).

In pattern IV, expression level is associated with the number of *cis*-acting element of transcription factor, ART1. In an accession of yorkshire fog (*Holcus lanatus*) grown on highly acidic soils, the expression of *HlALMT1* was twice as high as in the accession grown on neutral soil (Chen et al. 2013). The number of *cis*-acting elements of *HlART1* in the promoter region of *HlALMT1* was more in the accession grown on acid soil due to nucleotide substitution (Fig. 3D), indicating that the adaptation of *H. lanatus* to acidic soils may be achieved by increasing number of *cis*-acting elements for *ART1* in the promoter region of the *HlALMT1* gene, enhancing the expression of *HlALMT1* and the secretion of malate.

Genotypic difference in the expression of *Nrat1* was also found in rice (Li et al. 2014; Xia et al. 2014). Promotor analysis detected five unique SNPs, which are not related to the *cis*-acting element of ART1. It remains to examine whether these SNPs are involved in the transcriptional regulation of *Nrat1* in rice.

There is also a positive correlation between *OsFRDL4* expression level and the amount of citrate secretion in rice cultivars that are differing in Al tolerance (Yokosho et al. 2011), but the mechanism for regulating the expression of *OsFRDL4* is also unknown.

## References

Arenhart RA, Lima JC, Pedron M, Carvalho FE, Silveira JA, Rosa SB, Caverzan A, Andrade CM, Schünemann M, Margis R, Margis-Pinheiro M (2013) Involvement of *ASR* genes in aluminum tolerance mechanisms in rice. Plant Cell Environ 36:52–67

Arenhart RA, Bai Y, Valter de Oliveira LF, Neto LB, Schunemann M, Maraschin Fdos S, Mariath J, Silverio A, Sachetto-Martins G, Margis R, Wang ZY, Margis-Pinheiro M (2014) New insights into aluminum tolerance in rice: the ASR5 protein binds the *STAR1* promoter and other aluminum-responsive genes. Mol Plant 7:709–721

Chen ZC, Yamaji N, Motoyama R, Nagamura Y, Ma JF (2012) Up-regulation of a magnesium transporter gene *OsMGT1* is required for conferring aluminum tolerance in rice. Plant Physiol 159:1624–1633

Chen ZC, Yokosho K, Kashino M, Zhao FJ, Yamaji N, Ma JF (2013) Adaptation to acidic soil is achieved by increased *cis*-acting element numbers regulating *ALMT1* expression in *Holcus lanatus*. Plant J 76:10–23

Collins NC, Shirley NJ, Saeed M, Pallotta M, Gustafson JP (2008) An *ALMT1* gene cluster controlling aluminum tolerance at the *Alt4* locus of rye (*Secale cereale L.*). Genetics 179: 669–682

Delhaize E, Ma JF, Ryan PR (2012) Transcriptional regulation of aluminium tolerance genes. Trends Plant Sci 17:341–348

Ding ZJ, Yan JY, Xu XY, Li GX, Zheng SJ (2013) WRKY46 functions as a transcriptional repressor of *ALMT1*, regulating aluminum-induced malate secretion in Arabidopsis. Plant J 76:825–835

Fujii M, Yokosho K, Yamaji N, Saisho D, Yamane M, Takahashi H, Sato K, Nakazono M, Ma JF (2012) Acquisition of aluminium tolerance by modification of a single gene in barley. Nat Commun 3:713

Furukawa J, Yamaji N, Wang H, Mitani N, Murata Y, Sato K, Katsuhara M, Takeda K, Ma JF (2007) An aluminum activated citrate transporter in barley. Plant Cell Physiol 48:1081–1091

Garcia-Oliveira AL, Benito C, Prieto P, Menezes RA, Rodrigues-Pousada C, Guedes-Pinto H, Martins-Lopes P (2013) Molecular characterization of TaSTOP1 homoeologues and their response to aluminium and proton ($H^+$) toxicity in bread wheat (*Triticum aestivum* L.). BMC Plant Biol 13:134

Garcia-Oliveira AL, Martins-Lopes P, Tolrá R, Poschenrieder C, Tarquis M, Guedes-Pinto H, Benito C (2014) Molecular characterization of the citrate transporter gene *TaMATE1* and expression analysis of upstream genes involved in organic acid transport under Al stress in bread wheat (*Triticum aestivum*). Physiol Plant 152:441–452

Hoekenga OA, Maron LG, Cancado GMA, Piñeros MA, Shaff J, Kobayashi Y, Ryan PR, Dong B, Delhaize E, Sasaki T (2006) *AtALMT1*, which encodes a malate transporter, is identified as one of several genes critical for aluminum tolerance in Arabidopsis. Proc Natl Acad Sci U S A 103:9734–9743

Huang CF, Yamaji N, Mitani N, Yano M, Nagamura Y, Ma JF (2009) A bacterial-type ABC transporter is involved in aluminum tolerance in rice. Plant Cell 21:655–667

Huang CF, Yamaji N, Ma JF (2010) Knockout of a bacterial-type ATP-binding cassette transporter gene, *AtSTAR1*, results in increased aluminum sensitivity in Arabidopsis. Plant Physiol 153: 1669–1677

Huang CF, Yamaji N, Chen ZC, Ma JF (2012) A tonoplast-localized half-size ABC transporter is required for internal detoxification of aluminum in rice. Plant J 69:857–867
Iuchi S, Koyama H, Iuchi A, Kobayashi Y, Kitabayashi S, Kobayashi Y, Ikka T, Hirayama T, Shinozaki K, Kobayashi M (2007) Zinc finger protein STOP1 is critical for proton tolerance in Arabidopsis and coregulates a key gene in aluminum tolerance. Proc Natl Acad Sci U S A 104: 9900–9905
Kobayashi Y, Ohyama Y, Kobayashi Y, Ito H, Iuchi S, Fujita M, Zhao CR, Tanveer T, Ganesan M, Kobayashi M, Koyama H (2014) STOP2 activates transcription of several genes for Al- and low pH-tolerance that are regulated by STOP1 in Arabidopsis. Mol Plant 7:311–322
Kochian LV, Piñeros MA, Liu J, Magalhaes JV (2015) Plant adaptation to acid soils: the molecular basis for crop aluminum resistance. Annu Rev Plant Biol 66:23.1–23.28
Larsen PB, Geisler MJ, Jones CA, Williams KM, Cancel JD (2005) *ALS3* encodes a phloem-localized ABC transporter-like protein that is required for aluminum tolerance in Arabidopsis. Plant J 41:353–363
Larsen PB, Cancel J, Rounds M, Ochoa V (2007) Arabidopsis *ALS1* encodes a root tip and stele localized half type ABC transporter required for root growth in an aluminum toxic environment. Planta 225:1447–1458
Li XF, Ma JF, Matsumoto H (2000) Pattern of aluminum-induced secretion of organic acids differs between rye and wheat. Plant Physiol 123:1537–1543
Li JY, Liu J, Dong D, Jia X, McCouch SR, Kochian LV (2014) Natural variation underlies alterations in Nramp aluminum transporter (*Nrat1*) expression and function that play a key role in rice aluminum tolerance. Proc Natl Acad Sci U S A 111:6503–6508
Liang C, Piñeros M, Tian J, Yao Z, Sun L, Liu J, Shaff J, Coluccio A, Kochian LV, Liao H (2013) Low pH, aluminum and phosphorus coordinately regulate malate exudation through *GmALMT1* to improve soybean adaptation to acid soils. Plant Physiol 161:1347–1361
Ligaba A, Katsuhara M, Ryan PR, Shibasaka M, Matsumoto H (2006) The *BnALMT1* and *BnALMT2* genes from rape encode aluminum-activated malate transporters that enhance the aluminum resistance of plant cells. Plant Physiol 142:1294–1303
Liu J, Magalhaes JV, Shaff J, Kochian LV (2009) Aluminum activated citrate and malate transporters from the MATE and ALMT families function independently to confer Arabidopsis aluminum tolerance. Plant J 57:389–399
Ma JF, Chen ZF, Shen RF (2014) Molecular mechanisms of Al tolerance in gramineous plants. Plant Soil 381:1–12
Magalhaes JV, Liu J, Guimarães CT, Lana UG, Alves VM, Wang YH, Schaffert RE, Hoekenga OA, Piñeros MA, Shaff JE, Klein PE, Carneiro NP, Coelho CM, Trick HN, Kochian LV (2007) A gene in the multidrug and toxic compound extrusion (MATE) family confers aluminum tolerance in sorghum. Nat Genet 39:1156–1161
Maron LG, Piñeros MA, Guimarães CT, Magalhaes JV, Pleiman JK, Mao C, Shaff J, Belicuas SN, Kochian LV (2010) Two functionally distinct members of the MATE (multi-drug and toxic compound extrusion) family of transporters potentially underlie two major aluminum tolerance QTLs in maize. Plant J 61:728–740
Maron LG, Guimarães CT, Kirst M, Albert PS, Birchler JA, Bradbury PJ, Buckler ES, Coluccio AE, Danilova TV, Kudrna D, Magalhaes JV, Piñeros MA, Schatz MC, Wing RA, Kochian LV (2013) Aluminum tolerance in maize is associated with higher *MATE1* gene copy number. Proc Natl Acad Sci U S A 110:5241–5246
Melo JO, Lana UG, Piñeros MA, Alves VM, Guimarães CT, Liu J, Zheng Y, Zhong S, Fei Z, Maron LG, Schaffert RE, Kochian LV, Magalhaes JV (2013) In complete transfer of accessory loci influencing *SbMATE* expression underlies genetic background effects for aluminum tolerance in sorghum. Plant J 73:276–288
Ohyama Y, Ito H, Kobayashi Y, Ikka T, Morita A, Kobayashi M, Imaizumi R, Aoki T, Komatsu K, Sakata Y, Iuchi S, Koyama H (2013) Characterization of *AtSTOP1* orthologous genes in tobacco and other plant species. Plant Physiol 162:1937–1946

Ryan PR, Raman H, Gupta S, Horst WJ, Delhaize E (2009) A second mechanism for aluminum resistance in wheat relies on the constitutive efflux of citrate from roots. Plant Physiol 149: 340–351

Ryan PR, Raman H, Gupta S, Sasaki T, Yamamoto Y, Delhaize E (2010) The multiple origins of aluminium resistance in hexaploid wheat include *Aegilops tauschii* and more recent *cis* mutations to *TaALMT1*. Plant J 64:446–455

Sasaki T, Yamamoto Y, Ezaki B, Katsuhara M, Ahn SJ, Ryan PR, Delhaize E, Matsumoto H (2004) A wheat gene encoding an aluminum-activated malate transporter. Plant J 37:645–653

Sasaki T, Ryan PR, Delhaize E, Hebb DM, Ogihara Y, Kawaura K, Noda K, Kojima T, Toyoda A, Matsumoto H, Yamamoto Y (2006) Sequence upstream of the wheat (*Triticum aestivum* L.) *ALMT1* gene and its relationship to aluminum resistance. Plant Cell Physiol 47:1343–1354

Sawaki Y, Iuchi S, Kobayashi Y, Kobayashi Y, Ikka T, Sakurai N, Fujita M, Shinozaki K, Shibata D, Kobayashi M, Koyama H (2009) STOP1 regulates multiple genes which protect *Arabidopsis* from proton and aluminum toxicities. Plant Physiol 150:281–294

Sawaki Y, Kihara-Doi T, Kobayashi Y, Nishikubo N, Kawazu T, Kobayashi Y, Koyama H, Sato S (2013) Characterization of Al-responsive citrate excretion and citrate-transporting MATEs in *Eucalyptus camaldulensis*. Planta 237:979–989

Tokizawa M, Kobayashi M, Saito T, Kobayashi M, Iuchi S, Nomoto M, Tada Y, Yamamoto YY, Koyama H (2015) SENSITIVE TO PROTON RHIZOTOXICITY1, CALMODULIN BINDING TRANSCRIPTION ACTIVATOR2, and other transcription factors are involved in ALUMINUM-ACTIVATED MALATE TRANSPORTER1 expression. Plant Physiol 167: 991–1003

Tovkach A, Ryan PR, Richardson AE, Lewis DC, Rathjen TM, Ramesh S, Tyerman SD, Delhaize E (2013) Transposon-mediated alteration of *TaMATE1B* expression in wheat confers constitutive citrate efflux from root apices. Plant Physiol 161:880–892

Tsutsui T, Yamaji N, Ma JF (2011) Identification of a *cis*-acting element of ART1, a C2H2-type zinc-finger transcription factor for aluminum tolerance in rice. Plant Physiol 156:925–931

Wu X, Li R, Shi J, Wang J, Sun Q, Zhang H, Xing Y, Qi Y, Zhang N, Guo YD (2014) *Brassica oleracea MATE* encodes a citrate transporter and enhances aluminum tolerance in *Arabidopsis thaliana*. Plant Cell Physiol 55:1426–1436

Xia JX, Yamaji N, Kasai T, Ma JF (2010) Plasma membrane-localized transporter for aluminum in rice. Proc Natl Acad Sci U S A 107:18381–18385

Xia JX, Yamaji N, Ma JF (2013) A plasma membrane-localized small peptide is involved in rice aluminum tolerance. Plant J 76:345–355

Xia JX, Yamaji N, Che J, Shen RF, Ma JF (2014) Differential expression of *Nrat1* is responsible for Al-tolerance QTL on chromosome 2 in rice. J Exp Bot 65:4297–4304

Yamaji N, Huang CF, Nagao S, Yano M, Sato Y, Nagamura Y, Ma JF (2009) A zinc finger transcription factor ART1 regulates multiple genes implicated in aluminum tolerance in rice. Plant Cell 21:3339–3349

Yokosho K, Yamaji N, Ueno D, Mitani N, Ma JF (2009) OsFRDL1 is a citrate transporter required for efficient translocation of iron in rice. Plant Physiol 149:297–305

Yokosho K, Yamaji N, Ma JF (2010) Isolation and characterization of two *MATE* genes in rye. Funct Plant Biol 37:296–303

Yokosho K, Yamaji N, Ma JF (2011) An Al-inducible MATE gene is involved in external detoxification of Al in rice. Plant J 69:1061–1069

Zhu XF, Shi YZ, Lei GJ, Fry SC, Zhang BC, Zhou YH, Braam J, Jiang T, Xu XY, Mao CZ, Pan YJ, Yang JL, Wu P, Zheng SJ (2012) *XTH31*, encoding an *in vitro* XEH/XET-active enzyme, regulates aluminum sensitivity by modulating in vivo XET action, cell wall xyloglucan content, and aluminum binding capacity in Arabidopsis. Plant Cell 24:4731–4747

Zhu XF, Wan JX, Sun Y, Shi YZ, Braam J, Li GX, Zheng SJ (2014) Xyloglucan endotransglucosylase-hydrolase17 interacts with xyloglucan endotransglucosylase- hydrolase31 to confer xyloglucan endotransglucosylase action and affect aluminum sensitivity in Arabidopsis. Plant Physiol 165:1566–1574

# Aluminum-Dependent Root Growth Inhibition as Mediated by DNA-Damage Responses

**Caroline A. Sjogren and Paul B. Larsen**

**Abstract** Aluminum (Al) toxicity is a global agricultural problem that reduces crop yields primarily due to root growth inhibition. Several advances in our knowledge of Al resistance mechanisms have recently been made through studies of Al exclusion, yet due to the complicated nature of studying internalized Al, it has proven difficult to determine the biochemical basis of Al toxicity and tolerance. Recent studies show that Al triggers a DNA-damage response mediated by the cell cycle checkpoint *ATAXIA TELANGIECTASIA MUTATED AND RAD3-RELATED* (*ATR*). This is an active process that forces terminal differentiation of the root meristem and is at least in part the cause of stoppage of root growth following chronic exposure to Al. Interestingly, unlike well-studied stressors like IR or gamma (γ) radiation, Al is a real world genotoxic stress that represents a novel system for analysis of DNA damage in biological systems under environmentally relevant conditions. Understanding DNA-damage response and repair pathways following Al treatment in plant systems can offer more effective and safer agricultural practices but also begs to serve as a beacon of caution about Al as a genotoxic stress in other organisms since the cell cycle checkpoint machinery that has been attributed to Al responses is universally found in eukaryotes.

## 1 Aluminum Resistance

Responses to the toxic effects of Al soils have been extensively documented from a wide range of plant species. The primary consequence of Al exposure is severe inhibition of growth of the primary root (Foy et al. 1978). Al exposure has been found to impede cell division and cell elongation, nutrient uptake, $IP_3$ and hormone signaling, cytoskeletal structure, $Ca^{2+}$ homeostasis, vesicle trafficking, plasma membrane integrity, and chromatin structure (Kochian 1995). Plants have adopted two distinctly different strategies for preventing Al-dependent growth inhibition.

C.A. Sjogren • P.B. Larsen (✉)
Department of Biochemistry, University of California, Riverside, CA 92521, USA
e-mail: paul.larsen@ucr.edu

 
S.K. Panda, F. Baluška (eds.), *Aluminum Stress Adaptation in Plants*, Signaling and Communication in Plants 24, DOI 10.1007/978-3-319-19968-9_3

Exclusion of Al from roots, which is considered to be an Al resistance mechanism, is an effective and simple strategy for increasing root growth in an Al toxic environment largely because Al does not come into contact with its biochemical targets. Al exclusion is a distinctly different mechanism from true Al tolerance, in which roots cope with internalized Al. The study of Al tolerance has generally been considered to be an intractable problem largely because of the predicted complexity of Al toxicity, with the expectation being that changes in any one target of Al would have little positive impact on growth due to the sheer number of biochemical targets of Al. Consequently, while Al resistance mechanisms have received a tremendous amount of attention in recent years due to their relative simplicity, this chapter will focus on our emerging understanding of Al toxicity and Al tolerance mechanisms, which have largely been associated with Al acting as a genotoxin that activates the DNA-damage checkpoint to halt cell cycle progression.

## *1.1 Aluminum Exclusion*

Al is the most abundant metal in the earth's crust, existing mostly in primary and secondary mineral compounds (FitzPatrick 1986). Under acidic conditions, Al-containing minerals in soils can dissolve, releasing Al into the soil solution as a trivalent cation ($Al^{3+}$) where it may contribute to soil acidity (Reynolds et al. 2001). In this form, Al can be taken up into plant tissues through the roots. Clearly, an effective and straightforward strategy for increasing root growth in Al toxic environments is Al exclusion, where plants prevent the internalization of Al. Certainly, based on the predicted complexity of Al-binding sites once it enters the cellular environment, prevention of Al uptake is by far the simplest approach for reducing Al toxicity.

Release of Al-chelating organic acids has been documented as the primary Al resistance mechanism in multiple plant species including wheat, maize, and barley. Organic acids excreted from roots chelate Al in the rhizosphere to form nontoxic complexes that in some way prevent Al uptake. By not internalizing Al, the root tip is protected from Al-related damage that can be severe enough to cause terminal differentiation of the quiescent center (QC) in the root apical meristem (Ma 2000; Kochian et al. 2005). Al chelation commonly occurs through exudation of malate, citrate, or oxalate to render Al insoluble (Kochian et al. 2005). Strangely, it has been argued that in animal systems, an $Al^{3+}$-citrate complex is the form that is most readily transported across a cellular membrane (Macdonald and Martin 1988), thus making it unclear as to why such complexes prevent internalization in plants. Regardless, besides preventing internalization into the symplast, these organic acid-Al complexes also reduce the capability of $Al^{3+}$ to directly interact with the negative charges of the apoplast such as polygalacturonic acids and other components of the cell wall, which would normally increase wall rigidity and cause gross physical damage upon cell elongation (Horst et al. 1999).

Organic acid-dependent Al exclusion was first reported in Al resistant snapbean cultivars (Miyasaka et al. 1991) and subsequently studied intensively in an Al resistant wheat cultivar that has roots that secrete malic acid into the rhizosphere in response to Al (Delhaize et al. 1993). Following characterization of the role of the wheat *Alt1* locus in Al exclusion, it was found that the Al resistance associated with it was dependent on increased expression levels of *ALMT1*, which encodes an Al-activated root malate efflux transporter that has subsequently been reported in several plant species (Sasaki et al. 2004; Hoekenga et al. 2006). It has been argued based on these studies that Al exclusion must be a rapid response to minimize Al uptake and subsequent Al-dependent stress. Interestingly, much of the findings on Al exclusion mechanisms have arisen from studies that move roots from a no $Al^{3+}$ environment to one that has highly toxic levels with the research focused on the immediate responses to Al. It is hard to imagine a real world situation in which roots go from an environment with little to no $Al^{3+}$ to one that has highly inhibitory concentrations. Therefore, it is arguable whether the approach of studying immediate responses to Al is necessarily relevant to Al toxicity in soils since stoppage of root growth in such an environment is likely due to chronic long-term exposure to $Al^{3+}$. Consequently, it is of critical importance to determine the toxic effects of Al as it accumulates within plant tissue.

## *1.2 Aluminum Toxicity*

Since Al can exist as a trivalent cation, Al will bind to a vast array of biochemical sites once internalized because it has a higher affinity for anionic targets in comparison to ions such as $Mg^{2+}$ and $Ca^{2+}$. Al is proposed to enter the symplast via an Nramp-like transporter, Nrat1 (Xia et al. 2010). Al accumulation within root tissue can be rapid upon exposure. In soybean roots, Al can accumulate in the symplasm of the three outer cortical cell layers after only 30 min of treatment (Barceló and Poschenrieder 2002). Rapid uptake is also observed in Arabidopsis roots, where significant levels of Al accumulate within 1 h of exposure (Barceló and Poschenrieder 2002). This accumulation of Al is not only very rapid but also disrupts cellular activities and induces stress signaling.

Internalized Al has both apoplastic- and symplastic-binding sites. The trivalent cation has the capability of indiscriminately binding to a wide range of negatively charged biomolecules within the cell including sites within the nucleus. $Al^{3+}$ can displace other cations like $Ca^{2+}$, $Mg^{2+}$, and $K^{+}$, thus inhibiting or altering the function of the structures with which Al associates (Silva et al. 2000). For example, $Al^{3+}$ has been found to bind $10^7$ times more tightly to ATP than $Mg^{2+}$, suggesting that $Al^{3+}$ is generally inhibitory to ATP requiring enzymatic reactions that depend on an Mg-ATP complex to function (Macdonald and Martin 1988). Specific toxic effects include inhibition of $Ca^{2+}$ influx across this membrane through blockage of $Ca^{2+}$ channels, disruption of $H^{+}$ homeostasis and neutralization of the zeta potential at the membrane surface, and inhibition of $H^{+}$ flux mediated by the $H^{+}$-ATPase (Degenhardt et al. 1998; Kochian et al. 2005).

In response to the cellular damage caused by internalization of Al, plants must activate signal transduction pathways in order to cope with the stress. It is well documented that Al induces transcriptional programs in order to facilitate detoxification and redistribution of Al within the plant body. For example, *ALS3* is an Al-inducible gene that encodes an ABC-like transporter proposed to redistribute Al away from the most Al sensitive tissues such as the root apical meristem in order to maintain root growth (Larsen et al. 2005). Other Arabidopsis genes that are Al inducible and have a protective role include those involved in oxidative stress such as *AtBCB* (blue copper-binding protein), peroxidases, glutathione-S-transferases, and superoxide dismutases (Richards et al. 1998). Overexpression of several of these types of genes in Arabidopsis reduces Al toxicity symptoms in conjunction with lowering Al-dependent oxidative damage (Ezaki et al. 2001). Such findings clearly point to ROS as being critical to manifestation of Al toxicity symptoms, suggesting that amelioration of Al-dependent oxidative stress may be a useful approach for improving crop growth in an Al toxic environment.

Another well-documented hallmark of Al-inducible response is rapid deposition of callose in the plasmodesmata, which can be observed in plants within a few hours of treatment with Al (Wissemeir et al. 1987). Callose is a β-1,3-glucan that is normally found very rarely within plant cells and only known to be involved in a few specific developmental processes, such as pollen tube growth and as part of p-protein plugs in sieve tube elements in response to wounding. Callose deposition in plasmodesmata is a regulated response to stress, likely to isolate affected or damaged cells from healthy ones. Because its deposition is so tightly linked to Al exposure, callose accumulation has been a useful marker for assessing manifestation of Al toxicity (Larsen et al. 1996; Ezaki et al. 2001).

These Al-inducible responses represent just a few processes that are related to the hundreds of genes that have been found to be upregulated following Al exposure (Chandran et al. 2008; Kumari et al. 2008). As of now, it has been nearly impossible to differentiate between those genes that are of central importance to Al toxicity and stoppage of root growth and those that encode peripheral secondary factors that are only tangentially related. Determining whether various factors are related to the primary or secondary effects of Al-dependent root growth inhibition has proven to be a great challenge to understanding the nature of Al toxicity and the molecular basis of Al-dependent stoppage of root growth, although recent results suggest that Al-dependent DNA damage may be of paramount importance.

## 2 Tolerance to Long-Term Aluminum Exposure

Plants are exposed to a variety of stresses with which they must constantly cope. Their sessile nature requires that they adopt effective strategies to sense and respond to a wide range of stresses for survival. In the case of internalized Al, responses can be subdivided into two categories: short term and long term with the latter likely being more relevant to the natural situation. While Al toxicity causes root growth inhibition by both inhibition of cell elongation and cell division, reduced cell elongation is likely a short-term or immediate response to Al. In contrast, arrest of cell division is argued to be a long-term or chronic symptom of Al toxicity (Kochian 1995). As previously argued, plants growing in Al toxic soils will reside in that environment for the entirety of the life cycle, and it is likely even their progeny will reside in the same Al toxic soils, thus suggesting that the cumulative effects of Al toxicity may have consequences not only for the plant itself but also for its progeny when one considers that Al may act as a DNA-damage agent. While many studies on Al report toxic effects from transient exposures to highly inhibitory levels of Al, it is unlikely that plants growing in Al toxic soils will experience a similar regimen of short, intermittent exposures to Al. Therefore, it is of great importance to study the effects of and response to long-term, chronic Al exposure as it most accurately replicates the true nature of the biological problem.

In addition to the importance of studying the effects of Al toxicity following chronic exposure, it is also necessary to study plant responses to physiologically relevant Al concentrations. This of course is complicated since Al speciation is pH dependent, meaning that Al toxicity is not only related to concentration but also the species that is present as a result of solution pH. For example, at or below pH 5.5, Al speciates in soil solutions into the trivalent cation form (Macdonald and Martin 1988). This is especially true when solution pH changes from 4.5 to 4.0. This small change in solution acidity will cause profound increases in the amount of Al that is found in the toxic $Al^{3+}$ form (Tyler et al. 1987). Therefore, in order to best understand how Al affects a plant in its environment, one must consider factors such as length of exposure, Al concentrations, and the pH of the environment, all of which are key determinants of the type of damage caused by Al and the severity of response to Al by the root.

## 2.1 Aluminum Tolerance Mechanisms in Plants

The mechanisms of Al resistance have been intensively studied on crop species using natural genetic variation within and across species, such as wheat and maize. While clearly an insightful approach that has given extensive knowledge on Al exclusion mechanisms, this work is limited based on currently existing variability with regard to growth in the presence of Al. Mutational approaches using the model plant species Arabidopsis have become an important complement to these studies. Beyond the obvious advantages of using a model species, Arabidopsis has a similar Al toxic threshold to many agriculturally relevant crop species making it a valuable system for investigating how plants sense and respond to Al through the identification of mutants with altered growth capabilities in the presence of Al (Larsen et al. 1996). This has been particularly true with regard to identification of *Al sensitive* (*als*) Arabidopsis mutants, which have reduced root growth in the presence of Al likely due to defects in mechanisms required for Al exclusion, Al detoxification, or response to Al-dependent damage. By screening for *Al sensitive*

Arabidopsis mutants, eight complementation groups were identified indicating that Al toxicity is complex, which is to be expected considering the likely number of factors involved in mechanisms of Al resistance and tolerance (Larsen et al. 1996). Most importantly, as will be discussed later, identification of these *als* mutants has allowed for use of a suppressor mutagenesis approach that has resulted in identification of factors that are important for Al-dependent stoppage of root growth.

From this original mutagenesis screen, *als3* was found to be a mutant with extreme Al hypersensitivity. Analysis of the recessive *als3-1* loss-of-function mutant showed that in the absence of Al, it was indistinguishable from Col-0 wt, thus indicating that it is at best only tangentially required for normal plant growth and development. In contrast, growth in the presence of levels of Al that have only a limited effect on Col-0 wt roots results in *als3-1* roots that are severely stunted, which is a consequence of the mutant roots being forced to undergo terminal differentiation at a level of Al that does not visibly impact wild-type growth. Clearly, based on the severity of the *als3-1* phenotype in the presence of Al, it likely represents a key factor in detoxification or redistribution of internalized Al.

Map-based cloning of *als3-1* showed that it represents a defect in a gene that encodes an ABC-like transporter homologous to bacterial ybbM, which is a metal resistance protein from *Escherichia coli*. Based on this similarity and the localization pattern of ALS3, which shows it predominantly at the plasma membrane of root cortical cells and cells of the vasculature, it was proposed that it redistributes Al away from the most sensitive plant tissues in order to maintain cell division (Larsen et al. 1997, 2005). Loss of *ALS3* as in the *als3-1* mutant would result in inappropriate accumulation of Al in vulnerable areas such as the root tip and would consequently cause growth arrest at levels of Al that have no measurable effect on wild type.

Consistent with the importance of *ALS3* to Al tolerance, an *ALS3* homolog was identified in rice, called *STAR2*. Although both *ALS3* and *STAR2* are required for plant Al tolerance, the expression patterns and cellular localization differ. *STAR2* is only expressed in roots upon Al treatment and is located in all cell types except for the epidermal cells in the mature root zone (Huang et al. 2009). In contrast, *ALS3* is expressed at a basal level in the vasculature throughout the plant and its expression is dramatically increased in the Arabidopsis root tip following exposure to Al (Larsen et al. 2005). Similar to ALS3, STAR2 contains several transmembrane domains that likely form a pore or channel that is involved in substrate movement. Both ALS3 and STAR2 lack an ATPase domain, making them unusual with regard to ABC transporters that often have the transmembrane domains and ATPase domain all as part of one protein. While a separate ATPase domain containing protein partner has not been found for ALS3 (Larsen et al. 2005), rice STAR2 was shown to interact with another protein, STAR1, which contains an ATPase domain (Huang et al. 2009). The STAR1/STAR2 complex functions together as a bacterial-type ABC transporter that is speculated to transport UDP-glucose, although it is currently unclear as to how the transport of UDP-glucose by STAR1/2 is responsible for rice Al tolerance (Huang et al. 2009).

Other Arabidopsis mutants that have been identified with altered responses to Al include *als1* and *als7*. Both *als1* and *als7* were identified in the original screen for Arabidopsis mutants with Al hypersensitivity. *ALS1* encodes a half type ABC transporter that was localized to the vacuolar membrane of root tip cells, suggesting that it may be important for compartmentalization of internalized Al (Larsen et al. 2007). *ALS7/SLOWWALKER* encodes a transcription factor that among other things is required for regulation of expression of genes whose products participated in the production of polyamines such as spermine (Nezames et al. 2012b). Since *als7-1* has Al hypersensitive roots, it is expected that reduced production of polyamines lowers the protective effect of these multicharged cations because of reduced capacity to compete with $Al^{3+}$ for binding to anionic sites within the root tip. Anionic targets of $Al^{3+}$ are expected to include negative charges in the plant cell wall as well as symplastic targets such as genomic DNA, which directly binds to polyamines such as spermine and spermidine.

## 2.2 als3-1 *as a Tool for Dissecting the Basis of Al Toxicity*

In Arabidopsis, *als3-1* has an extreme Al sensitivity phenotype in the presence of low to moderate concentrations of Al. Its roots are stunted and swollen at the apex due to terminal differentiation of the meristem, with root hairs and lateral roots originating at or close to the tip (Rounds and Larsen 2008). This Al sensitivity phenotype is easily distinguished from wild-type in the presence of Al and has made a fantastic system for identification of secondary suppressor mutations, since these suppressors result in long healthy roots in the presence of Al despite the mutational loss of *ALS3*. This is because these Al tolerant secondary suppressor mutants do not undergo the characteristic terminal differentiation displayed by Al-treated *als3-1* and sustain cell division and maintain root growth in the presence of Al. Most important to the success of this approach, the extreme differential in growth phenotypes between Al-treated *als3-1* roots and roots of *als3-1* suppressor mutants has allowed for the successful use of map-based cloning to actually identify the nature of the suppressor mutations.

As previously discussed, *ALS3* encodes an ABC-like transporter that is proposed to redistribute Al away from the most Al sensitive tissues such as the root apical meristem in order to maintain cell division (Larsen et al. 2005). Consequently, it was expected that a suppressor mutagenesis screen would result in identification of gain-of-function mutations that enhance the activity of factors involved in mechanisms such as Al exclusion. Surprisingly, screening for *als3-1* suppressor mutations has resulted in identification of two loss-of-function mutations that affect genes that act as cell cycle checkpoints that are activated in response to DNA damage. This strongly suggests that, since loss of either increases root growth even in comparison to Col-0 wt, stoppage of root growth following Al exposure is an active process mediated by a DNA-damage response pathway.

The first Al tolerant suppressor of *als3-1* identified was a loss-of-function mutation in *ATAXIA TELANGEICTASIA MUTATED AND RAD3-RELATED (ATR)*, a highly conserved factor universal to all eukaryotes that encodes a cell cycle checkpoint responsible for detecting and responding to DNA damage by stopping cell division and activating repair mechanisms (Rounds and Larsen 2008). ATR is highly related to ATM (ATAXIA TELANGIECTASIA MUTATED), with both functioning together to monitor genome integrity albeit in different capacities. Whereas ATM is activated by accumulation of double strand breaks in DNA (DSBs), ATR is responsive to persistent single strand DNA that accumulates for example when the replication fork stalls (Culligan et al. 2006). Interestingly, loss-of-function mutations for *ATR* or *ATM* result in seedlings that are highly sensitized to various genomic stresses yet in the case of Al, loss of *ATR* function confers substantive increases in Al tolerance. This intriguing conundrum will be discussed later.

The second mutation capable of suppressing the *als3-1* sensitivity phenotype was found in *ALUMINUM TOLERANT 2/TANMEI* (*ALT2*), which also encodes a cell cycle checkpoint that monitors and actively responds to Al toxicity through promotion of root growth inhibition (Nezames et al. 2012a). *ALT2* encodes a DWD (DDB1-binding WD-40) motif containing protein (Lee et al. 2008) homologous to Arabidopsis Cockayne Syndrome type A protein (AtCSA), which is an integral component of the mechanism required for detection of UV-B-dependent DNA damage. AtCSA is a recently described WD-40 protein that is a component of the DNA-Damage-Binding (DDB) machinery (Biedermann and Hellmann 2010), which is important to assess changes in DNA integrity and conformation. CSA was originally identified in humans as being a critical factor for detection of UV-related damage that results in DNA crosslinks (Biedermann and Hellmann 2010). At least in animals, CSA works in conjunction with CSB to scan DNA for areas that halt the progression of RNA polymerase II, which subsequently activates Transcription Coupled Nucleotide Excision Repair (TCNER) to remove the crosslinked nucleotides (Saijo 2013). While it is not clear exactly how AtCSA participates in DNA repair in Arabidopsis, the identification of a closely related DWD motif containing WD40 protein that is required for stoppage of root growth following Al treatment indicates that Al-dependent DNA damage is an important theme to Al responsive root growth inhibition.

## 2.3 Aluminum Tolerance Genes ATR and ALT2

In higher eukaryotes, *ATR* encodes a cell cycle checkpoint that senses DNA damage as part of a signal transduction pathway capable of halting the cell cycle in order to repair damaged foci (Siede et al. 2006). In Arabidopsis, ATR is known to signal repair of single-stranded DNA breaks and replication fork stalls that result in persistence of single-stranded DNA (Culligan and Britt 2008). *alt1-1* (*atr-4*) is a point mutation in *ATR* conferring a single amino acid substitution of G1098E in the highly conserved yet uncharacterized UME domain, which is speculated to function in protein-protein interactions, whereas *alt1-2* results from a L2553F substitution in the conserved phosphatidylinositol 3- and 4-kinase domain of ATR (Rounds and Larsen 2008). Both of these dominant-negative mutant alleles were originally found because of their capability to suppress the hypersensitivity phenotype of *als3-1* (Gabrielson et al. 2006). Further work has shown that a T-DNA insertion allele, *atr-2*, is also capable of fully suppressing *als3-1*, thus supporting the argument that *alt1-1* (*atr-4*) and *alt1-2* are mutations that reduce the function of ATR even though they are dominant.

This is particularly evident when one considers the response of *atr-2* and *atr-4* to other DNA-damage agents, which only adds to the conundrum of why Al activates this ATR-dependent pathway. While these mutants are Al tolerant, as demonstrated by their capacity to maintain root growth due to failure to arrest cell cycle progression and force QC differentiation, *atr-2* and *atr-4* roots exposed to different DNA-damage agents such as the replication fork poison hydroxyurea (HU) and the DNA crosslinkers cisplatin (CDDP) and Mitomycin C (MMC) exhibit extreme sensitivity (Rounds and Larsen 2008; Nezames et al. 2012a). From these results, ATR is absolutely necessary to repair stalled replication forks as well as DNA crosslinks yet loss of this repair factor confers Al tolerance. It is difficult to reconcile increased Al tolerance with loss of such a key DNA-damage response factor, although it could be argued that $Al^{3+}$ may bind to DNA in a manner similar to covalent crosslinkers but without the extreme detrimental effects of these crosslinkers. This argument is based on the likelihood that $Al^{3+}$ would interact with DNA electrostatically in a reversible manner, with binding likely holding DNA in an unfavorable conformation that subsequently is perceived as a replication fork stall by ATR. Interestingly, even though treatment with Al has been shown to result in DSBs and micronuclei (Matsumoto 1988; Karlik et al. 1980; Achary et al. 2013), a loss-of-function *atm* mutant was incapable of suppressing the Al hypersensitivity phenotype of *als3-1* (Rounds and Larsen 2008) suggesting that DSBs are not important to activating the DNA-damage checkpoint following Al treatment.

The second *als3-1* suppressor mutant identified is *alt2-1*, which is a loss-of-function mutation in the cell cycle checkpoint *TANMEI/ALT2*. *ALT2* encodes a WD-40 motif containing protein homologous to an integral component of the mechanism required for response to DNA damage, AtCSA, which is part of the DNA-damage-binding (DDB) machinery (Nezames et al. 2012a; Biedermann and Hellmann 2010). WD-40 proteins are commonly found in many different biochemical pathways, serving as scaffolds for protein complexes. In some cases, such proteins participate in SCF ubiquitin ligase complexes in order to tag target proteins for degradation. Such a role for ALT2 in response to Al is consistent with the observation that CULLIN4, which is a key component of SCF ubiquitin ligases, interacts directly with DWD motif containing proteins (Lee et al. 2008; Biedermann and Hellmann 2010). Such a function is distinctly different from what was previously described for human CSA, which functions in TCNER to monitor for conformational changes in DNA as assessed by blockage of transcription (Saijo 2013). Cooperation of a WD-40 motif containing protein with CULLIN4 is characteristic of Global Genome Nucleotide Excision repair (GGNER) and is independent of RNA polymerase II (Siede et al. 2006). At this point, it is unclear in which if either NER response pathway ALT2 may participate although the potential linkage of ALT2 to either is consistent with Al acting as a genotoxin.

Like *atr-2* and *atr-4*, *alt2-1* falls into the same conundrum since it fails to halt the cell cycle and force differentiation of QC in the presence of a normally inhibitory concentration of Al yet it is highly sensitive to DNA crosslinkers (Nezames et al. 2012a). A double mutant representing the loss-of-function of both *ATR* and *ALT2* showed no additive Al tolerance compared with the single mutants, suggesting that ATR and ALT2 act together to detect Al-dependent damage and actively halt root growth. Interestingly, *alt2-1* does not exhibit hypersensitivity to the replication fork poison HU while it does show extreme sensitivity to CDDP and MMC, which is consistent with it being a key regulator of cell cycle progression following exposure to DNA-damage agents (Nezames et al. 2012a). Therefore, at this point the genotoxic nature of Al has yet to be defined. However, it is clear from the unbiased *als3-1* suppressor screens, in which DDR factors were identified as Al tolerance mutations, an ATR- and ALT2-dependent cell cycle checkpoint pathway is key to stoppage of root growth and promotion of terminal differentiation following Al treatment.

## 2.4 Aluminum as a Cumulative, Genotoxic Stress

Based on the results with ATR and ALT2, one could argue that Al-dependent DNA damage is a critical determinant of root growth inhibition, yet the effects of Al on the nucleus are just beginning to be elucidated. Al rapidly accumulates to high levels in root meristem nuclei (Silva et al. 2000) and is especially concentrated around interphase chromatin as well as mitotic figures (De Boni et al. 1974). It is hypothesized that Al binds to the phosphate backbone of DNA (Andersson 1988), which would be expected to result from an electrostatic attraction of $Al^{3+}$ to the negative charges of the phosphodiester bonds. Such an association could increase the rigidity of euchromatin and relax supercoiled heterochromatin destabilizing genome topology through an ever-fluxing torsional tug-of-war. Binding of Al to DNA or chromatin could condense DNA molecules and inhibit cell division by reducing its capacity to provide a viable template for replication, mitotically relevant transcriptional events, and even proper DNA separation (Matsumoto 1988). It is not unreasonable to predict that such conformational changes could be perceived by ATR as being deleterious to replication fork progression, thus activating this cell cycle checkpoint.

It should be noted that chromosomal aberrations resulting in DNA breakage and intra-strand crosslinking are a reported consequence of chronic exposure to Al that lead to micronuclei formation (Matsumoto 1988; Karlik et al. 1980; Achary et al. 2013) although it is not clear how relevant these are to stoppage of root growth since ATM does not appear to have a role in this process (Rounds and

Larsen 2008). Clearly, further studies are necessary to define if and how Al interacts with genomic DNA, especially since it is the $Al^{3+}$ species that is predicted to bind to the negatively charged DNA backbone, yet intracellular pH is not expected to favor the formation of this species. In the end, utilizing Al toxicity as a real world model to study DNA damage in plant systems presents a novel system to study DDR without the use of rare chemotherapy drugs or types of radiation not found in earthly environments.

Ultimately, it seems counterintuitive that a plant gains tolerance to an agent that causes DNA damage by reducing the function of a factor necessary for DNA-damage detection. This certainly begs the question—if *atr* and *alt2* mutant roots can maintain root growth even in the presence of Al, what actual Al-dependent damage is detected by these cell cycle checkpoints? It cannot be ruled out that inappropriate activation of the cell cycle checkpoint machinery and the repair mechanisms that they regulate may actually be the cause of the damage such as DSBs and micronuclei. Consequently, one explanation could be that failure to activate this pathway prevents the program-dependent accumulation of the damage and results in roots that can grow in normally inhibitory levels of Al.

## 3 DNA-Damage Responses

Determination of the nature of Al genotoxicity is currently at the initial stages of investigation. While defining Al responsive mechanisms is suggestive of the type of damage Al causes, it is crucial to define the genotoxic consequence of Al that triggers this response, whether real or perceived. Additionally, unanswered questions also persist regarding which factors participate in conjunction with ATR and ALT2 to respond to Al-dependent damage since clearly Al responsive stoppage of root growth is a multistep process progressing through cell cycle arrest to terminal differentiation associated with endoreduplication or DNA replication without cytokinesis. Therefore, not only is further identification of Al tolerance factors crucial to our understanding of Al response signaling, it is also of critical importance to determine how these factors function together to promote Al-dependent cell cycle arrest causing terminal differentiation and subsequent plant growth inhibition. While the genotoxic consequences of Al in plants have yet to be elucidated, it is clear from our *als3-1* suppressor mutagenesis approach that components of the DDR machinery are mediating root growth inhibition in response to the likely negative impact of Al on DNA structure, integrity, or conformation.

Based on evidence gained within the fields of Al toxicity and plant DNA-damage response and repair, it can be hypothesized that ATR may trigger Al-dependent root growth inhibition through the p53-like transcription factor SUPPRESSOR OF GAMMA RESPONSE 1 (SOG1) to evoke a transcriptional response upregulating a suite of yet to be discovered genes involved in the Al stress response. In other eukaryotic species such as yeast, mice, and human, there are two related phosphotidylinositol-3-kinase (PI3K)-related protein kinases, ATAXIA

TELANGEICTASIA MUTATED (ATM) and ATR (Siede et al. 2006). Together they mediate cell cycle arrest and coordinate repair proteins through the transcription factors Chk1, Chk2, and p53 in order to regulate accurate mitotic growth (Sancar et al. 2004). While no Chk1 or Chk2 homologues have been identified in Arabidopsis (Yoshiyama et al. 2013b), both ATM and ATR play distinct roles in the DNA-damage response and mediate transcriptional responses through SOG1, which is predicted to function similarly to mammalian p53 and is required for transition from an active cell division to endoreduplication (Culligan et al. 2006; Yoshiyama et al. 2009).

## 3.1 Persistent Single-Stranded DNA Regulated by ATR

As noted earlier, in higher eukaryotes, ATR is a cell cycle checkpoint that senses DNA damage as part of a signal transduction pathway capable of arresting cell cycle progression in order to repair DNA damage (Siede et al. 2006). In Arabidopsis, ATR is known to detect single-strand DNA breaks and stalled replication fork structures that interfere with DNA replication (Culligan and Britt 2008), and this function is conserved throughout other model eukaryotic species like yeast and metazoa (Siede et al. 2006). As part of the mechanism to locate damaged DNA foci, ATR binds to a partner protein that is also highly conserved among eukaryotes, *ATR INTERACTING PROTEIN (ATRIP)* (Cortez et al. 2001; Siede et al. 2006).

The Arabidopsis ATRIP homologue was identified from two distinct genetic screens for mutants hypersensitive to UVB and hydroxyurea termed *SENSITIVE TO UV 2* (*SUV2*) and *HYDROXYUREA SENSITIVE 2* (*HUS2*) respectively (Sweeney et al. 2009; Sakamoto et al. 2009). *SUV2/HUS2* is the homologue of *ATR interacting protein* (*ATRIP*) from vertebrates, *Lcd1* from *S. cerevisiae*, and *Rad26* from *S. pombe* (Sweeney et al. 2009, Sakamoto et al. 2009; Rouse and Jackson 2000; Edwards et al. 1999). While the sequence of this factor is not conserved among eukaryotes, ATRIP function is. All factors exhibit functional conservation acting as DNA-binding proteins that partner with ATR/Mec1/Rad3 and target their respective kinase substrates to sites of DNA damage (Siede et al. 2006). Based on the evolutionarily conserved relationship between ATR homologues and the respective ATRIP homologues from each species, it could be speculated that Arabidopsis *atrip* loss-of-function mutants will display characteristics similar to that of the *atr* loss-of-function mutants and could prove to be Al tolerant. This would be consistent with ATRIP being required for recruitment of ATR to sites of Al-dependent DNA damage.

Based on the established DNA repair mechanisms in bacterial and mammalian systems, ATRIP may interact with Arabidopsis homologues of the heterotrimeric RPA complex, which is well known to bind to single-stranded DNA, as well as RAD17 and the "9-1-1" complex (RAD9, RAD1 and HUS1), and has been shown to participate in ATR-dependent repair mechanisms (Behailu et al. 2013; Siede et al. 2006). While no homologues in Arabidopsis have been found of downstream effectors of ATR such as Chk1 and Chk2 in vertebrates, Rad53 in *S. cerevisiae*, or Cds1 in *S. pombe*, SOG1 is a transcription factor known to be capable of acting downstream of ATM (Yoshiyama et al. 2009, 2013a, b). It is interesting to speculate that SOG1 may also function downstream of ATR in response to Al, since Al treatment causes hallmarks of endoreduplication in Al-treated roots of *als3-1* and even wild-type Arabidopsis.

### 3.2 *Double-Stranded Breaks Regulated by ATM*

In all eukaryotes studied that have both ATM and ATR homologues, each of the two related PI kinases have divergent functions as they recognize different DNA-damage lesions. While ATR recognizes persistent single-stranded DNA, single-strand DNA breaks, stalled replication forks, and subsequently signals for their repair, ATM mainly responds to double-strand DNA breaks (Yoshiyama et al. 2013b). SOG1 acts to link ATM with induction of responses following DNA damage (Preuss and Britt 2003) and because of this, it can be speculated that SOG1 is also a downstream effector of ATR. Further work on SOG1 has shown that following DNA damage, it is responsible for entrance into endocycling in an ATM-dependent manner (Culligan et al. 2006; Culligan and Britt 2008; Adachi et al. 2011) that results in induction of transcription targets (Yoshiyama et al. 2009). For example, following treatment with $\gamma$-radiation, SOG1 upregulates a suite of genes including those responsible for repair of DNA damage such as *BRCA1*, *AtRAD17*, and *AtRAD51* (Yoshiyama et al. 2009). However, as discussed earlier, there is no evidence that Al-responsive root growth inhibition is ATM-dependent, as an *atm* loss-of-function mutant does not suppress the *als3-1* Al hypersensitivity phenotype (Rounds and Larsen 2008). Based on the functional similarities between ATR and ATM, it is intriguing to consider that since Al-dependent root growth inhibition is largely an ATR-dependent phenomenon and Al leads to terminal differentiation coupled with endoreduplication, ATR may function through SOG1 to regulate this response.

### 3.3 *The Future of Aluminum-Inducible DNA-Damage Responses*

The identification of ATR and ALT2 as mediators of Al-dependent root growth inhibition presents new strategies and opportunities to engineer crop species capable of growing in Al toxic soils. Based on the critical roles of ATR and ALT2 in mediating Al-dependent root growth inhibition and the current knowledge gained from the field of plant DNA damage, further studies need to be done to identify mediators and effectors within the ATR- and ALT2-mediated Al response pathway, especially those responsible for controlling cell cycle arrest, damage repair, and subsequent promotion of endocycling at the root apical meristem to stop root growth. Our understanding of the genomic consequences caused by Al is still in the beginning stages, and more work is needed. Continued testing of DDR mutant responses to Al can give us the opportunity to elucidate further how genomic maintenance factors are involved in this biological problem. In addition to the value of gaining a better understanding the role of DDR factors and cell cycle checkpoints in mediating Al-dependent DNA damage, Al toxicity represents a novel and biologically relevant model to study ATR-dependent mechanisms in the DDR in general.

## References

Achary VM, Parinandi NL, Panda BB (2013) Calcium channel blockers protect against aluminium-induced DNA damage and block adaptive response to genotoxic stress in plant cells. Mutat Res 751(2):130–138

Adachi S, Minamisawa K, Okushima Y, Inagaki S, Yoshiyama K, Kondou Y, Kaminuma E, Kawashima M, Toyoda T, Matsui M, Kurihara D, Matsunaga S, Umeda M (2011) Programmed induction of endoreduplication by DNA double-strand breaks in Arabidopsis. Proc Natl Acad Sci U S A 108:10004–10009

Aklilu BB, Soderquist RS, Culligan KM (2013) Genetic analysis of the Replication Protein A large subunit family in Arabidopsis reveals unique and overlapping roles in DNA repair, meiosis and DNA replication. Nucleic Acids Res 42:1–15

Andersson M (1988) Toxicity and tolerance of aluminum in vascular plants: a literature review. Water Air Soil Pollut 39:439–462

Barceló J, Poschenrieder C (2002) Fast root growth responses root exudates and internal detoxification as clues to the mechanisms of aluminum toxicity and resistance: a review. Environ Exper Bot 48:75–92

Biedermann S, Hellmann H (2010) The DDB1a interacting proteins ATCSA-1 and DDB2 are critical factors for UV-B tolerance and genomic integrity in *Arabidopsis thaliana*. Plant J 62:404–415

Chandran D, Sharopova N, Ivashuta S, Gantt JS, Vandenbosch KA, Samac DA (2008) Transcriptome profiling identified novel genes associated with aluminum toxicity, resistance and tolerance in *Medicago truncatula*. Planta 228:151–166

Cortez D, Guntuku S, Qin J, Elledge SJ (2001) ATR and ATRIP: partners in checkpoint signaling. Science 294:867–870

Culligan KM, Britt AB (2008) Both ATM and ATR promote the efficient and accurate processing of programmed meiotic double-strand breaks. Plant J 4:629–638

Culligan KM, Robertson CE, Foreman J, Doerner P, Britt AB (2006) ATR and ATM play both distinct and additive roles in response to ionizing radiation. Plant J 48:947–961

De Boni U, Scott JW, Crapper DR (1974) Intracellular aluminum binding; a histochemical study. Histochemistry 40:31–37

Degenhardt J, Larsen PB, Howell SH, Kochian LV (1998) Aluminum resistance in the Arabidopsis mutant *alr-104* is caused by an aluminum-induced increase in rhizosphere pH. Plant Physiol 117:19–27

Delhaize E, Ryan PR, Randall PJ (1993) Aluminum tolerance in wheat (Triticum *aestivum* L.) II. Aluminum-stimulated excretion of malic acid from root apices. Plant Physiol 103:695–702

Edwards RJ, Bently NJ, Carr AM (1999) A Rad3-Rad26 complex responds to DNA damage independently of other checkpoint protein. Nat Cell Biol 1:393–398

Ezaki B, Katsuhara M, Kawamura M, Matsumoto H (2001) Different mechanisms of four aluminum (Al)-resistant transgenes for Al toxicity in Arabidopsis. Plant Physiol 127:918–927

FitzPatrick EA (1986) An introduction to soil science. Longman Scientific and Technical, Essex, 255 pp

Foy CD, Chaney RL, White MC (1978) The physiology of metal toxicity in plants. Annu Rev Plant Physiol 29:511–566

Gabrielson KM, Cancel JD, Morua LF, Larsen PB (2006) Identification of dominant mutations that confer increased aluminium tolerance through mutagenesis of the Al-sensitive Arabidopsis mutant, als3-1. J Exp Bot 57:943–951

Hoekenga OA, Maron LG, Piñeros MA, Cançado GM, Shaff J, Kobayashi Y, Ryan PR, Dong B, Delhaize E, Sasaki T, Matsumoto H, Yamamoto Y, Koyama H, Kochian LV (2006) *AtALMT1*, which encodes a malate transporter, is identified as one of several genes critical for aluminum tolerance in *Arabidopsis*. Proc Natl Acad Sci U S A 103:9738–9743

Horst WJ, Schmohl N, Kollmeier M, Baluska F, Sivaguru M (1999) Does aluminum inhibit root growth of maize through interaction with the cell wall, plasma membrane, cytoskeleton continuum? Plant Soil 192:23–30

Huang CF, Yamaji N, Mitani N, Yano M, Nagamura Y, Ma JF (2009) A bacterial type ABC transporter is involved in aluminum tolerance in rice. Plant Cell 21:655–667

Karlik SJ, Eichhorn GL, Lewis PN, Crapper DR (1980) Interaction of aluminum species with deoxyribonucleic acid. Biochemistry 19:5991–5998

Kochian LV (1995) Cellular mechanisms of aluminum toxicity and resistance in plants. Annu Rev Plant Physiol Plant Mol Biol 46:237–260

Kochian LV, Piñeros MA, Hoekenga OA (2005) The physiology, genetics and molecular biology of plant aluminum resistance and toxicity. Plant Soil 274:175–195

Kumari M, Taylor GJ, Deyholos MK (2008) Transcriptomic responses to aluminum stress in roots of *Arabidopsis thaliana*. Mol Genet Genom 279:339–357

Larsen PB, Tai C-Y, Kochian LV, Howell SH (1996) Arabidopsis mutants with increased sensitivity to aluminum. Plant Physiol 110:743–751

Larsen PB, Kochian LV, Howell SH (1997) Al inhibits both shoot development and root growth in *als3*, an al sensitive Arabidopsis mutant. Plant Physiol 114:1207–1214

Larsen PB, Geisler MJB, Jones CA, Williams KM, Cancel JD (2005) *ALS3* encodes a phloem-localized ABC transporter-like protein that is required for aluminum tolerance in Arabidopsis. Plant J 41:353–363

Larsen PB, Cancel J, Rounds M, Ochoa V (2007) Arabidopsis *ALS1* encodes a root tip and stele localized half type ABC transporter required for root growth in an aluminum toxic environment. Planta 225:1447–1458

Lee JH, Terzaghi W, Gusmaroli G, Charron JB, Yoon HJ, Chen H, He YJ, Xiong Y, Deng XW (2008) Characterization of Arabidopsis and rice DWD proteins and their roles as substrate receptors for CUL4-RING E3 ubiquitin ligases. Plant Cell 20:152–167

Ma JF (2000) Role of organic acids in detoxification of aluminum in higher plants. Plant Cell Physiol 4:383–390

Macdonald TL, Martin RB (1988) Aluminum ion in biological systems. Trends Biochem Sci 13:15–19

Matsumoto H (1988) Inhibition of proton transport activity of microsomal membrane vesicles of barley roots by aluminum. Soil Sci Plant Nutr 34:499–506

Miyasaka SC, Buta JG, Howell RK, Foy CD (1991) Mechanism of aluminum tolerance in snapbeans: root exudation of citric acid. Plant Physiol 96:737–743

Nezames CD, Sjogren CA, Barajas JF, Larsen PB (2012a) The Arabidopsis cell cycle checkpoint regulators TANMEI/ALT2 and ATR mediate the active process of aluminum-dependent root growth inhibition. Plant Cell 24:608–621

Nezames CD, Ochoa V, Larsen PB (2012b) Mutational loss of *Arabidopsis SLOW WALKER2* results in reduced endogenous spermine concomitant with increased aluminum sensitivity. Funct Plant Biol 40:67–78

Preuss SB, Britt AB (2003) A DNA-damage-induced cell cycle checkpoint in Arabidopsis. Genetics 164:323–334

Reynolds MP, Ortiz-Monasterio JI, McNab A (eds) (2001) Application of physiology in wheat breeding. CIMMYT, Mexico

Richards KD, Schott EJ, Sharma YK, Davis KR, Gardner RC (1998) Aluminum induces oxidative stress genes in *Arabidopsis thaliana*. Plant Physiol 116(1):409–418

Rounds MA, Larsen PB (2008) Aluminum dependent root growth inhibition results from AtATR dependent cell cycle arrest and loss of the quiescent center in Arabidopsis. Curr Biol 18:1495–1500

Rouse J, Jackson SP (2000) An essential gene involved in checkpoint control and regulation of the *MEC1* signaling pathway in *Saccharomyces cerevisiae*. EMBO J 19:5801–5812

Saijo M (2013) The role of Cockayne syndrome group A (CSA) protein in transcription-coupled nucleotide excision repair. Mech Ageing Dev 134:196–201

Sakamoto AN, Lan VT, Puripunyavanich V, Hase Y, Yokota Y, Shikazono N, Nakagawa M, Narumi I, Tanaka A (2009) A UVB-hypersensitive mutant in *Arabidopsis thaliana* is defective in the DNA damage response. Plant J 60:509–517

Sancar A, Lindsey-Boltz LA, Unsal-Kacmaz K, Linn S (2004) Molecular mechanisms of mammalian DNA repair and the DNA damage checkpoints. Annu Rev Biochem 73:39–85

Sasaki T, Yamamoto Y, Ezaki B, Katsuhara M, Ahn SJ, Ryan PR, Delhaize E, Matsumoto H (2004) A wheat gene encoding an aluminum-activated malate transporter. Plant J 37:645–653

Siede W, Kow YW, Doetsch PW (2006) DNA damage recognition. Taylor & Francis, New York, NY

Silva IR, Smyth TJ, Moxley DF, Carter TE, Allen NS, Rufty TW (2000) Aluminum accumulation at nuclei of cells in the root tip. Fluorescence detection using lumogallion and confocal laser scanning microscopy. Plant Physiol 123:543–552

Sweeney PR, Britt AB, Culligan KM (2009) The Arabidopsis ATRIP ortholog is required for a programmed response to replication inhibitors. Plant J 60:518–526

Tyler G, Berggren D, Bergkvist B, Falkengren-Grerup U, Folkeson L, Rühling Å (1987) In: Hutchinson TC, Meema KM (eds) Effects of atmospheric pollutants on forests, wetlands and agricultural ecosystems. NATO ASI series G16. Springer, Berlin, 347 pp

Wissemeir AH, Klotz F, Horst WJ (1987) Aluminum induced callose synthesis in roots of soybean (*Glycine max* L.). J Plant Physiol 129:487–492

Xia J, Yamaji N, Kasai T, Ma JF (2010) Plasma membrane-localized transporter for aluminum in rice. PNAS USA 107:18381–18385

Yoshiyama K, Conklin PA, Huefner ND, Britt AB (2009) Suppressor of gamma response 1 (SOG1) encodes a putative transcription factor governing multiple responses to DNA damage. PNAS USA 106:12843–12848

Yoshiyama KO et al (2013a) ATM-mediated phosphorylation of SOG1. EMBO Rep 14(9):817–822

Yoshiyama KO, Sakaguchi K, Kimura S (2013b) DNA damage response in plants: conserved and variable response compared to animals. Biology 2:1338–1356

# Signaling Pathways of Aluminum-Induced Programmed Cell Death in Higher Plants

**Hu-Yi He, Ming-Hua Gu, and Long-Fei He**

**Abstract** Aluminum (Al) is the most abundant metals in the earth's crust. Al stress triggers the production of nitric oxide (NO) and hydrogen peroxide ($H_2O_2$). The homeostasis between NO and $H_2O_2$ may be a key decision point of cell survival or cell death. Al toxicity can break NO and $H_2O_2$ homeostasis and induce programmed cell death (PCD) in plants, which is characterized by nucleus condensation and crescent-shaped, marginalized chromatin aggregation, and DNA Ladder. This multiple programming and damaged process is mediated via two signaling pathways. One is mitochondria-dependent pathway. The excess Al toxicity-generated production of ROS leads to lipid peroxidation and induces the opening of MPTP, and then causes the release of Cyt c and finally results in PCD. Another is multi-organelle and nucleus-guided mitochondria-independent pathway, which is executed by regulating gene expressions of PCD promoter and suppressor. The promoters include senescence-associated gene (*SAG*), vacuole processing enzyme (*VPE*), poly (ADP ribose) polymerase (*PARP-1*), and *PDCD5*. Bax inhibitor-1 (*BI-1*), ACCELERATED CELL DEATH2 (*ACD2*), and LESION SIMULATING DISEASE1 (*LSD1*) all belong to the suppressor. There is a negative relationship between the occurrence of PCD and Al resistance, so the negative regulation of Al-induced PCD may be an important mechanism of Al tolerance. In this review, we highlight the newest advances about Al-induced PCD in the context of the relevant literature and enlarge our knowledge on cell death signaling pathways in plants under Al stress.

H.-Y. He
College of Agronomy, Guangxi University, Nanning 530004, China

M.-H. Gu • L.-F. He (✉)
College of Agronomy, Guangxi University, Nanning 530004, China

National Key Laboratory of Conservation and Utilization of Subtropical Agrobioresources, Guangxi University, Nanning 530004, China
e-mail: lfhe@gxu.edu.cn



S.K. Panda, F. Baluška (eds.), *Aluminum Stress Adaptation in Plants*, Signaling and Communication in Plants 24, DOI 10.1007/978-3-319-19968-9_4

# 1 Introduction

Aluminum (Al) is a major limiting factor for crop production in acid soil. When soil pH drops below 5.0, the soluble $Al^{3+}$ is toxic to plants. The inhibition of root elongation is the initial symptom of Al toxicity. Al initially reduces cell elongation, thus decreasing root growth and causing damage to epidermal and cortical cells (Blamey et al. 2004). As a DNA-damaging agent in vivo, Al halts cell cycle progression and forces differentiation of the quiescent center. The cell cycle checkpoint regulators TANMEI/ALT2 and ATR mediate the active process of Al-dependent root growth inhibition (Nezames et al. 2012).

Extensive efforts have been made; plant species have evolved diverse mechanisms of Al tolerance, including the secretion of Al-induced organic acids, immobilization of Al at cell wall, and increasing in rhizosphere pH (He et al. 2012), but the detailed mechanisms of Al toxicity and tolerance are still poorly understood. Programmed cell death (PCD) is defined as a form of cell death involving a series of orderly processes mediated by intracellular death programs, regardless of the triggers or the hallmarks of its exhibits (Zhang and Xing 2008). PCD is a foundational cellular process in plant development and elimination of damaged cells under environmental stresses. Recently, there are some reports on Al-induced PCD in plants, such as rice (Meriga et al. 2004), barley (Pan et al. 2001; Tamas et al. 2005), tobacco (Yamaguchi et al. 1999), peanut (Zhan et al. 2009), onion (Achary et al. 2008; Andrade-Vieira et al. 2011), soybean (Rath and Barz 2000), tomato (Yakimova et al. 2007), and maize (Boscolo et al. 2003). Interestingly, Al-induced PCD process may be controlled by different signaling pathway. The manipulation of the negative regulation process of PCD may provide a novel mechanism for conferring Al tolerance (Zheng et al. 2007). To elucidate the regulatory mechanisms of Al toxicity and tolerance, herein we discuss cell death pathways during Al-induced PCD in plants by combining relevant literature.

# 2 Al-Induced PCD in Plants

Recent studies have described some apoptotic hallmarks that appeared upon Al treatment in plant cells (Table 1). For example, Al promoted $Fe^{2+}$-induced lipid peroxidation and caused death of tobacco suspension cells (Yamamoto et al. 1997). Under $Al^{3+}$ treatment with $Fe^{2+}/Fe^{3+}$ together, the plasma membrane integrity of tobacco suspension cells was destructed, resulting in the inhibition of cell growth (Ikegawa et al. 1998). In tobacco cultured cells, Al promoted $Fe^{2+}$-mediated lipid peroxidation and caused cell death, which required high concentrations of cytoplasmic $Ca^{2+}$ and protease activities. This type of cell death-generated DNA fragmentation belonged to PCD (Yamamoto et al. 2002). When tobacco cells were treated with 50 μmol/L $AlCl_3$ for 18 h, a large number of superoxide anion and $H_2O_2$ arose from mitochondria. Subsequently, the membrane potential and

**Table 1** The reports on Al-induced PCD in plants

| Plant species | Al concentration (μmol/L) | Treatment time (h) | Characteristics | Relevant signal molecule | Reference |
|---|---|---|---|---|---|
| *Nicotiana tabacum* | 120 | 18 | – | $H_2O_2$ | Yamamoto et al. (1997) |
| | – | 8–12 | – | ROS | Ikegawa et al.(1998) |
| | 100 | 24 | DNA degradation | ROS | Yamamoto et al. (2002) |
| | 50 | 18 | Cytosolic shrinkage, nucleus fragmentation | ROS | Panda et al. (2008) |
| | 100 | 6 | DNA degradation | ROS | Wang et al. (2009) |
| *Hordeum vulgare* | 100–1000 | 8 | DNA degradation | ROS | Pan et al. (2001) |
| *Barley* | 2000 | 20–24 | – | $H_2O_2$ | Tamas et al. (2005) |
| *Glycine max* | 15 | 4 | Loss of cell viability | ROS | Rath and Barz (2000) |
| *Zea mays* | 36 | 48 | DNA degradation | ROS | Boscolo et al. (2003) |
| *Oryza sativa* | 80 | 8–56 | DNA degradation | – | Meriga et al. (2004) |
| *Solanum lycopersicum* | 100 | 24 | Cytosolic shrinkage, nuclear condensation | ROS | Yakimova et al. (2007) |
| *Allium cepa* | 50–200 | 4 | DNA damage | ROS | Achary et al. (2008) |
| *Arachis hypoganea* | 400 | 96 | Nucleus shrinkage, apoptotic body | – | Zhan et al. (2009, 2013, 2014) |
| | 100 | 4 | DNA cleavage, DAPI staining, gene expression, cytochrome C release | $H_2O_2$ | Huang et al. (2014a, b) |
| *Arabidopsis* | 500 | 1 | – | ROS | Li and Xing (2011) |

*Note*: ROS means reactive oxygen species

ATP content were declined. The release of cytochrome c (Cyt c) from mitochondria caused PCD (Panda et al. 2008). When tobacco was exposed to 100 μmol/L $Al^{3+}$ for 6 h, the genomic DNA of wild-type and non-transgenic plants were degraded. Overexpression of the Ced-9 gene can inhibit Al-induced PCD in tobacco (Wang et al. 2009). After barley was treated with 0.1–1 mmol/L Al for 8 h, root tip cells generated DNA fragmentation but did not produce apoptotic bodies. Al-induced cell death of barley root tip cells may be a PCD process (Pan et al. 2001). When barley root border cells were treated with 2 mmol/L Al for 20–24 h, apoptosis-like (AL) phenomenon occurred (Tamas et al. 2005). It was showed a distinct and longtime increase in lipid peroxidation within 4 h upon transfer to an Al-containing culture medium with a calculated Al activity of 15 μM soybean cells (Rath and Barz 2000). Maize root tips were treated with 36 μmol/L $Al^{3+}$ for 48 h; the result of TUNEL detection is positive (Boscolo et al. 2003). When rice was treated with 80 μmol/L Al stress for 8–56 h, DNA breakage occurred in root tip cells (Meriga et al. 2004). Tomato suspension cells were treated with 100 μmol/L Al for 24 h; only 67.5 % cells emitted fluorescence by FDA staining, indicating 32.5 % cells had died (Yakimova et al. 2007). When onion root cells were treated with 50–200 μmol/L Al, distinct trailing emerged from comet assay (Achary et al. 2008). Therefore, the negative regulation of Al-induced PCD may be an important mechanism of Al tolerance.

Al induced caspase-3-like activation and PCD, which provided new insight into the signaling cascades that modulate Al phytotoxicity mechanism (Li and Xing 2011). Al induced obvious PCD morphological characteristics, including nucleus condensation, crescent-shaped or oval-shaped, and similar apoptotic bodies. The difference of Al-induced PCD has a negative correlation with Al tolerance of peanut root tips (Zhan et al. 2009).

As described above, it can be seen that Al stress induces morphological changes of plant cells significantly, exhibiting distinct characteristics corresponding to PCD such as nucleus condensation and crescent-shaped, marginalized chromatin aggregation, DNA Ladder, cytochrome C release, special gene expression, etc. And apoptotic bodies are formed in some cases. The physiological aspects of Al-induced PCD also are altered, including severe damage of the mitochondrial respiratory functions, changes of the redox status and the internal structure, and tardy responses to environmental stress.

## 3 NO and $H_2O_2$ Homeostasis

Reactive oxygen species (ROS) and NO are highly reactive and diffusible molecules, and they are known to play key signaling roles in both animal and plant cells, regulating many physiological responses. NO has a strong relationship with another reactive species: hydrogen peroxide. ROS are not only toxic by products of aerobic metabolism with strictly controlled cellular levels, but they also function as signaling agents regulating many biological processes and producing pleiotropic effects.

Al treatments induced cell death possibly via a ROS-activated signal transduction pathway (Pan et al. 2001). Roots are the major sites of Al localization, and accumulation of Al promoted oxygen free radicals mediated peroxidation of membranes (Meriga et al. 2004).

ROS have become recognized as important modulators of plant PCD with emphasis on $H_2O_2$ (Gadjev et al. 2008). Root growth inhibition by Al is probably caused by cell death due to peroxidase-mediated $H_2O_2$ production (Simonovicova et al. 2004). Al-induced cell death of barley-root border cells is correlated with peroxidase- and oxalate oxidase-mediated $H_2O_2$ production (Tamas et al. 2005). It has established that $H_2O_2$ is a key player in stress and PCD responses (Gechev and Hille 2005). Low concentrations of Al stimulate the production of ROS and subsequent cell death (Yakimova et al. 2007).

Our results showed that Al stress induced ROS burst, upregulated Rboh and COX gene expression, increased mitochondrial permeability transition pore (MPTP) opening, decreased inner mitochondrial membrane potential ($\Delta\psi m$), released cytochrome c from mitochondria to cytoplasm, activated caspase 3-like protease activity. Exogenous $H_2O_2$ aggravated the changes caused by Al and accelerated PCD occurrence, but ROS scavenger CAT and AsA reversed the changes caused by Al and inhibited PCD production (Huang et al. 2014b). Al inhibited catalase (CAT) activity and enhanced the activities of superoxide dismutase (SOD), guaiacol peroxidase (GPX), and ascorbate peroxidase (APX) significantly in a dose-response manner (Achary et al. 2008). Through reactive oxygen intermediates (ROI), the biphasic (hormetic) mode of action of Al that at high doses-induced DNA damage and at low nontoxic doses-conferred genomic protection was mediated (Achary and Panda 2010). Al-induced mitochondrial ROS possibly originated from complex I and III damage in the respiratory chain through the interaction between Al and iron–sulfur (Fe–S) protein (Li and Xing 2011). The specificity of the biological responses to ROS depends on the chemical identity of ROS, intensity of the signal, sites of production, plant developmental stage, previous stresses encountered, and interactions with other signaling molecules such as NO, lipid messengers, and plant hormones (Gechev et al. 2006). Hydrogen peroxide works synergistically with NO to stimulate or delay programmed cell death and assist in defense responses to pathogens (de Pinto et al. 2006; Besson-Bard et al. 2008).

NO is a freely diffusible, gaseous free radical and an important signaling molecule in animals. In plants, NO influences growth and development, and can affect plant responses to stress. Al affects mitochondrial functions, which leads to ROS production, probably the key critical event in Al inhibition of cell growth (Yamamoto et al. 2002). NO is often generated contemporaneously with $H_2O_2$. The effects of NO are the results of its interaction with ROS in some cases, and these interactions can be cytotoxic or protective. The presence of NO donors delays the loss of CAT and SOD that metabolize ROS, speculating that NO may be an endogenous modulator of PCD in barley aleurone cells (Beligni et al. 2002). NO plays an important role in protecting the plant against Al-induced oxidative stress (Wang and Yang 2005). The reduction of endogenous NO concentrations resulting from inhibition of nitric oxide synthase (NOS) activity could underpin Al-induced arrest of root elongation in *H. moscheutos* (Tian et al. 2007). However, by alleviating Al-induced oxidative stress in red kidney bean roots, nitrate reductase (NR)-dependent NO production plays an important role in providing protection against Al toxicity (Wang et al. 2010). $Al^{3+}$ induced an increase of NO in rice seedlings, proposing that NO activated multiple pathways that enhance rice adaptation to $Al^{3+}$ toxicity (Yang et al. 2013). It is indicated that the contribution of NOS or NR-mediated NO production is dependent on plant species and environmental stimuli. NO maybe controls PCD by regulating the expression of PCD-related genes (Zhan et al. 2011). NO was first seen as punctuate foci at the cell surface, and subsequent NO was an intercellular signal that functions in cell-to-cell spread of the HR (Zhang et al. 2003).

Owing to their mobility, NO and $H_2O_2$ may act as signal transmission mediator of oxidative and nitrosative stress. Elevated NO levels are sufficient to induce PCD in Arabidopsis cells independent of ROS (Clarke et al. 2000). The simultaneous increase of NO and ROS activated a process of death with the typical cytological and biochemical features of hypersensitive PCD and a remarkable rise in PAL activity. Under the simultaneous generation of NO and ROS, the cellular antioxidant capabilities were also suppressed (de Pinto et al. 2002). SOD accelerates $O_2^-$ dismutation to $H_2O_2$ to minimize the loss of NO by reaction with $O_2^-$ and to trigger hypersensitive cell death through NO/$H_2O_2$ cooperation. The rates of production and dismutation of $O_2^-$ generated during oxidative burst play a crucial role in the modulation and integration of NO/$H_2O_2$ signaling in hypersensitive reaction (HR) (Delledone et al. 2001). Al exposure caused rapid depolarization of the plasma membrane. The extent of depolarization in cells of the distal was much more extensive than in the proximal portion of the transition zone. Cells of the distal portion of the transition zone emitted large amounts of NO, and this was blocked by Al treatment (Illes et al. 2006).

There is a convergence between NO and $H_2O_2$ signaling, which functions at the center of cellular stress responses. In the process of normal development, plants maintain a tight NO and $H_2O_2$ homeostasis. When plants are subjected to environmental stress, if the balance between NO and $H_2O_2$ production is in favor of NO, plants show favorable stress tolerance. If the balance is in favor of $H_2O_2$, plants will be easily injured and even die. Al toxicity can break NO and $H_2O_2$ homeostasis and induce PCD in plants. Conserved negative regulators of PCD are involved in integrated regulation of cell survival and Al-induced PCD (Wang et al. 2009). Eukaryotic cells have to constantly cope with environmental cues, and cell survival or death is the only possible outcome (Cacas 2010). The homeostasis between NO and $H_2O_2$ is key decision point of cell survival or cell death. Meanwhile, alternative oxidase (AOX), the unique respiratory terminal oxidase in plants, not only alleviated excessive ROS accumulation but also suppressed NO concentration. So AOX plays a central role in NO and ROS homeostasis in mitochondria (Gupta et al. 2012) and was also demonstrated to play protective roles in Al-induced protoplast death (Li and Xing 2011).

## 4 Transcription Factors Related to Al-Induced PCD

Genetic and functional genomic studies have shown that many transcription factors (TFs) play essential roles in developmental PCD and abiotic stress PCD. Three basic helix-loop-helix transcription factors, UDT1 (bHLH164), TDR1 (bHLH5), and EAT1/DTD1 (bHLH141), are known to function in rice pollen development. bHLH142 acts downstream of UDT1 and GAMYB but upstream of TDR1 and EAT1 in pollen development. In vivo and in vitro assays demonstrated that bHLH142 and TDR1 proteins interact. Transient promoter assays demonstrated that regulation of the EAT1 promoter requires bHLH142 and TDR1. EAT1 positively regulates the expression of AP37 and AP25, which induce tapetal programmed cell death. The bHLH142 transcription factor coordinates with TDR1 to modulate the expression of EAT1 and regulate tapetal programmed cell death and pollen development (Ko et al. 2014).

It is identified that a glyoxal oxidase (GLOX1), a pectin methylesterase (VANGUARD1), and an A1 aspartic protease (UNDEAD) are direct targets of MYB80. TUNEL assays showed that when UNDEAD expression was silenced using small interfering RNA, premature tapetal and pollen programmed cell death occurred, resembling the myb80 mutant phenotype. UNDEAD possesses a mitochondrial targeting signal and may hydrolyze an apoptosis-inducing protein(s) in mitochondria (Phan et al. 2011).

WRKY transcription factors have been implicated in various transcriptional programs, including biotic and abiotic stress responses, growth, and development (Pandey and Somssich 2009; Rushton et al. 2010, 2012; Van Aken et al. 2013). As the most widely discussed $H_2O_2$-inducible representative of the family, WRKY52 is a senescence-related factor and its overexpression leads to accelerated senescence (Miao et al. 2004). ORESARA1 SISTER1 (ORS1), a member of the NAC transcription factor (TF) family, triggers expression of senescence-associated genes through a regulatory network that may involve cross-talk with $H_2O_2$-dependent signaling pathways (Balazadeh et al. 2011). A C2H2-type zinc finger transcription factor ART1 (for Al resistance transcription factor 1), which specifically regulates the expression of genes related to Al tolerance in rice (*Oryza sativa*), was identified. ART1 regulates 31 genes implicated in Al tolerance in both internal and external detoxification of Al at different cellular levels, including STAR1 and 2 in rice (Yamaji et al. 2009). It had been successful in identification of cis-acting element of ART1, which is present in the promoter regions of 29 genes out of 31 genes regulated by ART1 (Tsutsui et al. 2011). It is regret that there are no reports on the transcription factors related to Al-induced PCD.

# 5 The Genes Related to Al-Induced PCD

The genes controlling the genotypic variation in $Al^{3+}$ tolerance have been cloned such as ALMTs (Aluminum-activated malate transporter) and MATEs (multidrug and toxic compound extrusion), which have been successfully expressed in plants (wheat, barley, Arabidopsis, and rice) as well as tobacco suspension cells (Ryan et al. 2011). STOP1 (sensitive to proton rhizotoxicity1) and ART1 ($Al^{3+}$ resistance transcription factor 1) share significant sequence similarity and appear to act as transcription factors to enhance the expression of a range of genes in $Al^{3+}$-treated roots. STOP1 is a Cys2His2-type zinc-finger protein belonging to a family of transcription factors and localizes to the nucleus. The stop1 mutant is also sensitive to $Al^{3+}$ (but not other metal ions). STOP1 likely functions as a transcription factor that regulates the expression of proton and $Al^{3+}$ responsive genes. ART1, similar to STOP1, belongs to the family of Cys2His2-type zinc-finger transcription factors. ART1 regulates the expression of multiple $Al^{3+}$-tolerance genes in rice such as *OsFRDL4*, *STAR1/2*, *Nrat1* (Nramp aluminum transporter 1), and *OsALS1* (Delhaize et al. 2012).

Natural senescence is a genetically determined cell death progress, characterized by upregulation of many senescence-associated genes (SAGs) (Rosenvasser et al. 2006). A hypothesis was proposed that SAGs can serve as integrators of different signaling pathways that control environmental responses (Balazadeh et al. 2010). We isolated *AhSAG* (a senescence-associated gene) from cDNA library of Al-stressed peanut with PCD, which Open reading frame (ORF) of AhSAG is 474 bp, encoding a SAG protein composed of 157 amino acids. The *AhSAG* was transferred into tobacco. Compared to the control and the antisense transgenic tobacco plants, the fast development and blossom of the sense transgenic plants happened to promote senescence. The ability of Al tolerance in sense transgenic tobacco was lower than in antisense transgenic tobacco according to root elongation and Al content analysis. The expression of AhSAG-GFP was higher in sense transgenic tobacco than in antisense transgenic tobacco. It showed that *AhSAG* can induce or promote the occurrence of PCD in plants (Zhan et al. 2013).

In animal, one group of cysteine proteinases, the cysteine-dependent aspartate-specific proteinases (caspases), are involved in a proteolytic signaling cascade that controls apoptosis. The similar apoptotic caspase cascade has not been uncovered in plants, but other proteolytic enzymes involved in PCD had been found, which are localized in different compartments of plant cells: the cytoplasm (metacaspases), the vacuoles (VPE), and the intercellular fluid (phytaspases). Vacuolar processing enzyme (VPE) is a cysteine-dependent protease responsible for caspase-1 activity in plant and is localized in plant cell vacuoles (see Fig. 1), where it participates in the processing of vacuolar proteins, and its physiological role has been most extensively investigated (Hatsugai et al. 2006). The Arabidopsis genome has four *VPE* homologues traditionally distributed into seeds, $\beta$*VPE* and $\delta$*VPE*, and vegetative tissues, $\alpha$*VPE* and $\gamma$*VPE*. $\gamma$*VPE* is a vacuolarlocalised cysteine protease with a caspase-1 like activity involved in the activation and maturation of downstream

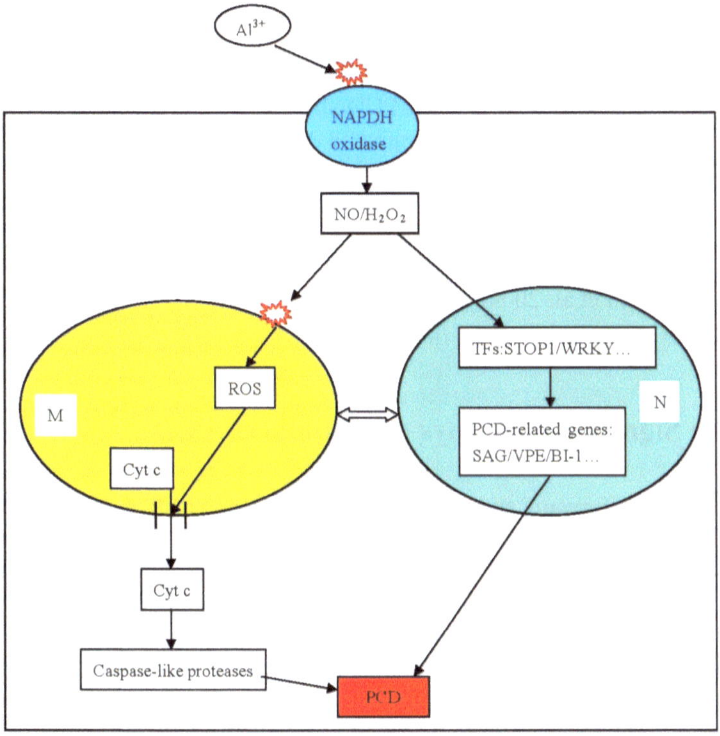


**Fig. 1** A schematic illustration of possible signaling pathways of aluminum-induced programmed cell death in higher plants

vacuolar hydrolytic enzymes that trigger hypersensitive cell death and tissue senescence. This work provides evidence that *γVPE* is strongly expressed in *Arabidopsis* guard cells and is involved in water stress response (Albertini et al. 2014). VPE functions as executioner of plant PCD through disrupting the vacuole in pathogenesis, seed development, and abiotic stress such as heat shock (Li et al. 2012). Real-time quantitative RT-PCR indicated that four VPE genes (*NtVPE-1a*, *NtVPE-1b*, *NtVPE-2*, *NtVPE-3*) were more or less enhanced by the Al exposure over the control levels. Especially, the expressions of the NtVPE-1a and the NtVPE-1b were significantly enhanced, by 2.5-fold under the Al stress. It is concluded that an enhancement of VPE activity by Al is controlled at transcriptional level and is a key factor leading to a loss of integrity of the plasma membrane and a loss of growth capacity (Kariya et al. 2013).

*Caenorhabditis elegans* apoptotic suppressor Ced-9, a Bcl-2 homologue, inhibited both the Al-induced PCD and Al-induced activity of caspase-like vacuolar processing enzyme (VPE) in tobacco. Furthermore, Ced-9 significantly alleviated Al inhibition of root elongation, decreased Al accumulation in the root tip, and greatly inhibited Al-induced gene expression in early response to Al, leading to enhancing the tolerance of tobacco plants to Al toxicity. It suggests that Ced-9 promotes Al tolerance in plants via inhibition of Al-induced PCD, indicating that conserved negative regulators of PCD are involved in integrated regulation of cell survival and Al-induced PCD (Wang et al. 2009).

PpBI-1 can attenuate Al-induced PCD and enhance Al tolerance in transgenic yeast (Zheng et al. 2007). The programmed cell death 5 (*PDCD5*) gene encodes a protein that shares significant homology with the corresponding proteins of species ranging from yeast to mice (Liu et al. 1999). Overexpression of OsPDCD5 genes induces PCD in rice (Attia et al. 2005). As a molecular chaperone, mitochondrial HSP70 may be involved in PCD initiation by reducing $\Delta\psi m$ in mitochondrial outer membrane (Chen et al. 2009).

## 6 The Signaling Pathways of Al-Induced PCD

Over the years, forward and reverse genetic screens have uncovered numerous regulators of PCD in plants. However, to date, molecular networks are far from being deciphered (Cacas 2010). Inside the cell, the compartments that produce the highest amounts of ROS and NO are chloroplasts and mitochondria. The mitochondrial electron transport chain harbors electrons with sufficient free energy to directly reduce $O_2$ which is considered the unavoidable primary source of mitochondrial ROS generation. It was suggested that the mitochondrial transmembrane potential loss and the changes in distribution and mobility of mitochondria, as well as the production of ROS, play important roles during UV-induced plant PCD (Gao et al. 2008).

Based on understanding of related knowledge and NO signaling network proposed by us (He et al. 2014), a new mechanism of Al-induced PCD is proposed in Fig. 1. Al-induced PCD may be mediated via two divergent signaling pathways. One is mitochondria-dependent pathway. Al stress provokes the activity of NADPH oxidase, triggers ROS burst, ROS burst works as a signal of PCD production, opens MPTP, releases cytochrome c, activates caspase 3-like protease, and then promotes PCD occurrence (Huang et al. 2014b). Through $NO/H_2O_2$ cooperation, SOD accelerates $O_2^-$ dismutation to $H_2O_2$ to minimize the loss of NO by reaction with $O_2^-$ to trigger hypersensitive cell death (Delledone et al. 2001).

Another is multi-organelle-participated and nucleus-guided mitochondria-independent pathway, which is executed by regulating gene expressions of PCD promoter and suppressor, then vacuolar collapse, a loss of plasma membrane integrity, and eventually reaching to a loss of growth capacity. The promoter includes senescence-associated gene (*SAG*), vacuole processing enzyme (*VPE*), poly (ADP ribose) polymerase (*PARP-1*), and programmed cell death 5 (*PDCD5*). Bax inhibitor-1 (*BI-1*), ACCELERATED CELL DEATH2 (*ACD2*), and LESION SIMULATING DISEASE1 (*LSD1*) all belong to the suppressor. Al is able to not only generate a signal cascade but also modulates other signal cascades generated by other types of stress in plants (Poot-Poot and Hernandez-Sotomayor

2011). The final output of the cascade depends on the intensity of Al stress, $NO/H_2O_2$ signaling, and two-way communication between two signaling pathways.

As a stress sensor of death signals and a dispatcher of PCD, mitochondria can serve in plant and animal cell death (Jones 2000). In contrast, the part played by mitochondria in the death of plant cells has little attention. High $Al^{3+}$ concentration treatment induced mitochondrial permeability transition pore (MPTP) opening, increased mitochondrial membrane permeability, Cyt c released into the cytoplasm, activated caspase 3-like protease, which might induce PCD in root tip (Zhan et al. 2009; Huang et al. 2014b). Al treatment and oxidative stress in the sensitive maize line induced cell death in root tips cells (Boscolo et al. 2003). Al enhances ferrous ion ($Fe^{2+}$)-mediated lipid peroxidation which is the primary factor leading to cell death in nutrient medium in tobacco cells (Yamaguchi et al. 1999). Spent pot liner (SPL) is solid waste from the Al industry. This toxic agent, consisting of cyanides, fluorides, organics, and metals, leads to cell damage and disturbance (Andrade et al. 2010). SPL induces apoptosis-like PCD in root meristem cells of *Allium cepa* (Andrade-Vieira et al. 2011). NO can ameliorate remarkably mitochondrial respiratory dysfunction resulted from Al stress (He et al. 2006). Al induced oxidative burst at the cell surface through up- or downregulation of some of the key enzymes of oxidative metabolism ultimately resulting in oxidative stress leading to DNA damage and cell death in root cells of *Allium cepa* (Achary et al. 2008).

Mitochondria are the main target for oxidative damage to proteins under well-irrigated and drought conditions (Bartoli et al. 2004). As a semiautonomous organelle, mitochondrion is a common factor that integrates $NO/H_2O_2$ signaling. Mitochondria constitute a major source of ROS and have been proposed to integrate the cellular responses to stress. Oxidative stress increased mitochondrial electron transport, resulting in amplification of $H_2O_2$ production and cell death. The increased generation of $H_2O_2$ also caused the opening of the mitochondrial transmembrane potential (MTP) and the release of Cyt c from mitochondria (Tiwari et al. 2002; Huang et al. 2014b). Exposure to $H_2O_2$ caused the opening of permeability transition pores in the inner mitochondrial membrane. Cytosolic Cyt c plays an essential role in the execution of apoptosis (Takeyama et al. 2002). We found that $\Delta\Psi m$ loss is a common early marker in plant PCD; mitochondrial Cyt c release is an obligatory step in PCD control also (Huang et al. 2014b). Mitochondrial swelling and MTP loss, as well as the generation of mitochondrial ROS, play important roles in Al-induced PCD (Li and Xing 2011). Al toxicity affects severely the mitochondrial respiratory functions and alters the redox status studied in vitro and also the internal structure, which seems to cause finally cell death in tobacco cells (Panda et al. 2008).

However, it has recently been shown that PCD can still occur even when the mitochondria are removed, revealing that there is a mitochondria-independent signaling pathway in nucleus. Proteolytic cleavage of nuclear lamin was conserved in plant PCD (Sun et al. 1999). The nuclear matrix largely remained intact during the course of apoptosis, maintaining the integrity of apoptotic cells and connecting the apoptotic bodies and apoptotic nucleus (Zhao et al. 2001). As one of the hallmarks of apoptosis, chromatin condensation is regulated by nucleoplasmin (Lu et al. 2005). NO and $H_2O_2$ cause an induction of caspase-like proteases previously characterized in physiological nucellar PCD (Lombardi et al. 2010). Using physiological, biochemical, and genetic approaches, we recently demonstrated that *AhSAG* could induce or promote Al-induced PCD (Zhan et al. 2013). Although VPE is structurally unrelated to caspases, plants have evolved a regulated cellular suicide strategy that is mediated by VPE and the cellular vacuole (Hatsugai et al. 2004). Al induced the activity of caspase-like VPE, a crucial executioner of PCD in tobacco (Wang et al. 2009). *Bcl-2* overexpression suppresses $H_2O_2$-induced PCD via *OsVPE2* and *OsVPE3*, but not via *OsVPE1* and *OsVPE4*, in rice (Deng et al. 2011).

The Arabidopsis *PARP-1* shows high homology to human *PARP-1*, and its activity is inhibited by the caspase-3 inhibitor (Ac-DEVD-CHO). By regulating synthesis of PAR, *PARP-1* processes diverse signals and directs cells to specific fates (DNA repair, energy depletion, or cell death) (Luo and Kraus 2012). Because the PDCD5 protein can translocate rapidly to the nucleus in cells undergoing apoptosis, overexpression of the *OsPDCD5* gene induces PCD in rice (Attia et al. 2005). *PpBI-1* (*Phyllostachys praecox*) inhibits Al-induced PCD and promotes Al tolerance in yeast (Zheng et al. 2007). The C-terminal hydrophilic region of BI-1 is essential for the inhibition of cell death. $H_2O_2$-mediated cell death was suppressed in tobacco BY-2 cells overexpressing *AtBI-1* (Kawai-Yamada et al. 2004). The *Arabidopsis* ACD2 protein protects cells from PCD caused by endogenous porphyrin-related molecules like red chlorophyll catabolite or exogenous protoporphyrin IX (Pattanayak et al. 2012).

LSD1 is an important negative regulator of PCD in Arabidopsis. The loss-of-function mutations in LSD1 cause run away cell death triggered by ROS (Li et al. 2013). Although caspases are proteases that act as key components of animal apoptosis, plants have no orthologous caspase sequences in their genomes. Metacaspase-8 is part of an evolutionary conserved PCD pathway activated by oxidative stress, so metacaspases may be the functional homologues of animal caspases in these organisms (He et al. 2008). The prolonged activation of the mitogen-activated protein kinase (MAPK) pathway in cells could disrupt the redox balance, which leads to the generation of ROS and eventually cell death (Ren et al. 2002). The PCD-related genes are mediated by TFs, redox changes, MAPK cascades, microRNAs, and their interactions with each other.

Moreover, Cyt c induced in vitro apoptosis of carrot nucleus, indicating there is a signal communication between mitochondria and nucleus (Zhao et al. 1999). Chloroplasts may be involved in mediating certain types of plant PCD (Chen and Dickman 2004). Doyle et al. (2010) found that chloroplasts can play a significant role in Al-PCD regulation. Distinct organelles sense a broad range of stimuli, if necessary, engage cell death signaling pathways. The endomembrane system (ES) seems to harbor a significant number of cell death mediators (Cacas 2010). *AtLrgB*, which encodes a homolog of the bacterial membrane protein LrgB, functions against cell death (Yamaguchi et al. 2012).

## 7 Conclusions and Perspectives

In conclusion, Al stress not only triggers the production of NO and $H_2O_2$ but also induces PCD by breaking their homeostasis. Al-induced PCD is characterized by nucleus condensation and crescent-shaped, marginalized chromatin aggregation, and DNA Ladder. In the light of relevant literature, Al toxicity initiates PCD via two signaling pathways. One is mitochondria-dependent pathway. The excess Al toxicity-generated production of ROS lead lipid peroxidation, induce the opening of MPTP, cause the release of Cyt c, activate caspase 3-like protease, and finally result in PCD. Another is mitochondria-independent pathway existing in nucleus. It is a multiple organelle-participated and nucleus-guided process, which is executed by regulating expressions of PCD-related genes, such as *SAG*, *VPE*, *BI-1*, *ACD2*, *PDCD5*, and *LSD1*. Since there is a negative relationship between the occurrence of PCD and Al-resistance in peanut (Zhan et al. 2013), the negative regulation of Al-induced PCD may be an important mechanism of Al tolerance.

Although researches on signaling molecules, related proteins, and genes of Al-induced PCD in plants have made some progress, its precise mechanism is still unclear. For example, how is the relationship between PCD occurrence and Al tolerance in different plants? Whether the mitochondria lie in the control center of Al-induced PCD? What kinds of species are signaling factors related to Al-induced PCD? Whether common regulatory pathway or mechanism exists? What are the similarities and differences of Al-induced PCD mechanism at the molecular level? Which kinds of transcription factors are related to Al-induced PCD? The role of nuclease and specific protease in Al-induced PCD is still unknown. Deep research on molecular mechanism and regulatory pathways of Al-induced PCD help to elucidate the mechanisms of Al toxicity and Al tolerance in plants, providing opportunities for enhancing the $Al^{3+}$ resistance of plants by marker-assisted breeding and through biotechnology.

**Acknowledgements** We apologize to many of our colleagues for not being able to cite many exceptional articles due to space limitations. This work is supported by grants from the National Natural Science Foundation of China (No. 31260296 and 30960181) and 2011 Guangxi Innovation Program for Graduates (GXU11T31076).

## References

Achary VMM, Panda BB (2010) Aluminum-induced DNA damage and adaptive responses to genotoxic stress in plant cells are mediated through reactive oxygen intermediates. Mutagenesis 25:201–209

Achary VMM, Jena S, Panda KK, Panda BB (2008) Aluminum induced oxidative stress and DNA damage in root cells of *Allium cepa* L. Ecotoxicol Environ Saf 70:300–310

Albertini A, Simeoni F, Galbiati M, Bauer H, Tonelli C, Cominelli E (2014) Involvement of the vacuolar processing enzyme γVPE in response of *Arabidopsis thaliana* to water stress. Biol Plant 58:531–538

Andrade LF, Davide LC, Gedraite LS (2010) The effects of cyanide compounds, fluorides, aluminum, and inorganic oxides present in spent pot liner on germination and root tip cells of *Lactuca sativa*. Ecotoxicol Environ Saf 73:626–631

Andrade-Vieira LF, Gedraite LS, Campos JM, Davide LC (2011) Spent pot liner (SPL) induced DNA damage and nuclear alterations in root tip cells of *Allium cepa* as a consequence of programmed cell death. Ecotoxicol Environ Saf 74:882–888

Attia K, Li KG, Wei C, He GM, Su W, Yang JS (2005) Overexpression of the *OsPDCD5* gene induces programmed cell death in rice. J Integr Plant Biol 47:1115–1122

Balazadeh S, Siddiqui H, Allu AD, Matallana-Ramirez LP, Caldana C, Mehrnia M, Zanor MI, Kohler B, Mueller-Roeber B (2010) A gene regulatory network controlled by the NAC transcription factor ANAC092/AtNAC2/ORE 1 during salt-promoted senescence. Plant J 62:250–264

Balazadeh S, Kwasniewski M, Caldana C, Mehrnia M, Zanor MI, Xue GP, Mueller-Roeber B (2011) ORS1, and $H_2O_2$-responsive NAC transcription factor, controls senescence in *Arabidopsis thaliana*. Mol Plant 4:346–360

Bartoli CG, Gomez F, Martinez DE, Guiamet JJ (2004) Mitochondria are the main target for oxidative damage in leaves of wheat (*Triticum aestivum* L.). J Exp Bot 55:1663–1669

Beligni MV, Fath A, Bethake PC, Lamattina L, Jones RL (2002) Nitric oxide acts as an antioxidant and delays programmed cell death in barley aleurone layers. Plant Physiol 129:1642–1650

Besson-Bard A, Pugin A, Wendehenne D (2008) New insights into nitric oxide signaling in plants. Annu Rev Plant Biol 59:21–39

Blamey FPC, Nishizawa NK, Yoshimura E (2004) Timing, magnitude, and location of initial soluble aluminum injuries to mungbean roots. Soil Sci Plant Nutr 50:67–76

Boscolo PRS, Menossi M, Jorge RA (2003) Aluminum-induced oxidative stress in maize. Phytochemistry 62:181–189

Cacas JL (2010) Devil inside: does plant programmed cell death involve the endomembrane system? Plant Cell Environ 33:1453–1473

Chen SR, Dickman MB (2004) Bcl-2 family members localize to tobacco chloroplasts and inhibit programmed cell death induced by chloroplast-targeted herbicides. J Exp Bot 55:2617–2623

Chen X, Wang Y, Li J, Jiang A, Cheng Y, Zhang W (2009) Mitochondrial proteome during salt stress-induced programmed cell death in rice. Plant Physiol Biochem 47:407–415

Clarke A, Desikan R, Hurst RD, Hancock JT, Neill SJ (2000) No way back: nitric oxide and programmed cell death in *Arabidopsis thaliana* suspension cultures. Plant J 24:667–677

de Pinto MC, Tommasi F, De Gara LD (2002) Changes in the antioxidant systems as part of the signaling pathway responsible for the programmed cell death activated by nitric oxide and reactive oxygen species in tobacco bright-yellow cells. Plant Physiol 130:698–708

de Pinto MC, Paradiso A, Leonetti P, De Gara L (2006) Hydrogen peroxide, nitric oxide and cytosolic ascorbate peroxidase at the crossroad between defence and cell death. Plant J 48:784–795

Delhaize E, Ma JF, Ryan PR (2012) Transcriptional regulation of aluminium tolerance genes. Trends Plant Sci 17:341–348

Delledone M, Zeier J, Marocco A, Lamb C (2001) Signal interactions between nitric oxide and reactive oxygen intermediates in the plant hypersensitive disease resistance response. Proc Natl Acad Sci USA 98:13454–13459

Deng MJ, Bian HW, Xie YK, Kim YH, Wang WZ, Lin EP, Zeng ZH, Guo F, Pan JW, Han N, Wang JH, Qian Q, Zhu MY (2011) *Bcl-2* suppresses hydrogen peroxide-induced programmed cell death via *OsVPE2* and *OsVPE3*, but nor via *OsVPE1* and *OsVPE4*, in rice. FEBS J 278:4797–4810

Doyle SM, Diamond M, McCabe PF (2010) Chloroplast and reactive oxygen species involvement in apoptotic-like programmed cell death in *Arabidopsis* suspension cultures. J Exp Bot 61:473–482

Gadjev I, Stone JM, Gechev TS (2008) Programmed cell death in plants: new insights into redox regulation and the role of hydrogen peroxide. Int Rev Cell Mol Biol 270:87–144

Gao C, Xing D, Li L, Zhang L (2008) Implication of reactive oxygen species and mitochondrial dysfunction in the early stages of plant programmed cell death induced by ultraviolet-C overexposure. Planta 227:755–767

Gechev TS, Hille J (2005) Hydrogen peroxide as a signal controlling plant programmed cell death. J Cell Biol 168:17–20

Gechev TS, Van Breusegem F, Stone JM, Denev L, Laloi C (2006) Reactive oxygen species as signals that modulate plant stress responses and programmed cell death. Bioessays 28:1091–1101

Gupta KJ, Igamberdiev AU, Mur LAJ (2012) NO and ROS homeostasis in mitochondria: a central role for alternative oxidase. New Phytol 195:1–3

Hatsugai N, Kuroyanagi M, Yamada K, Meshi T, Tsuda S, Kondo M, Nishimura M, Hara-Nishimura I (2004) A plant vacuole protease, VPE, mediates virus-induced hypersensitive cell death. Science 305:855–858

Hatsugai N, Kuroyanagi M, Nishimura M, Hara-Nishimura I (2006) A cellular suicide strategy of plants: vacuole-mediated cell death. Apoptosis 11:905–911

He HY, He LF, Li XF, Gu MH (2006) Effects of sodium nitroprusside on mitochondrial function of rye and wheat root tip under aluminum stress. J Plant Physiol Mol Biol 32:239–244

He R, Drury GE, Rotari VI, Gordon A, Willer M, Farzaneh T, Woltering EJ, Gallois P (2008) Metacaspase-8 modulates programmed cell death induced by ultraviolet light and $H_2O_2$ in Arabidopsis. J Biol Chem 283:774–783

He HY, Zhan J, He LF, Gu MH (2012) Nitric oxide signaling in Al stress in plants. Protoplasma 249:483–492

He HY, Gu MH, He LF (2014) The role of nitric oxide in programmed cell death in higher plants. In: Khan MN et al (eds) Nitric oxide in plants: metabolism and role in stress physiology. Springer, Heidelberg, pp 281–296

Huang WJ, Oo TL, He HY, Wang AQ, Zhan J, Li CZ, Wei SQ, He LF (2014a) Al induces rapidly mitochondriadependent programmed cell death in Al-sensitive peanut root tips. Bot Stud 55, e67

Huang WJ, Yang XD, Yao SC, Oo TL, He HY, Wang AQ, Li CZ, He LF (2014b) Reactive oxygen species burst induced by Al stress triggers mitochondria-dependent programmed cell death in peanut root tip cells. Plant Physiol Biochem 82:76–84

Ikegawa H, Yamamoto Y, Matsumoto H (1998) Cell death caused by a combination of aluminum and iron in cultured tobacco cells. Physiol Plant 104:474–478

Illes P, Schlicht M, Pavlovkin J, Lichtscheidl I, Baluska F, Ovecka M (2006) Aluminum toxicity in plants: internalization of aluminum into cells of the transition zone in Arabidopsis root apices related to changes in plasma membrane potential, endosomal behaviour, and nitric oxide production. J Exp Bot 57:4201–4213

Jones A (2000) Does the plant mitochondrion integrate cellular stress and regulate programmed cell death? Trends Plant Sci 5:225–230

Kariya K, Demiral T, Sasaki T, Tsuchiya Y, Turkan I, Sano T, Hasezawa S, Yamamoto Y (2013) A novel mechanism of aluminium-induced cell death involving vacuolar processing enzyme and vacuolar collapse in tobacco cell line BY-2. J Inorg Biochem 128:196–201

Kawai-Yamada M, Ohori Y, Uchimiya H (2004) Dissection of *Arabidopsis* Bax inhibitor-1 suppressing Bax-, hydrogen peroxide-, and salicyclic acid-induced cell death. Plant Cell 16:21–32

Ko SS, Li MJ, Ku MSB, Ho YC, Lin YJ, Chuang MH, Hsing HX, Lien YC, Yang HT, Chang HC, Chan MT (2014) The bHLH142 transcription factor coordinates with TDR1 to modulate the expression of EAT1 and regulate pollen development in Rice. Plant Cell 26:2486–2504

Li Z, Xing D (2011) Mechanistic study of mitochondria-dependent programmed cell death induced by aluminium phytotoxicity using fluorescence techniques. J Exp Bot 62:331–343

Li Z, Yue H, Xing D (2012) MAP kinase 6-mediated activation of vacuolar processing enzyme modulates heat shock-induced programmed cell death in Arabidopsis. New Phytol 195:85–96

Li YS, Chen LC, Mu JY, Zuo JR (2013) LESION SIMULATING DISEASE1 interacts with catalases to regulate hypersensitive cell death in Arabidopsis. Plant Physiol 163:1059–1070

Liu HT, Wang YQ, Zhang YM (1999) TFAR19, a novel apoptosis-related gene cloned from human leukemia cell line TF-1 could enhance apoptosis of some tumor cells induced by growth factor withdraw. Biochem Biophys Res Commun 245:203–210

Lombardi L, Ceccarelli N, Picciarelli P, Sorce C, Lorenzi R (2010) Nitric oxide and hydrogen peroxide involvement during programmed cell death of *Sechium edule* nucellus. Physiol Plant 140:89–102

Lu Z, Zhang C, Zhai Z (2005) Nucleoplasmin regulates chromatin condensation during apoptosis. Proc Natl Acad Sci U S A 102:2778–2783

Luo X, Kraus WL (2012) On PAR with PARP: cellular stress signaling through poly (ADP-ribose) and PARP-1. Genes Dev 26:417–432

Meriga B, Reddy BK, Rao KR, Reddy LA, Kishor PB (2004) Aluminum-induced production of oxygen radicals, lipid peroxidation and DNA damage in seedlings of rice (*Oryza sativa* L.). J Plant Physiol 161:63–68

Miao Y, Laun T, Zimmermann P, Zentgraf U (2004) Targets of the WRKY53 transcription factor and its role during leaf senescence in Arabidopsis. Plant Mol Biol 55:853–867

Nezames CD, Sjogren CA, Barajas JF, Larsen PB (2012) The Arabidopsis cell cycle checkpoint regulators TANMEI/ALT2 and ATR mediate the active process of aluminum-dependent root growth inhibition. Plant Cell 24:608–621

Pan JW, Zhu MY, Chen H (2001) Aluminum-induced cell death in root-tip cells of barley. Environ Exp Bot 46:71–79

Panda SK, Yamamoto Y, Kondo H, Matsumoto H (2008) Mitochondrial alterations related to programmed cell death in tobacco cells under aluminum stress. CR Biol 331:597–610

Pandey SP, Somssich IE (2009) The role of WRKY transcription factors in plant immunity. Plant Physiol 150:1648–1655

Pattanayak GK, Venkataramani S, Hortensteiner S, Kunz L, Christ B, Moulin M, Smith AG, Okamoto Y, Tamiaki H, Sugishima M, Greenberg JT (2012) ACCELERATED CELL DEATH2 suppresses mitochondrial oxidative bursts and modulates cell death in Arabidopsis. Plant J 69:589–600

Phan HA, Iacuone S, Li SF, Parish RW (2011) The MYB80 transcription factor is required for pollen development and the regulation of tapetal programmed cell death in *Arabidopsis thaliana*. Plant Cell 23:2209–2224

Poot-Poot W, Hernandez-Sotomayor SMT (2011) Aluminum stress and its role in the phospholipid signaling pathway in plants and possible biotechnological applications. IUBMB Life 63:864–872

Rath I, Barz W (2000) The role of lipid peroxidation in aluminum toxicity in soybean cell suspension cultures. Z Naturforsch 55:957–964

Ren DT, Yang HP, Zhang SQ (2002) Cell death mediated by MAPK is associated with hydrogen peroxide production in Arabidopsis. J Biol Chem 277:559–565

Rosenvasser S, Mayak S, Friedman H (2006) Increase in reactive oxygen species (ROS) and in senescence-associated gene transcript (SAG) levels during dark-induced senescence of Pelargonium cuttings, and effect of gibberellic acid. Plant Sci 170:873–879

Rushton PJ, Somssich IE, Ringler P, Shen QJ (2010) WRKY transcription factors. Trends Plant Sci 15:247–58

Ryan PR, Tyerman SD, Sasaki T, Furuichi T, Yamamoto Y, Zhang WH, Delhaize E (2011) The identification of aluminium-resistance genes provides opportunities for enhancing crop production on acid soils. J Exp Bot 62:9–20

Simonovicova M, Huttova J, Mistrik I, Siroka B, Tamas L (2004) Root growth inhibition by aluminum is probably caused by cell death due to peroxidase-mediated hydrogen peroxide production. Protoplasma 224:91–98

Sun Y, Zhu H, Zhou J, Dai Y, Zhai Z (1999) Menadione-induced apoptosis and the degradation of lamin-like proteins in tobacco protoplasts. Cell Mol Life Sci 55:310–316

Takeyama N, Mike S, Hirakawa A, Tanaka T (2002) Role of mitochondrial permeability transition and cytochrome c release in hydrogen peroxide-induced apoptosis. Exp Cell Res 274:16–25

Tamas L, Budlkova S, Huttova J, Mistrik I, Simonovicovicova M, Sirka B (2005) Aluminum-induced cell death of barley-root border cells is correlated with peroxidase- and oxalate oxidase-mediated hydrogen peroxide production. Plant Cell Rep 24:189–194

Tian QY, Sun DH, Zhao MG, Zhang WH (2007) Inhibition of nitric oxide synthase (NOS) underlies aluminum-induced inhibition of root elongation in *Hibiscus moscheutos*. New Phytol 174:322–331

Tiwari BS, Belenghi B, Levine A (2002) Oxidative stress increased respiration and generation of reactive oxygen species, resulting in ATP depletion, opening of mitochondrial permeability transition and programmed cell death. Plant Physiol 128:1271–1281

Tsutsui T, Yamaji N, Ma JF (2011) Identification of a cis-acting element of ART1, a $C_2H_2$-type zinc-finger transcription factor for aluminum tolerance in Rice. Plant Physiol 156:925–931

van Aken O, Zhang B, Law S, Narsai R, Whelan J (2013) AtWRKY40 and AtWRKY63 modulate the expression of stress-Responsive nuclear genes encoding mitochondrial and chloroplast proteins. Plant Physiol 162:254–271

Wang YS, Yang ZM (2005) Nitric oxide reduces aluminum toxicity by preventing oxidative stress in the roots of *Cassia tora* L. Plant Cell Phyisol 46:1915–1923

Wang WZ, Pan JW, Zheng K, Chen H, Shao HH, Guo YJ, Bian HW, Han N, Wang JH, Zhu MY (2009) Ced-9 inhibits Al-induced programmed cell death and promotes Al tolerance in tobacco. Biochem Biophys Res Commun 383:141–145

Wang HH, Huang JJ, Bi YR (2010) Nitrate reductase-dependent nitric oxide production is involved in aluminum tolerance in red kidney bean roots. Plant Sci 179:281–288

Yakimova ET, Kapchina-Toteva VM, Woltering EJ (2007) Signal transduction events in aluminum-induced cell death in tomato suspension cells. J Plant Physiol 164:702–708

Yamaguchi Y, Yamamoto Y, Matsumoto H (1999) Cell death process initiated by a combination of aluminum and iron in suspension-cultured tobacco cells (*Nicotiana tabacum*): apoptosis-like cell death mediated by calcium and proteinase. Soil Sci Plant Nutr 15:647–657

Yamaguchi M, Takechi K, Myouga F, Imura S, Sato H, Takio S, Shinozaki K, Takano H (2012) Loss of the plastid envelop protein AtLrgB causes spontaneous chlorotic cell death in *Arabidopsis thaliana*. Plant Cell Physiol 53:125–134

Yamamoto Y, Hachiya A, Matsumoto H (1997) Oxidative damage to membrane by a combination of aluminum and iron in suspension-cultured tobacco cells. Plant Cell Physiol 38:1333–1339

Yamamoto Y, Kobayashi Y, Devi SR, Rikiishi S, Matsumoto H (2002) Aluminum toxicity is associated with mitochondrial function and the production of reactive oxygen species in plant cells. Plant Physiol 128:63–72

Yang L, Tian D, Todd CD, Luo Y, Hu X (2013) Comparative proteome analysis reveal that nitric oxide is an important signal molecule in the response of rice to aluminum toxicity. J Proteom Res 12:1316–1330

Zhan J, Kou RJ, Li CZ, He HY, He LF (2009) Effects of aluminum on physiological characteristics of mitochondrial membrane in peanut root tips. Acta Agron Sin 35:1059–1067

Zhan J, Wang TJ, He HY, Li CZ, He LF (2011) Effects of SNP on *AhSAG* and *AhBI-1* genes expression and amelioration of aluminum stress of peanut (*Arachis hypoganea* L.). Acta Agron Sin 37:459–468

Zhan J, He HY, Wang TJ, Wang AQ, Li CZ, He LF (2013) Aluminum-induced programmed cell death promoted by *AhSAG*, asenescence-associated gene in *Arachis hypoganea* L. Plant Sci 210:108–117

Zhan J, Li W, Hy H, Li CZ, He LF (2014) Mitochondrial alterations during Al-induced PCD in peanut root tips. Plant Physiol Biochem 75:105–113

Zhang L, Xing D (2008) Methyl jasmonate induces production of reactive oxygen species and alterations in mitochondrial dynamics that precede photosynthetic dysfunction and subsequent cell death. Plant Cell Physiol 49:1092–1111

Zhang C, Czymmek KJ, Shapiro AD (2003) Nitric oxide does not trigger early programmed cell death events but may contribute to cell-to-cell signaling governing progression of the Arabidopsis hypersensitive responses. MPMI 16:962–972

Zhao Y, Sun Y, Jiang Z, Zhai Z (1999) Cytochrome c induces *in vitro* apoptosis of carrot nucleus. Chin Sci Bull 44:1181–1185

Zhao Y, Wu M, Shen Y, Zhai Z (2001) Analysis of nuclear apoptotic process in a cell-free system. Cell Mol Life Sci 58:298–306

Zheng K, Pan JW, Ye L, Fu Y, Peng HZ, Wan BY (2007) Programmed cell death-involved aluminum toxicity in yeast alleviated by antiapoptotic members with decreased calcium signals. Plant Physiol 143:38–49

Yamaji N, Huang CF, Nagao S, Yano M, Sato Y, Nagamura Y, Ma JF (2009) A zinc finger transcription factor ART1 regulates multiple genes implicated in aluminum tolerance in rice. Plant Cell 21:3339–3349

# Mechanisms of Hyper-resistance and Hyper-tolerance to Aluminum in Plants

Charlotte Poschenrieder, Roser Tolrà, Roghieh Hajiboland, Catalina Arroyave, and Juan Barceló

**Abstract** As a widespread, permanent stress factor in acid soils, Aluminum toxicity has driven the evolution of different mechanisms that allow plants to colonize these adverse environments. Even more, Al-induced stimulation of growth has frequently been observed in highly adapted plants. Plant strategies for handling excess Al span from highly efficient exclusion (hyper-resistance) to the tolerance of extremely high Al accumulation within leaf tissues (hyper-tolerance). This chapter, after considering potential mechanisms for Al-induced growth stimulation, gives an overview of the current knowledge on Al hyper-resistance and Al hyper-tolerance mechanisms in plants with special focus on both the highly efficient excluder species of the genus *Urochloa* (former *Brachiaria*) and the most studied Al accumulators, tea and buckwheat.

## 1 Introduction

Aluminum toxicity affects plants that are not- or only poorly adapted to acid soil conditions. Especially in tropical areas, crop production on acid mineral soils can be severely affected. In contrast, the natural vegetation that has evolved under these conditions perfectly handles Al toxicity, in addition to other adverse factors associated with these soils, such as low P and Ca availability, high concentrations of $H^+$ and of soluble Mn and Fe, and drought stress (Fageria and Baligar 2008; Yang et al. 2013; Kabaz-Saberi et al. 2014).

C. Poschenrieder (✉) • R. Tolrà • J. Barceló
Lab. Fisiología Vegetal, Facultad Biociencias, Universidad Autónoma de Barcelona, Bellaterra, Spain
e-mail: charlotte.poschenrieder@uab.es

R. Hajiboland
Plant Science Department, University of Tabriz, Tabriz, Iran

C. Arroyave
Facultad Ciencias Exactas y Naturales, Inst. Biología, Universidad Antioquia, Medellin, Colombia



S.K. Panda, F. Baluška (eds.), *Aluminum Stress Adaptation in Plants*, Signaling and Communication in Plants 24, DOI 10.1007/978-3-319-19968-9_5

While our knowledge about Al toxicity and tolerance mechanisms in major crop plants has made excellent progress in recent years (see reviews, e.g., by Kochian et al. 2005; Delhaize et al. 2007, 2012; Poschenrieder et al. 2008; Ma et al. 2014), the research efforts addressing the natural vegetation exposed to high Al availability is still in its exploratory phase mainly focusing systematics and Al accumulation patterns (Jansen et al. 2002, 2003), while mechanistic aspects are just emerging. Most information is available on some highly adapted species of important commercial interest, such as the tea shrub (*Camellia sinensis*) and signal grass (*Urocloa decumbens* formerly *Brachiaria decumbens*).

These species have extremely contrasting strategies for managing excess Al. While *C. sinensis* and other Theaceae species accumulate huge concentrations of Al in the shoots reaching hyperaccumulation levels of more than 1000 mg $kg^{-1}$ (Matsumoto et al. 1976; Jansen et al. 2002; Carr et al. 2003; Osawa et al. 2013), signal grass is a highly efficient Al excluder (Wenzl et al. 2001; Arroyave et al. 2011). Applying Levitt's concepts of stress tolerance and stress resistance (Levitt 1980), tea is an extremely Al-tolerant species. In contrast, signal grass is extremely Al resistant due to efficient avoidance of Al accumulation, even under conditions of very high Al availability. Nonetheless, in highly resistant species or varieties that are naturally adapted to acid or metalliferous soils, the ability of efficient metal exclusion is further combined with higher tissue tolerance than in non-adapted plants (Llugany et al. 2003; Arroyave et al. 2013). Multiple mechanisms (i.e., genes) therefore must cooperate to achieve hyper-resistance.

Metal hyperaccumulation and metal tolerance can be genetically independent traits (Bert et al. 2003). However, hyper-tolerance traits are required for hyperaccumulators to survive in their natural habitats. In this sense, hyper-tolerance can be clearly defined as a function of the high internal metal concentration achieved in the shoots of the species without a significant growth reduction. In the case of Al hyperaccumulators, shoot concentrations higher than 1000 mg $kg^{-1}$ should correspond to hyper-tolerance. According to this criterion, a few cultivated species can be considered as Al hyperaccumulators: tea, buckwheat, and *Hydrangea*. However, most hyperaccumulators are wild tropical shrubs and trees, among others, from the Melastomataceae, Rubiaceae, Proteaceae, Vochysiaceae families (Haridasan 1982; Jansen et al. 2002; Metali et al. 2012). Aluminum hyperaccumulation has also been reported in some Pteridophyte species growing on acid soils (Olivares et al. 2009).

(Hyper)-resistance is to be defined as function of the external concentration available to the plant causing a standardized effect. Under field conditions, the percent Al saturation of the cation exchange sites or the concentrations of exchangeable Al are commonly used for characterizing the Al stress intensity, while in hydroponic studies the activity of $Al^{3+}$ or the sum of soluble monomeric Al species defines the stress treatment. Hyper-resistant species like *U. decumbens* can grow on soils with up to 80 % Al saturation and suffer 50 % inhibition of root elongation at soil solution $Al^{3+}$ activities higher than 30 μM (Wenzl et al. 2001; Bitencourt et al. 2011). Similar hyper-resistance can be found in some legume species of tropical origin such as *Mucuna nivea*, *M. deeringiana*, *M. aterrima*, and *Vigna unguiculata* (Meda and Furlani 2005). Also certain upland rice varieties can be considered as hyper-resistant to Al suffering less than 10 % yield reduction in soils with around 85 % Al saturation (Sarkarung 1986). Stimulation of root elongation by 160 μM $Al^{3+}$ has been observed in some extremely Al-resistant rice varieties (Famoso et al. 2011). In comparison, a wheat variety considered to be highly Al resistant, e.g., *Triticum aestivum* cv Atlas, suffers a 50 % growth inhibition at 2–5 μM $Al^{3+}$ activity (Wheeler et al. 1992; Poschenrieder et al. 2008).

## 2 Growth Stimulation by Al

Up-to-date no essential biological role is recognized for Al in any organism. Nonetheless, Al-induced stimulation of growth has frequently been reported. Multiple mechanisms may account for such hormetic responses caused by low levels of potentially toxic metal ions (Poschenrieder et al. 2013). Two main patterns in the stimulation response to Al should be distinguished. (1) A short, transient increase of growth mainly observed in lab studies when plants are exposed to Al in low ionic strength nutrient solutions with low pH and (2) a permanent Al-induced increase in productivity of Al hyper-resistant or hyper-tolerant plants.

Inhibition of root elongation is often used as a fast, reliable indicator of Al sensitivity in genotype screening in single salt or low-ionic-strength nutrient solutions with low pH. In such a system, acid soil sensitive varieties can exhibit substantial Al-induced enhancement of root elongation for several minutes or hours due to amelioration of $H^+$ toxicity by $Al^{3+}$ (Llugany et al. 1995). According to the Gouy–Chapman–Stern Model, $Al^{3+}$ can ameliorate $H^+$ toxicity due to electrostatic displacement from the cell membrane surface (Kinraide et al. 1992; Kinraide 1998). Although the positive effect in sensitive plant is only transitory because $Al^{3+}$ gets toxic to the roots.

However, not in all cases the Al-induced growth stimulation is transitory. Al and proton tolerance are two independent factors, and long-term growth stimulation can be observed in Al-tolerant species. The suboptimal growth of a *Betula pendula* race at pH 4.2 was ameliorated by low Al concentrations even after 28 days of exposure (Kidd and Proctor 2000). Contrastingly, the growth of a race that grew optimal at this pH was not enhanced by Al. Birch is considered an Al-tolerant species. However, detailed growth studies under different $H^+$ and $Al^{3+}$ concentrations using birch races from different locations showed that the tolerance to $H^+$ and $Al^{3+}$ was strictly in accordance with the prevailing soil conditions at the sites of the plants' origin (Kidd and Proctor 2000). There seems to be a specific proportion of $H^+$ and $Al^{3+}$ for optimal growth. This optimal proportion reflects the soil solution composition at the site where the plant had evolved and is most probably conditioned by the $Ca^{2+}$ concentration in the solution. In certain species, local adaptation to acid soil conditions and Al toxicity can be a relatively fast process (Gould et al. 2014).

Besides proton toxicity, other adverse inorganic factors may also drive Al-induced growth stimulation. Due to its high prooxidant activity, cationic iron both in the form of $Fe^{3+}$ and $Fe^{2+}$ is more toxic than $Al^{3+}$ (Kinraide et al. 2011). Plants can suffer from iron toxicity, visible in the form of leaf bronzing, especially in acid soils under reducing conditions (Shabala et al. 2014). For economic reasons, this phenomenon has mostly been investigated in rice (Becker and Asch 2005; Mongon et al. 2014). However, few data on the interaction between Al and Fe under acid soil conditions are available (Ayeni et al. 2014). As in the case of Al, resistance to Fe toxicity seems mainly based on more efficient Fe exclusion, especially from the shoots. However, shoot tolerance mechanisms are also operating in some rice lines (Wu et al. 2014). Interestingly, in a comparative study on wheat varieties differing in resistance to Al, Fe, and Mn toxicity, the Al-resistant varieties accumulated as low or even lower shoot Fe concentrations than the Fe-resistant variety (Khabaz-Saberi et al. 2012). Interference between Al and Fe transport seems possible and Al may stimulate growth by alleviating Fe toxicity. In fact, under lab conditions leaf bronzing in hydroponically grown tea plants was accompanied by high leaf Fe concentration. Addition of 200 μM Al to hydroponically grown tea plants reduced hematoxylin stainable Fe in the roots, alleviated the leaf bronzing (Fig. 1a, b), and reduced root and leaf Fe concentrations, while enhancing plant growth (Hajiboland et al. 2013b). However, even under conditions without Fe toxicity or other apparent stress conditions, Al supply enhances growth in tea plants (Fig. 1c). This sustained Al-induced stimulation of growth is accompanied by enhanced photosynthesis, increased antioxidant defenses, and less cell wall lignification (Hajiboland et al. 2013c). This could be related to the interaction of Al with cell wall-bound phenolics and boron (see also Sect. 4.4). In fact, Al stimulates growth not only in B-sufficient but also in B-deficient tea plants (Fig. 1c). A further mechanism proposed for Al-induced growth stimulation in plants highly adapted to acid soils is the favorable influence of Al on phosphorus acquisition (Osaki et al. 1997; Watanabe and Osaki 2002) (see also Sect. 4.2).

## 3 Mechanisms of Al Hyper-resistance

Root tips are the primary target for Al phytotoxicity and, most probably, also the site of Al sensing (Liu et al. 2014). Cell division, cell elongation, and root cell patterning are highly sensitive to Al (Doncheva et al. 2005; Amenós et al. 2009). Interactions of Al with root tip cell walls, plasmalemma, cytoskeleton, vesicle transport, mitochondria, and nucleoli have been observed after seconds to minutes of Al exposure. The root transition zone has been proposed as most sensitive site. Fast induction of ethylene in response to Al stress in this zone mediates auxin-regulated root growth inhibition (Massot et al. 2002; Yang et al. 2014). Resistance mechanisms have to avoid the accumulation of toxic Al in the root apex. Moreover, mechanisms efficient in the exclusion of Al from the sensitive shoots must operate.

**Fig. 1** *Camellia sinensis* plants grown in hydroponics. (**a**) Leaf of plant grown without Al showing leaf bronzing, a symptom of iron toxicity. Leaf iron concentration 900 mg $kg^{-1}$ dry weight. (**b**) Plants grown in solution with 300 μM Al; no toxicity symptoms, leaf iron concentration 300 mg $kg^{-1}$ dry weight. (**c**) Tea plants cultivated in solution with sufficient (+B) or deficient (−B) boron supply and either with (+Al) or without (−Al)

## *3.1 Molecular Basis for Al Exclusion*

Al-induced exudation of organic acids, mostly citrate, malate, or oxalate is the best-characterized mechanism responsible for Al exclusion from root tips. A single dominant gene is responsible for this in most cereals (wheat, barley, oat), while multiple genes seem responsible in maize and rice (Guimaraes et al. 2014; Ma et al. 2014). As Al-induced organic acid exudation is covered in other chapters of this book, here we will focus mainly on other mechanisms that seem relevant for hyper-resistance to Al. In fact, there is increasing experimental evidence that in several species, organic acid exudation is not the only mechanism contributing to

Al (hyper)-resistance (Vázquez et al. 1999; Kidd et al. 2001; Wenzl et al. 2001, 2002).

The extraordinary high Al resistance in rice, especially upland rice, is only in part due to Al-induced organic acid exudation. Genome-wide association mapping and QTL analysis revealed three regions involved in natural variation of Al resistance: *ART1, STAR2,* and *Nrat1* (Famoso et al. 2011). ART1 is a transcription factor that regulates at least 31 genes (Yamaji et al. 2009). Several are related to Al resistance (Ma et al. 2014; see more detailed explanations also in chapter "Transcriptional Regulation of Al Tolerance in Plants"). Among those are genes for Al-induced citrate exudation (*OsFRDL4*), magnesium transport (*OsMGT1*), and vacuolar Al storage (*Nrat1*). STAR1/STAR2 coding for UDP-glucose release to the cell wall seems involved in avoidance of Al binding in the apoplast. Taken together these results indicate that in rice, hyper-resistance to Al implies less binding of Al in the apoplast, due to both alteration in the composition of cell walls and chelation of Al by Al-induced citrate exudation, enhanced Al transport into root vacuoles, and enhanced Mg uptake. Avoidance of Al-induced Mg deficiency is crucial for the maintenance of ATP availability in the heterotrophic root cells (Gout et al. 2014), facilitating resistance strategies based on active transport mechanisms.

## 3.2 Hyper-resistance in *Urochloa* sp

*Urochloa decumbens* (former *Brachiaria decumbens*; see Torres González and Morton (2005), for phylogenetic relations), a pasture grass of African origin, is among the most Al-resistant Poaceae. The mechanisms behind this extraordinary stress resistance are still not established. Amongst the *Urochloa* species, both Al accumulation in the roots and Al sensitivity followed the order *U. decumbens* < *U. brizantha* < *U. ruziziensis* (Arroyave et al. 2013). *U. ruzizensis* is a relative sensitive species within the *Urochloa* genus. However, comparison of Al-induced inhibition of root elongation among different species revealed that growth of *U. ruzizensis* was not affected at $Al^{3+}$ activities that inhibited growth of buckwheat, which has been described as "highly Al resistant" (Wenzl et al. 2001).

The most Al-resistant *Urochloa* species *U. decumbens* and *U. brizanta* are tetraploids with a complex genome of more than 1600 Mbp (Silva et al. 2013). Available genome data are mainly focused on marker-assisted breeding of the fast growing, less Al-resistant *U. ruziziensis* with a small genome (600 Mbp). In the more Al-resistant *U. brizantha*, some gene information is available. However, this is mainly restricted to studies on differential gene expression related to apomixis versus sexual reproduction and the development of floral organs (Silveira et al. 2012; Lazerda et al. 2013), a topic of high interest for breeding of valuable pasture varieties. This scarcity of genome information for *U. decumbens* hampers the investigation of the genetic basis of Al hyper-resistance in these species and, up-to-date, mainly physiological approaches have been performed.

Attempts to relate Al resistance in *U. decumbens* or *U. brizantha* with root exudation of organic acids failed (Wenzl et al. 2001; Ishikawa et al. 2000), while internal detoxification by Al binding to organic acids has been suggested as an Al tolerance mechanism inside the root tip cells of signal grass (Wenzl et al. 2002). A further characteristic of *U. decumbens* is that even when exposed to 200 μM Al (32 μM $Al^{3+}$ activity), the root tips do not stain with hematoxylin (Arroyave et al. 2011, 2013). Only a few hematoxylin stained spots were found scattered on the root tip surface. Morin staining revealed that Al accumulated in root hairs. Apparently, this Al accumulation occurred inside the hair cells rather than in the cell walls (Arroyave et al. 2011). A further unusual feature of this species is the formation of abundant root hairs very close to the tip (300–500 μm). Moreover, roots of signal grass present a multiseriate exodermis which apparently hampers apoplastic access to the inner cortex. Taken together, these results indicate that the Al resistance in *Urochloa* implies low apoplastic binding of Al in the root tips. Whether this is achieved by the exudation of a still uncharacterized Al-binding substance or by changes in the cell wall composition remains to be established. Besides the important Al exclusion from roots, *U. decumbens* also efficiently restricts Al transport from roots to shoots. To what extent active Al efflux, Al accumulation in the root hairs that have a fast turnover, and/or vacuolar storage contribute to this is still unclear and deserves further investigation.

## 4 Mechanisms of Hyper-tolerance

Maintenance of a plant's fitness even with a burden of several thousands of mg of Al per kilo dry weight in photosynthetically active leaves requires extremely efficient mechanisms of tolerance allowing detoxifying and compartmentalizing the toxic $Al^{3+}$. Species with such hyper-tolerance of Al share some common characteristics: (1) with some exceptions, these plants are tropical shrubs or trees, (2) they have a high tissue concentration of organic acids, especially oxalate and citrate, and (3) they contain a high concentration of phenolic substances.

### *4.1 Predominance of Shrubs and Trees Among Al Hyperaccumulator*

The question why Al hyper-tolerance has mainly evolved in shrubs and trees and only in a few herbaceous species is still open. The proposal of some speculative causes may stimulate further research efforts into this question. Main differences between the perennial shrubs or trees and herbaceous plants are size and life cycle length. Most herbaceous plants are relatively small annuals or bi-annuals. Their root systems mainly explore the superficial soil horizon where organic matter content is usually higher and, in consequence, $Al^{3+}$ activity is lower than in deeper soil. If the root system of these plants is challenged by high $Al^{3+}$ activity, organic acid exudation is an efficient way to prevent both injuries to the sensitive root tips and the excessive Al uptake and transport to the sensitive shoots. In shrubs or trees with longevity of more than decades, and a size that requires much deeper root systems, the challenge is much stronger. This especially gains relevance taking into account that most hyper-accumulating perennials have evolved in the tropics where subsoil acidity is much more common than in boreal and temperate regions (Brunner and Sperisen 2013). In the tropics shrubs and trees are photosynthetically active all over the year, and many plants have perennial leaves with large surface areas. In these plants no unfavorable, cold winter or dry hot summer period induces dormancy reducing metabolic activity and transpiration. Under these circumstances, the Al exclusion strategy may be less advantageous energetically than internal chelation and compartmentation. However, by far, not all tropical shrubs and trees hyperaccumulate Al in their shoots. For example, *Melaleuca cajuputi*, *Acacia mangium*, and *Leucaena leucocephala* efficiently exclude Al by root exudation of organic acids (Osaki et al. 1997). Also Eucalyptus species mainly exclude Al from the shoots while accumulating it in the roots (Ikka et al. 2013). Moreover, Al hyperaccumulation is not restricted to shrubs and trees of the tropics. The Mediterranean *Plantago almogravensis* is a notable exception. This woody cushion plant is an extremely rare endemism only occurring on Al- and Fe-rich hardpans of the eroded podzol in SW Portugal (Serrano et al. 2011). Buckwheat, *Fagopyrum esculentum*, an annual herbaceous species, is the best studied exception to the role that Al hyperaccumulators are shrubs or trees. Al hyperaccumulation seems to be a general characteristic of the genus because of both xylem sap Al concentrations in the range of 2–4 mM, and high Al leaf concentrations have been observed in different *F. tataricum and F. homotropicum* species (Wang et al. 2015).

Aluminum toxicity is a widespread inorganic stress of natural origin. Such stress factors (e.g., also salinity or nitrogen deficiency) certainly are working as efficient drivers of the evolution of different strategies that allow plants to colonize these adverse soils. The presence of Al hyperaccumulation in Pteridophytes suggests that the Al hyper-tolerance strategy is an ancient mechanism within the evolutionary time scale of terrestrial plants (Olivares et al. 2009). Moreover, phylogenetic analysis corroborates the view that this characteristic has appeared several times during plant evolution (Jansen et al. 2002). Tropical trees seem especially prone to evolution of different mechanisms for adaptation to soil with different nutrient stress factors. Recent investigation on *Geissois* sp indicate that originating from a single colonist, 13 different species evolved in a relative short time in New Caledonia. Also an Al hyperaccumulator, *Geissois polyphylla,* was found on non-ultramfic soil in this Island (Pillon et al. 2014). Three out of seven nickel hyperaccumulating *Geissois* species growing on ultramafic soil use different Ni chelating mechanisms. Such metabolic diversity may also operate in Al hyperaccumulators where large concentrations of potential Al-binding substance have been reported (Barceló and Poschenrieder 2002).

## 4.2 The Role of Rhizosphere in Al Hyperaccumulation

Organic acid exudation by root tips is not only a characteristic of Al-resistant crop varieties such as wheat, sorghum, or bean but also of Al hyperaccumulators. Protection against $Al^{3+}$ of the sensitive root tips where active cell division and elongation occurs is apparently also required in these hyper-tolerant species. However, in contrast to Al excluders, in Al hyperaccumulators the exudation of organic acids does not avoid, but favor Al uptake. A distinctive feature of species native and highly adapted to acid soils is that root tips mainly release oxalate and, to a lesser extent, citrate, but usually not malate (Barceló and Poschenrieder 2002). The stability constants for Al-organic acid complexes are higher for citrate and oxalate than for malate.

Fe and Al, together with Ca, may compete for binding to organic acids in the rhizosphere. Speciation studies in the rhizosphere of Al-hyperaccumulating plants are scarce and mostly refer to tea plantations. Cultivation of tea leads to intense soil acidification, enhancement of Al and Fe bioavailability, and depletion of Ca (Wang et al. 2010; Aleekseva et al. 2011). High oxalate and succinate concentrations in the rhizosphere soil are correlated to the enhanced Al and Fe bioavailability (Chen et al. 2006). In fact, exudation of oxalate by the roots of Al hyperaccumulators does not lead to restriction of Al uptake, but rather favors Al uptake (Watanabe and Osaki 2002). It has been proposed that oxalate exudation by the roots of hyperaccumulator *Melastoma malabathricum* is an important mechanism for increasing phosphate availability to the trees by releasing phosphate from insoluble Al-phosphate (Watanabe et al. 2002). This tree species is able to hyperaccumulate Al even from tropical peat soils with low Al availability (Osaki et al. 1998).

Besides oxalate exudation, young roots of tea plants also release considerable amounts of phenolic substances. In contrast to Al-induced organic acid exudation, however, the release of phenolics from roots of Al exposed tea plants is observed after weeks of exposure when also exudation of phenolics of plants without Al supply is enhanced (Hajiboland et al. 2015).

It has been proposed that the high volume of mucilage produced by the root tips of *M. malabathricum* concentrate considerable amounts of Al from low Al soil substrate. No organic acids were found in this mucilage (Watanabe et al. 2008). Due to high charge density, this mucilage has high affinity for trivalent cations, but its binding strength is weak because of a high degree of methylation. Therefore, mucilage-Al remains in bioavailable form and is taken up into the root. Contrastingly, in non-hyperaccumulating species, mucilage has a high binding strength and traps Al (Li et al. 2000) hampering its uptake. This may or may not contribute to enhanced Al resistance of Al excluding species (Horst et al. 1982; Li et al. 2000).

## 4.3 Root Uptake, Radial Transport, and Compartmentation of Al in Hyperaccumulators

Laboratory studies with *Camellia oleifera* seedlings growing in $CaCl_2$ solutions spiked with different Al forms revealed higher Al uptake when supplying Al as $Al^{3+}$, Al-malate, or Al-fluoride than in the form of Al citrate or Al oxalate (Zeng et al. 2011). Also in *F. esculentum* much more Al was accumulated from solutions with Al in the form of $AlCl_3$ than when exposed to Al-oxalate (1:3) complex (Ma and Hiradate 2000). This suggests that Al is mainly taken up in its ionic form as $Al^{3+}$. So while oxalate enhances bioavailability of Al in the rhizosphere of hyper-accumulators, it reduces uptake in comparison to the ionic $Al^{3+}$. Whether some dissociation of the Al-oxalate complex occurs in the root apoplast before Al enters into the symplasm has not yet been analyzed. The Al-oxalate (1:1) complex has also been proposed as the Al species taken up in *F. esculentum* (Klug and Horst 2010). However, after entrance into the root cells of buckwheat, Al is complexed in nontoxic form by oxalate with a metal ligand ratio of 1:3 (Ma et al. 1998).

The Al-induced release of oxalic acid from buckwheat roots occurs without lag time (Wang et al. 2015) and increases over time at least during the first 72 h after starting the Al exposure. During the first hours of exposure, Al induces a substantial enhancement of the expression of *FeAlS3*, while its expression decreases with increasing oxalate exudation (Reyna-Llorens et al. 2015). The *FeAlS3* has 75 % identity with *ALS3* of *A. thaliana* which encodes an ABC transporter known to function in Al tolerance by transporting either Al or a metabolite involved in Al tolerance (Larsen et al. 2005). Abscisic acid supply enhanced *FeALS3* expression in a similar way, and it has been suggested that the upregulation of this gene may be regulated by ABA, induced as an early stress response to Al (Reyna-Llorens et al. 2015). The recently published global transcription analysis of Al-induced genes in *F. esculentum* also revealed upregulation of, among others, *FeALS3*, *FeALS1*, and *FeMATE1* and *FeMATE2* genes (Yokosho et al. 2014).

The sites of Al uptake into the root have mainly been studied in *F. esculentum* and *C. sinensis*. In buckwheat, the 10 mm root tip is the main zone of Al uptake, while xylem loading occurs in the older subapical region 10–20 mm from the tip (Klug et al. 2011). In tea, staining with hematoxylin or morin showed that Al is highly accumulated in the root tip (root caps and adjacent meristematic cells). Towards the young subapical part, Al was found in root hairs and epidermal and cortical cells. In more basal regions, intense staining at the endodermal level suggests that Al entrance into the central cylinder is hampered in older root parts (Hajiboland and Poschenrieder 2015). As in buckwheat, also in tea the Al loaded into xylem seems mainly coming from the Al taken up in the root tips. This Al then moves symplastically to the more basal parts where it is loaded into the newly differentiated xylem vessels developing in this zone. The huge accumulation of Al in the abundant root hairs suggested that Al could also be transported radially from the root hair zone to the central cylinder. This radial transport in the subapical zone apparently only occurs in young parts of the root system where new branchings disrupt the suberized endodermal barrier. Moreover, it has to be taken into account that lateral roots of tea plants can present an exodermis as close as 1 mm to the apex. Such an apoplastic barrier may further contribute to inhibit uncontrolled apoplastic Al-access to the central part of the root (Tanimoto et al. 2004). This may be of special importance under conditions that favor plants' transpiration. Aluminum (hyper)accumulation is dependent on transpiration (Shen and Ma 2001). Studies on Al uptake kinetics in hyperaccumulators agree with the view that Al uptake is a passive process (Ma and Hiradate 2000; Watanabe et al. 2001; Ruan and Wong 2004). The high relative humidity during most time of the year in the tropical zones where Al hyperaccumulators have preferentially evolved reduces transpiration and may help to limit excessive Al accumulation.

Although huge amounts of Al are translocated out of the root system into the shoots of Al hyperaccumulators, the roots also accumulate considerable Al amounts. Therefore, besides chelation in nontoxic form, Al compartmentation in the roots is also an important issue. Cell walls and vacuoles are potential sites for Al storage in the roots. Al in cell wall of buckwheat roots was revealed by lumogallion staining (Reyna-Llorens et al. 2015), while morin mainly stained the Al-oxalate in the cytoplasmic ring surrounding the vacuoles (Klug et al. 2011).

In the root tips of tea, Al was mainly found in the cell walls (Hajiboland and Poschenrieder 2015). In the whole root system, however, cell wall-bound Al accounted for only 50 % of the total Al (Hajiboland et al. 2015) indicating the importance of other compartments, mainly cytosol and vacuoles. As in buckwheat, Al in the cytosol of tea root cells is chelated with oxalate (Morita et al. 2008), while the binding form of Al in the cell walls of tea roots is still not clearly established. Pectin and hemicellulose (Gao et al. 2014) as well as cell wall phenolic acids (Hajiboland et al. 2015) have been proposed. In tea plants, high concentration of cell wall-bound phenolics could act as a potential target for Al. Al binding to phenolics in tea roots could be the reason for the lack of any negative influence of Al on root elongation, which is to be expected if Al would cross-link cell wall pectins. Moreover, Al bound to cell walls of tea is stained with morin, while morin according to Eticha et al. (2005) is unable to stain pectin-bound Al. The increasing Al accumulation in a non-exchangeable Al fraction in tea roots has been related to the accumulation of phenolic compounds in the endodermis layer sequestering a considerable amount of Al in the cell walls (Ruan and Wong 2004). Al binding to the cell wall-bound phenolic acids would reduce their availability for subsequent enzymatic reactions and lower lignin content (Hajiboland et al. 2015). This may contribute to long-term Al-induced growth stimulation in the tea plants.

## *4.4 Long Distance Transport and Leaf Distribution of Al in Hyperaccumulators*

Aluminum in the xylem has been visualized using morin staining followed by fluorescence microscopy (Klug et al. 2011; Hajiboland and Poschenrieder 2015). In *F. esculentum*, Al in roots is chelated by oxalate, while during its path into the xylem a ligand exchange occurs and Al is transported in the xylem sap in the form of a citrate complex. Also in tea and *M. malabathricum* (Watanabe and Osaki 2001), citrate is the main ligand for Al transport in the xylem sap (Morita et al. 2004). The exact mechanisms determining how and where this ligand exchange occurs, and whether active loading and unloading mechanisms are operating in the xylem transport of Al, are still unknown. Transpiration influences the transport and in tea Al typically accumulates in the leaf margins close to the end of the xylem vessels in the leaf epidermal cells and in the leaf hairs (Tolrà et al. 2011; Hajiboland and Poschenrieder 2015). In the leaves of both buckwheat and *M. malabathricum*, Al is mainly bound to oxalate, so that also in the leaves a ligand exchange between citrate and oxalate must occur. Oxalate again is the major ligand, and vacuolar storage of Al–oxalate seems to be the main compartmentation strategy in buckwheat (Shen and Ma 2001). In tea, besides organic acids, the high concentration of phenolic substances has also been involved in Al detoxification. The $^{27}$Al-NMR technique identified a catechin–Al complex as main Al form in tea leaves (Nagata et al. 1992). The high number of hydroxyl groups in tea leaf phenolics (five in epicatechin and up to eight in the case of epigallocatechingallate) makes them extraordinarily strong ligands for Al (Barceló and Poschenrieder 2002; Tolrà et al. 2005). While in tea infusions, large amount of Al is bound to these flavonoids reducing the bioavailability of Al to consumers of tea leaf infusions; the in vivo role of these phenolics in Al compartmentation in the tea plant is still not clearly established. In vivo the phenolic substances seem mainly localized in the mesophyll vacuoles of the tea leaf, also in vascular bundles and in chloroplasts, but to a lesser extent in the epidermis. This distribution is not in agreement with the preferential accumulation of Al in the epidermal cells and the cuticle of tea leaves as shown by X-ray fluorescence spectro-microscopy (Tolrà et al. 2011). Compartmentation of Al in other hyperaccumulating species is less studied. A most striking Al distribution has been reported for leaves of *Qualea grandiflora* and *Callisthene major* (Vochysiaceae). These species of the Brazilian Cerrado vegetation apparently accumulated Al in their chloroplasts without damage (de Andrade et al. 2011). These results obtained with hematoxylin staining should be confirmed using different localization methods.

Within the tea plant, older leaves accumulate much higher Al concentrations than young leaves. This can already be observed in young 2-month-old plants (Hajiboland et al. 2013c), but is even more pronounced in older plants (Carr et al. 2003). Such a distribution suggests that Al has low mobility and is poorly mobilized from older to young leaves. However, the possibility of phloem transport of Al in hyperaccumulators is still under debate. Hematoxylin staining revealed the presence of Al in phloem of Al hyperaccumulators of the Brazilian Cerrado vegetation (Haridasan et al. 1986). More recently, X-ray fluorescence-microscopy also confirmed Al within the phloem tissue of tea leaves (Tolrà et al. 2011). Based on Al analysis in roots and leaves of buckwheat plants that after 12 days in Al-spiked medium were transferred to medium without Al, it has been proposed that Al is a phloem-immobile element. In these experiments, the Al content of the new leaves formed after the transfer was extremely low. In contrast, in old leaves the Al content continued to increase because of transport from the roots (Shen and Ma 2001). Contrasting results were obtained with *C. sinensis* plants using a shorter, 24 h, period of root loading with Al. Under these conditions, a clear decline in the Al content of old leaves during further growth in the Al free medium is observed. About 20–35 % of Al loaded in the mature leaves is remobilized after 2 weeks indicating that Al is highly mobile in this hyperaccumulator plant (Hajiboland et al. 2015). This view is also in line with results obtained by different experimental techniques with oil tea plants, *C. oleifera* (Zeng et al. 2013). Moreover, these authors revealed that Al can be retranslocated via phloem from the shoot to the roots and from leaves to seeds and backwards. Phloem transport thus could be an important mechanism for controlling the Al accumulation in the different organs avoiding excessive accumulation and damage. The form in which Al is translocated in the phloem is still not established. Analysis of carbohydrates in tea leaf phloem sap showed that main components are fructose, sucrose, and glucose, while no polyols were detected (Hajiboland et al. 2013a). Interestingly, in tea plants under boron deficiency, Al binding to cell walls is enhanced from 50 to 80 %, while the concentration of soluble Al as the readily re-translocable form in the phloem decreases (Hajiboland et al. 2015).

## 4.5 *Integrated View and Outlook of Al Hyperaccumulation in Tea*

Although Al uptake is considered a passive process and Al hyperaccumulation is enhanced by transpiration, this does not imply that the entire amount of Al of the soil solution circulates through these plants in a completely uncontrolled process. In the roots, uptake is mainly restricted to the apex. In the subapical root hair zone of the young roots, the entrance into the apoplastic pathway is hindered by a suberized exodermis that develops very close to the tip. The suberized endodermis is a second barrier for the apoplastic transport. However, apoplastic byflow could occur at sites of lateral root development. The symplastic pathway of chelated Al by sure is controlled by membrane transport proteins. Moreover, efflux channels analogue to those described, for example, for Zn xylem loading in Zn hyperaccumulators (Papoyan and Kochian 2004) may also operate in Al hyperaccumulation. Compartmentation of Al in the leaves, especially vacuolar storage and export into leaf hairs also must imply transport systems. In fact, recent studies with buckwheat leaves revealed 25 transporter genes upregulated upon Al exposure (Yokosho et al. 2014). There are now convincing data that Al can be transported in the phloem, a process that may be an important way to avoid excessive accumulation of Al to toxic levels in sensitive sites, e.g., young leaves or seed embryos. The form(s) of Al transported in the phloem and the regulatory mechanism(s) for this translocation need to be explored. As tea is a high-value crop, genomic information for this species is growing quickly. Although up to now most of these studies at the molecular genetics level are being focused on the metabolic pathways, substantial progress in the characterization of the molecular mechanisms underlying Al hyper-accumulation in tea are to be expected in the near future.

**Acknowledgments** Authors' cited work was supported by the Spanish MICINN (projects BFU2010-14873 and BFU2013-42839-R).

## References

Aleekseva T, Alekseev A, Xu R-K, Zhao A-Z, Kalinin P (2011) Effect of soil acidification by a tea plantation on chemical and mineralogical properties of Alfisols in eastern China. Environ Geochem Health 33:137–148

Amenós M, Corrales I, Poschenrieder C, Illéš P, Baluška F, Barceló J (2009) Different effects of aluminum on the actin cytoskeleton and brefeldin A-sensitive vesicle recycling in root apex cells of two maize varieties differing in root elongation rate and aluminium tolerance. Plant Cell Physiol 50:528–540

Arroyave C, Barceló J, Poschenrieder C, Tolrà R (2011) Aluminium-induced changes in root epidermal cell patterning, a distinctive feature of hyperresistance to Al in *Brachiaria decumbens*. J Inorg Biochem 105:1477–1483

Arroyave C, Tolrà R, Thuy T, Barceló J, Poschenrieder C (2013) Differential aluminum resistance in *Brachiaria* species. Environ Exp Bot 89:11–18

Ayeni O, Kambizi L, Laubscher C, Fatoki O, Olatunji O (2014) Risk assessment of wetland under aluminium and iron toxicities: a review. Aq Ecosyst Health Manag 17:122–128

Barceló J, Poschenrieder C (2002) Fast root growth responses, root exudates, and internal detoxification as clues to the mechanisms of aluminium toxicity and resistance: a review. Environ Exp Bot 48:75–92

Becker M, Asch F (2005) Iron toxicity in rice-conditions and management concepts. J Plant Nutr Soil Sci 168:558–573

Bert V, Meerts P, Saumitou-Laprade P, Salis P, Gruber W, Verbruggen N (2003) Genetic basis of Cd tolerance and hyperaccumulation in Arabidopsis halleri. Plant Soil 249:9–18

Bitencourt GA, Chiari L, Laura VA, do Valle CB, Jank L, Moro JR (2011) Aluminum tolerance on genotypes of signal grass. Rev Bras Zootec 40:245–250

Brunner I, Sperisen C (2013) Aluminum exclusion and aluminum tolerancein woody plants. Front Plant Sci 4:172

Carr HP, Lombi E, Kupper H, McGrath SP, Wong MH (2003) Accumulation and distribution of aluminium and other elements in tea (*Camellia sinensis*) leaves. Agronomie 23:705–710

Chen YM, Wang MK, Zhuang SY, Chiang PN (2006) Chemical and physical properties of rhizosphere and bulk soils of three tea plants cultivated in Ultisols. Geoderma 136:378–387

De Andrade LRM, Barros LMG, Echevarria GF, do Amaral LIV et al (2011) Al-hyperaccumulator Vochysiaceae from the Brazilian Cerrado store aluminum in their chloroplasts without apparent damage. Environ Exp Bot 70:37–42

Delhaize E, Gruber B, Ryan PR (2007) The roles of organic anion permeases in aluminium resistance and mineral nutrition. FEBS Lett 581:2255–2262
Delhaize E, Ma JF, Ryan PR (2012) Transcriptional regulation of aluminum tolerance genes. Trends Plant Sci 17:341–348
Doncheva S, Amenós M, Poschenrieder C, Barceló J (2005) Root cell patterning: a primary target for aluminium toxicity in maize. J Exp Bot 56:1213–1220
Eticha D, Stass A, Horst WJ (2005) Localization of aluminium in the maize root apex: can morin detect cell wall bound aluminium? J Exp Bot 56(415):1351–1357
Fageria NK, Baligar VC (2008) Ameliorating soil acidity of tropical oxisols by liming for sustainable crop production. Adv Agron 99:345–399
Famoso AN, Zhao K, Clark RT, Tung CW, Wright MH, Bustamante C, Kochian LV, McCouch SR (2011) Genetic architecture of aluminum tolerance in rice (*Oryza sativa*) determined through genome-wide association analysis and QTL mapping. PLoS Genet 7, e1002221
Gao HJ, Zhao Q, Zhang XC, Wan XC, Mao JD (2014) Localization of fluoride and aluminum in subcellular fractions of tea leaves and roots. J Agr Food Chem 62:2313–2319
Gould B, McCouch S, Geber M (2014) Variation in soil aluminium tolerance genes is associated with local adaptation to soils at the Park Grass Experiment. Mol Ecol 23:6058–6072
Gout E, Rébeillé F, Douce R, Bligny R (2014) Interplay of $Mg^{2+}$, ADP, and ATP in the cytosol and mitochondria: unraveling the role of $Mg^{2+}$ in cell respiration. Proc Natl Acad Sci U S A 111: E4560–E4567
Guimaraes CT, Simoes CC, Pastina MM, Maron LG et al (2014) Genetic dissection of Al tolerance QTLs in the maize genome by high density SNP scan. BMC Genom 15:153
Hajiboland R, Poschenrieder C (2015) Localization and compartmentation of Al in the leaves and roots of tea plants. Phyton In J Exp Bot (in press)
Hajiboland R, Bahrami-Rad S, Bastani S, Tolrà R, Poschenrieder C (2013a) Boron re-translocation in tea (*Camellia sinensis* (L.) O. Kuntze) plants. Acta Physiol Plant 35:2373–2381
Hajiboland R, Barceló J, Poschenrieder C, Tolrà R (2013b) Amelioration of iron toxicity: a mechanism for aluminum-induced growth stimulation in tea plants. J Inorg Biochem 128: 183–187
Hajiboland R, Rad SB, Barceló J, Poschenrieder C (2013c) Mechanisms of aluminum-induced growth stimulation in tea (*Camellia sinensis*). J Plant Nutr Soil Sci 176:616–625
Hajiboland R, Bastani S, Bahrami-Rad S, Poschenrieder C (2015) Interactions between aluminum and boron in tea (*Camellia sinensis*) plants. Acta Physiol Plant (in press)
Haridasan M (1982) Aluminum accumulation by some Cerrado native species of Central Brazil. Plant Soil 65:265–273
Haridasan M, Paviani TI, Schiavini I (1986) Localization of aluminum in the leaves of some aluminum-accumulating species. Plant Soil 94:435–437
Horst WJ, Wagner A, Marschner H (1982) Mucilage protects root-meristem from aluminum injury. J Plant Physiol 105:435–444
Ikka T, Ogawa T, Li D, Hiradate S, Morita A (2013) Effect of aluminum on metabolism of organic acids and chemical forms of aluminum in root tips of *Eucalyptus camaldulensis* Dehnh. Phytochemistry 94:142–147
Ishikawa S, Wagatsuma T, Sasaki R, Ofei-Manu P (2000) Comparison of the amount of citric and malic acids in Al media of seven plant species and two cultivars in five species. Soil Sci Plant Nutr 46:751–758
Jansen S, Broadley MR, Robbrecht E, Smets E (2002) Aluminum hyperaccumulation in angiosperms: a review of its phylogentic significance. Bot Rev 68:235–269
Jansen S, Watanabe T, Dessein S, Smets E, Robbrecht E (2003) A comparative study of metal levels in leaves of some Al-accumulating Rubiaceae. Ann Bot 91:657–663
Khabaz-Saberi H, Barker SJ, Rengel Z (2012) Tolerance to ion toxicities enhances wheat (Triticum aestivum L.) grain yield in waterlogged acidic soils. Plant Soil 354:371–381
Khabaz-Saberi H, Barker SJ, Rengel Z (2014) Tolerances to ion toxicities enhances wheat grain yield in acid soils prone to drought and transient waterlogging. Crop Pasture Sci 65:862–867

Kidd PS, Proctor J (2000) Effects of aluminium on the growth and mineral composition of *Betula pendula* Roth. J Exp Bot 51:1057–1066

Kidd PS, Llugany M, Poschenrieder C, Gunsé B, Barceló J (2001) The role of root exudates in aluminium resistance and silicon-induced amelioration of aluminium toxicity in three varieties of maize (*Zea mays* L.). J Exp Bot 52:1339–1352

Kinraide TB (1998) Three mechanisms for the calcium aññeviation of mineral toxicities. Plant Physiol 118:513–520

Kinraide TB, Ryan PR, Kochian LV (1992) Interactive effects of Al3+, H+, and other cations on root elongation considered in terms of cell-surface electrical potential. Plant Physiol 99:1461–1468

Kinraide TB, Poschenrieder C, Kopittke PM (2011) The standard electrode potential (E-theta) predicts the prooxidant activity and the acute toxicity of metal ions. J Inorg Biochem 105:1438–1445

Klug B, Horst WJ (2010) Oxalate exudation into the root tip water free space confers protection from aluminum toxicity and allows accumulation in the symplast in buckwheat (Fagopyrum esculentum). New Phytol 187:380–391

Klug B, Specht A, Horst WJ (2011) Aluminium localization in root tips of the aluminium-accumulating plant species buckwheat (*Fagopyrum esculentum* Moench). J Exp Bot 62:5453–5462

Kochian LV, Piñeros MA, Hoekenga OA (2005) The Physiology, genetics and molecular Biology of plant aluminum resistance and toxicity. Plant Soil 274:175–195

Larsen PB, Geisler J, Jones C, Williams K, Cancel J (2005) ALS3 encodes a phloem-localized ABC transporter-like protein that is required for aluminum tolerance in Arabidopsis. Plant J 41: 353–363

Lazerda AL, de Dusi DMA, Alves E, Rodrigues JCM et al (2013) Expression analysis of Brachiaria brizantha genes encoding ribosomal proteins *BbrizRPS8*, *BbrizRPS15a*, and *BbrizRPL41* during development of ovaries and anthers. Protoplasma 250:505–514

Levitt J (1980) Responses of plants to environmental stresses, vol 1, 2nd edn. Academic Press, New York, NY

Li XF, Ma JF, Hiradate S, Matsumoto H (2000) Mucilage strongly binds aluminum but does not prevent roots from aluminum injury in *Zea mays*. Physiol Plant 108:152–160

Liu J, Piñeros MA, Kochian LV (2014) The role of aluminum sensing and signaling in plant aluminum resistance. J Integr Plant Biol 56:221–230

Llugany M, Poschenrieder C, Barceló J (1995) Monitoring of aluminium-induced inhibition of root elongation in four maize cultivars differing in tolerance to aluminium and proton toxicity. Physiol Plant 93:265–271

Llugany M, Lombini A, Poschenrieder C, Dinelli E, Barceló J (2003) Different mechanisms account for enhanced copper resistance in *Silene armería* ecotypes from mine spoil and serpentine sites. Plant Soil 251:55–63

Ma JF, Hiradate S (2000) Form of aluminium for uptake and translocation in buckwheat (*Fagopyrum esculentum* Moench). Planta 211:355–360

Ma JF, Hiradate S, Matsumoto H (1998) High aluminum resistance in buckwheat. II Oxalic acid detoxifies aluminum internally. Plant Physiol 117:753–759

Ma JF, Xhen ZC, Shen RF (2014) Molecular mechanisms of Al tolerance in gramineous plants. Plant Soil 381:1–12

Massot N, Nicander B, Barceló J, Poschenrieder C, Tillberg E (2002) A rapid increase in cytokinin levels and enhanced ethylene production precede Al3+-induced inhibition of root growth in vean seedlings (*Phaseolus vulgaris* L.). Plant Growth Reg 37:105–112

Matsumoto H, Hirasawa E, Morimura S, Takahashi E (1976) Localization of aluminum in tea leaves. Plant Cell Physiol 17:627–631

Meda AR, Furlani PR (2005) Tolerance to aluminum toxicity by tropical leguminous plants used as cover crops. Braz Arch Biol Technol 48:309–317

Metali F, Salim KA, Burslem DFRP (2012) Evidence of foliar aluminium accumulation in local, regional and global dataset of wild plants. New Phytol 193:637–649

Mongon J, Konnerup D, Colmer TD, Rerkasem B (2014) Responses of rice to Fe2+ in aerated and stagnat conditions: growth, root porosity and radial oxygen barrier. Funct Plant Biol 9:922–929
Morita A, Horie H, Fujii Y, Takatsu S, Watanabe N, Yagi A, Yokota H (2004) Chemical forms of aluminum in xylem sap of tea plants (*Camellia sinensis* L.). Phytochemistry 65:2775–2780
Morita A, Yanagisawa O, Takatsu S, Maeda S, Hiradate S (2008) Mechanism for the detoxification of aluminum in roots of tea plant (*Camellia sinensis* (L.) Kuntze). Phytochemistry 69:147–153
Nagata T, Hayatsu M, Kosuge N (1992) Identification of aluminium forms in tea leaves by $^{27}$Al NMR. Phytochemistry 31:1215–1218
Olivares E, Peña E, Marcano E, Mostacero J, Aguiar G, Benítez M, Rengifo E (2009) Aluminum accumulation and its relationship with mineral nutrients in 12 pteridophytes from Venezuela. Environ Exp Bot 65:132–141
Osaki M, Watanabe T, Tadano T (1997) Beneficial effect of aluminium on growth of plants adapted to low pH soils. Soil Sci Plant Nutr 43:551–563
Osaki M, Watanabe T, Ishizawa T, Nilnod C, Nuyim T, Sittibush C, Tadano T (1998) Nutritional characteristics in leaves of native plants grown in acid sulfate, peat, sandy podzolic, and saline soils distributed in peninsular Thailand. Plant Soil 201:175–182
Osawa H, Ikeda S, Tange T (2013) The rapid accumulation of aluminum is ubiquitous in both evergreen and deciduous leaves of Theaceae and Ternstroemiaceae plants over a wide pH range in acidic soils. Plant Soil 363:49–59
Papoyan A, Kochian LV (2004) Identification of *Thlaspi caerulescens* genes that may be involved in heavy metal hyperaccumulation and tolerance. Characterization of a novel heavy metal transporting ATPase. Plant Physiol 136:3814–3823
Pillon Y, Hopkins HCF, Rigault F, Jaffré T, Stacy EA (2014) Cryptic adaptive radiation in tropical forest trees in New Caledonia. New Phytol 202:521–530
Poschenrieder C, Gunsé B, Corrales I, Barceló J (2008) A glance into aluminum toxicity and resistance in plants. Sci Total Environ 400:356–368
Poschenrieder C, Cabot C, Martos S, Gallego B, Barceló J (2013) Do toxic ions induce hormesis in plants? Plant Sci 212:15–25
Reyna-Llorens I, Corrales I, Poschenrieder C, Barceló J, Cruz-Ortega R (2015) Both aluminum and ABA induce the expression of an ABC-like transporter gene (FeALS3) in the Al-tolerant species *Fagopyrum esculentum*. Environ Exp Bot 111:74–82
Ruan JY, Wong MH (2004) Aluminium absorption by intact roots of the Al-accumulating plant *Camellia sinensis* L. Agronomie 24:137–142
Sarkarung S (1986) Screening upland rice for aluminum tolerance and blast resistance. In: Progress in upland rice research. Proceeding of the 1985 Jakarta Conference. The International Rice Research Institute, Manila, Philippines, pp 272–281
Serrano HC, Pinto MJ, Martins-Louçao MA, Branquinho C (2011) How does Al hyperaccumulator plant respond to a natural field gradient of soil phytoavailable Al. Sci Total Environ 409:3749–3756
Shabala S, Shabala L, Barceló J, Poschenrieder C (2014) Membrane transporters mediating root signaling and adaptive responses to oxygen deprivation and soil flooding. Plant Cell Environ 37:2216–2233
Shen R, Ma JF (2001) Distribution and mobility of aluminium in an Al-accumulating plant, *Fagopyrum esculentum* Moench. J Exp Bot 52:1683–1687
Silva PIT, Martins AM, Gouvea EG, Pessoa-Filho M, Ferreira ME (2013) Development and validation of microsatellite markers for *Brachiaria ruziziensis* obtained by partial genome assembly of Illumina single-end reads. BMC Genom 14:17
Silveira E, Guimaraes L, de Dusi DMA, Silva F et al (2012) Expressed sequence-tag analysis of ovaries of Brachiaria brizantha reveals genes associated with the early steps of embryo sac differentiation of apomitic plants. Plant Cell Rep 31:403–416
Tanimoto E, Homma T, Matsuo K, Hoshino T, Lux A, Luxova M (2004) Root structure and cell wall extensibility of adventitious roots of tea (*Camellia sinensis* cv. Yabukita). Biologia 59:57–66

Tolrà RP, Poschenrieder C, Luppi B, Barceló J (2005) Aluminium-induced changes in the profiles of both organic acids and phenolic substances underlie Al tolerance in *Rumex acetosa* L. Environ Exp Bot 54:231–238

Tolrà R, Vogel-Mikus K, Hajiboland R, Kump P et al (2011) Localization of aluminium in tea (*Camellia sinensis*) leaves using low energy X-ray fluorescence spectro-microscopy. J Plant Res 124:165–172

Torres González AM, Morton CM (2005) Molecular and morphological phylogenetic analysis of *Brachiaria* and *Urochloa* (Poaceae). Mol Phylogenet Evol 37:36–44

Vázquez MD, Poschenrieder C, Corrales I, Barceló J (1999) Change in apoplastic aluminum during the initial growth response to aluminum by roots of a tolerant maize variety. Plant Physiol 119:435–444

Wang H, Xu R-K, Wang N, Li X-H (2010) Soil acidification of alfisols as influenced by tea cultivation in China. Pedosphere 20:799–806

Wang H, Chen RF, Iwashita T, Shen RF, Ma JF (2015) Physiological characterization of aluminum tolerance and accumulation in tartary and wild buckwheat. New Phytol 205:273–279

Watanabe T, Osaki M (2001) Influence of aluminum and phosphorus on growth and xylem sap composition in *Melastoma malabathricum* L. Plant Soil 237:63–70

Watanabe T, Osaki M (2002) Role of organic acids in aluminum accumulation and plant growth in *Melastoma malabathricum*. Tree Physiol 22:785–792

Watanabe T, Osaki M, Tadano T (2001) Al uptake kinetics in roots of *Melastoma malabathricum* L. – an Al accumulator plant. Plant Soil 231:283–291

Watanabe T, Misawa S, Hiradate S, Osaki M (2008) Characterization of root mucilage from Melastoma malabathricum, with emphasis on its role in aluminum accumulation. New Phytol 178:581–589

Wenzl P, Patiño GM, Chaves AL, Mayer JE, Rao IM (2001) The high level of aluminum resistance in signal grass is not associated with known mechanisms of external aluminum detoxification in root apices. Plant Physiol 125:1473–1484

Wenzl P, Chaves A, Patiño GM, Mayer JE, Rao IM (2002) Aluminum stress stimulates the accumulation of organic acids in root apices of Brachiaria species. J Plant Nutr Soil Sci 165:582–588

Wheeler DM, Edmeades DC, Chriestie RA, Gardner R (1992) Effect of aluminum on the growth of 34 plant species – a summary of results obtained in low ionic-strength solution culture. Plant Soil 146:61–66

Wu LB, Shhadi MY, Gregorio G, Matthus E et al (2014) Genetic and physiological analysis of tolerance to acute iron toxicity in rice. Rice 7:8

Yamaji N, Huang CF, Nagao S, Yano M et al (2009) A zinc finger transcription factor ART1 regulates multiple genes implicated in aluminum tolerance in rice. Plant Cell 21:3339–3349

Yang ZB, Rao IM, Horst WJ (2013) Interaction of aluminium and drought stress on root growth and crop yield on acid soils. Plant Soil 372:3–25

Yang ZB, Geng XY, He CM, Zhang F et al (2014) TAA1-regulated local auxin biosynthesis in the root ápex transition zone mediates the aluminium-induced inhibition of root growth in Arabidopsis. Plant Cell 26:2889–2904

Yokosho K, Yamaji N, Ma JF (2014) Global transcriptome analysis of Al-induced genes in an Al-accumulating species, common buckwheat (*Fagopyrum esculentum* Moench). Plant Cell Physiol 55:2077–2091

Zeng QL, Chen RF, Zhao XQ, Wang HY, Shen RF (2011) Aluminium uptake and accumulation in the hyperaccumulator *Camellia oleifera* Abel. Pedosphere 21:358–364

Zeng QL, Chen RF, Zhao XQ, Shen RF, Noguchi A, Shinmachi F, Hasegawa I (2013) Aluminum could be transported via phloem in Camellia oleifera Abel. Tree Physiol 33:96–105

# Significant Role of the Plasma Membrane Lipid Bilayers in Aluminum Tolerance of Plants

**Tadao Wagatsuma, Eriko Maejima, Toshihiro Watanabe, Md. Shahadat Hossain Khan, and Satoru Ishikawa**

**Abstract** We propose a new aluminum (Al) tolerance mechanism which should be discriminated from the exclusion mechanism: plasma membrane (PM) lipid bi-layers barrier mechanism (Abbreviation: Plasma membrane lipid mechanism). It is defined as the retardation of Al permeation through the PM lipid bilayers ascribed to the specific composition of lipid molecules in the PM. The lipid phase separation response of the PM in the root tip portion of plants caused by the binding of phospholipids with $Al^{3+}$ is an important first step in Al toxicity. This response develops most strongly in the elongation zone of the root tip by increasing the distances between solid state lipid aggregates. Sterols are independent of the action of $Al^{3+}$ because of their electrical neutrality. Purified PMs from Al-tolerant cultivars showed less permeability than PMs from Al-sensitive cultivars without the participation of organic acid anion exudation. Lowering the phospholipid content and the increasing the sterol content of PMs from root tips produce PMs with less surface negativity and is a common strategy for Al tolerance in several plant species. *PAH* encoding phosphatidate phosphohydrolase and *HMG* encoding 3-hydroxy-3-methylglutaryl CoA reductase are speculated to be promising candidate genes for lowering the phospholipid content and for increasing the sterol content, respectively, to generate new Al-tolerant plants. Phenolics are present in high concentrations in plants, especially in hyper Al-tolerant plants; however, their existence within the lipid bilayers of root PMs needs to be clarified to understand

T. Wagatsuma (✉)
Faculty of Agriculture, Yamagata University, Tsuruoka, Japan
e-mail: wagatuma@tds1.tr.yamagata-u.ac.jp

E. Maejima • T. Watanabe
Graduate School of Agriculture, Hokkaido University, Sapporo, Japan
e-mail: ericom@chem.agr.hokudai.ac.jp; nabe@chem.agr.hokudai.ac.jp

M.S.H. Khan
HMD Science and Technology University, Dinajpur, Bangladesh
e-mail: khan_bely@yahoo.com

S. Ishikawa
National Agricultural Research Center, Tsukuba, Japan
e-mail: isatoru@affrc.go.jp



S.K. Panda, F. Baluška (eds.), *Aluminum Stress Adaptation in Plants*, Signaling and Communication in Plants 24, DOI 10.1007/978-3-319-19968-9_6

their potential to reduce PM permeability. Combined studies on the ionome and lipidome of plant PMs may provide insights useful for breeding plants with multiple tolerances to complex ionic environments.

## 1 Plasma Membrane Lipid Bilayers Barrier Mechanism as a Significant and Ubiquitous Aluminum Exclusion Mechanism

Aluminum (Al) stress responses have been considered to be controlled by two mechanisms: exclusion mechanism and internal tolerance mechanism (Taylor 1991; Kochian and Jones 1997; Matsumoto 2003; Kochian et al. 2005; Ma 2007). Although in many review articles and the related references, special interest has been focused on the exudation of organic acid anions, two reviews have pointed out the role of the plasma membrane (PM) in the Al exclusion mechanism. In one review, release of low-molecular-weight Al-chelating ligands into the rhizosphere, root-induced pH increase in the rhizosphere, increased binding of Al within the cell wall, decreased permeability of the PM to Al influx, and binding of Al within the mucigel associated with the root apex were proposed as being associated with the Al exclusion mechanism (Kochian and Jones 1997). In their review, Kochian and Jones (1997) also suggested that an investigation of the role of Al interactions with the lipid layer component of the PM in Al phytotoxicity may be a useful direction for future research. In another review, Al immobilization at the cell wall, selective permeability of the PM, a plant-induced pH barrier in the rhizosphere, exudation of chelate ligands, exudation of phosphate, and Al efflux were proposed as being associated with the Al exclusion mechanism (Taylor 1991). These two groups have published their PM-lipid related research work (Jones and Kochian 1997; Zhang et al. 1996, 1997); however, until now, further detailed research has not been reported although Jones and Kochian (1997) clearly showed that specific phospholipid (PL) molecules and nonenzymatic binding domains in the PM were most likely the sites of Al toxicity in plants.

Because the Al tolerance mechanism should be defined based on its site-specific function, we propose to divide the present definition of the Al exclusion mechanism into two compartments: an Al detoxification mechanism on the outside of the PM and a PM lipid bilayers barrier mechanism. The Al detoxification mechanism includes all the items described by Kochian and Jones (1997) except for the one related to the PM. The PM lipid bilayers barrier mechanism is defined as the retardation of Al permeation through the PM lipid bilayers based on the specific composition of the lipid molecules in the PM. Until now, the PM lipid bilayers barrier mechanism has not been regarded as a major contribution factor to Al resistance (Yermiyahu et al. 1997; Ahn et al. 2004); however, remarkable progress on this mechanism has been reported recently (Khan et al. 2009; Kobayashi et al. 2013; Maejima et al. 2014; Wagatsuma et al. 2015). Thus, we believe this is an excellent time to summarize and review the studies related to this mechanism.

# 2 Lipid Phase Separation by $Al^{3+}$ as an Important First Step of Al Toxicity

## *2.1 Mechanism of Lipid Phase Separation by Polyvalent Cations*

Enlarged cracks within the lipid bilayers form the hydrophilic route for the passive permeation of Al into the cytoplasm through the PM from rhizosphere; therefore, phase separation of the lipid bilayers is the causal risk for greater Al permeation, which is connected directly with high Al toxicity. To explore the neurotoxicity mechanism of $Al^{3+}$ and other polyvalent cations, PL model membranes have been used. Lipid phase separation of phosphatidylserine (PS)-containing lipid vesicles monitored by NBD (4-nitrobenz-2-oxa-1,3-diazole) fluorescence quenching was demonstrated at less than 30 μM $Al^{3+}$, whereas the effect of $Cd^{2+}$ and $Mn^{2+}$ on quenching was much less pronounced and was only demonstrated in the 0.1–1 mM range (Deleers et al. 1985). Increasing amounts of phosphatidylcholine (PC) or phosphatidylethanolamine (PE) in the vesicles decreased $Al^{3+}$-induced quenching. Maximal lipid phase separation was also demonstrated in the mixed PE-cholesterol vesicles at concentrations of $Al^{3+}$ between 87.5 and 125 μM, while millimolar concentrations of $Ca^{2+}$, $Mn^{2+}$, $Cd^{2+}$, and $Zn^{2+}$ had no effect (Deleers et al. 1987). The presence of acidic PS in the mixed phospholipid (PL) vesicles was not a prerequisite for the interaction of lipid phase separation by $Al^{3+}$. Further, when the PL vesicles contained only PE and cholesterol, $Al^{3+}$ was the only cation that provoked lipid phase separation. This was the first evidence that acidic PS or negative lipids were not required for cation-induced lipid phase separation. These findings are considered to be useful model system for the actual PM lipid bilayers, because in the plant root PM, acidic phosphatidylglycerol and phosphatidic acid are minor lipid classes despite of their existence mainly in the outer leaflet of the bilayers (Larsson et al. 1990). PS is also a minor acidic PL; however, it is known to exist in the cytoplasmic leaflet of the bilayers of normal animal cells and appears in the outer leaflet only in apoptosis. PC and PE are the major PL classes, which together with neutral phytosterols are the major lipid components of the PM lipid bilayers (Yoshida and Uemura 1986; Brown and DuPont 1989). The lipid bilayers are composed of many lipid molecules in fluid conditions that are controlled by various forces, including repulsive and attractive forces between the head groups in the lipid molecules, the molecular shape, the ratio between the volumes of the head group and the hydrocarbon region, and hydration forces (Boggs 1987). The repulsive forces are primarily electrostatic repulsive forces between similarly charged lipids, while the attractive forces are the electrostatic interactions between oppositely charged groups and intermolecular hydrogen bonding between charged or neutral lipids with hydrogen-donating and -accepting groups. Structural modifications to the lipid such as increased dehydration contribute to increases in the strength or probability of hydrogen bonding interactions (Boggs 1987). Hydrogen-donating groups in lipid molecules include $NH_3^+$, $NH_2$, POH, COH,

COOH, and HNC=O, while hydrogen accepting groups include some of these groups as well as PO, COO, OC=O, and COC with different strengths of the hydrogen bonds between them. For example, 3β-OH of sterol, a hydrogen-donating group, could form a hydrogen bond with a hydrogen acceptor, such as the ester oxygen atoms of glycerolipids or the amide oxygen of sphingolipids with the carbonyl of the saturated acyl chain of PC. The number of water molecules generally associated with PC, PE, and PS is 12, 12, and 23, respectively (Hauser and Phillips 1979) although different numbers of water molecules also have been reported (Cevc 1982). The creation of the surface water-flooded layer has been considered to be a consequence of direct hydrogen bonding; i.e., coulombic and dipolar forces between the water molecules and lipid polar residues. These water molecules are tightly hydrogen bonded to the phosphodiester group. Cations and water molecules compete for the same binding sites, and the expulsion of water as a result of the interaction of divalent metal ions with the negatively charged phosphate groups of PL has been demonstrated by relaxation enhancement studies using $Mn^{2+}$ (Hauser and Phillips 1979).

## 2.2 Removal of Hydrogen-Bonded Water Molecules from Hydrophilic Polar Groups by $Al^{3+}$ as First Step of Al Toxicity

The order of the effective ionic radius (pm) is $Al^{3+}$ (53.5) < $Yb^{3+}$ (98.5) < $Ca^{2+}$ (100.0) < $La^{3+}$ (121.6), and the order of ionic potential calculated from the ratio of valence to the effective ionic radius (nm) is $Al^{3+}$ (56.1) > $Yb^{3+}$ (30.5) > $La^{3+}$ (24.7) > $Ca^{2+}$ (20.0) (Ishikawa et al. 1996). $Al^{3+}$ has the smallest size and highest valence among these cations; therefore, the ionic potential of $Al^{3+}$ is higher than the ionic potential of the other three cations. When a cation with high ionic potential binds to a polarizable anion, the covalent bond that forms will have the highest binding energy among all chemical bonds. This highest binding has been ascribed to entropic stability after the release of water molecules from the coordinated waters of PL by the Eigen mechanism (Ishikawa and Wagatsuma 1998). Haller and Freiser (1976) found that the binding of divalent cations ($Ca^{2+}$, $Ba^{2+}$, $Mg^{2+}$, $Zn^{2+}$) or two orders of magnitude lower concentrations of trivalent cations ($La^{3+}$, $Ce^{3+}$) to acidic PL (PS) black lipid membranes reduced the electrostatic repulsion of the polar group causing the film to condense. The water concentration within the hydrophobic core of a lipid layer should be about 100 mM, assuming a volume of 250 $cm^3$ for a mole of hydrocarbon chain (Miller 1987; Meier et al. 1990). It has been postulated that the lipid bilayers are traversed by chains of water molecules held together by hydrogen bonds. These trapped water molecules are extruded from the hydrocarbon core of the bilayers by the condensation.

Based on these extensive studies of lipid membrane physics, the interaction of $Al^{3+}$ with the lipid bilaycrs can be summarized as follows. The binding affinity of

$Al^{3+}$ with the phosphate groups of PLs is strong. Therefore, $Al^{3+}$ can remove hydrogen-bonded water molecules from these surface hydrophilic polar groups as well as extrude the trapped water molecules from the hydrocarbon core of the bilayers. The extrusion of the water molecules disrupts the normal membrane fluidity, inducing the dispersed aggregated domains composed of several numbers of PL molecules, and the membrane, which is now in a crystal state, becomes rigid and gel-like. Finally, considerable phase separation occurs between each aggregated domain, and the newly formed phase separation spaces expand as the cells grow (Ishikawa and Wagatsuma 1998). This expansion induces greater Al permeation through the PM lipid bilayers into the cytoplasm. The remarkable PM destruction and Al accumulation observed primarily around the root elongation zone (Sivaguru and Horst 1998; Wagatsuma et al. 2005) support this scheme.

# 3 Less Al Permeation Through PM Lipid Bilayers as Determinant for Al Tolerance

## *3.1 Effects of DNP, CCCP, or $N_2$ on Membrane Permeability and Al Uptake*

The effects of anaerobic conditions ($N_2$) or 2,4-dinitrophenol (DNP) on the integrity of PMs have been investigated from when mineral physiological studies first began. Strontium uptake by the non-vacuolated sections of the primary root of maize (*Zea mays* L.) increased considerably under $N_2$ conditions compared with in normal aerobic conditions, which suggested that the PM was destroyed by anaerobiosis (Handley and Overstreet 1963). Excised roots of barley (*Hordeum vulgare* L.) lost organic acid anions (OA), amino acids, and $K^+$ and Cl ions when treated with $N_2$ or DNP, indicating that these non-metabolic conditions injured the PM (Hiatt and Lowe 1967). Uptake of Al by the excised roots of cabbage (*Brassica oleracea* L.), lettuce (*Lactuca sativa* L.), and kikuyu grass (*Pennisatum clandestinum* Chiov.) was also enhanced by DNP (Huett and Menary 1979). Various kinds of metabolic inhibitors, especially chloroform gas and DNP, significantly increased Al uptake by the excised roots of several plant species (Wagatsuma 1983). Potassium in the roots of Japanese radish (*Raphanus sativus* L.) decreased under $N_2$ gas treatment, indicating the destruction of the PM. The increase in the Al uptake caused by exposure to $N_2$ was in the order barley, edible burdock (*Arctium lappa* L.), and spinach (*Spinacia oleracea* L.) > pea (*Pisum sativum* L.) > rice (*Oryza sativa* L.). This order is similar to the levels of Al tolerance among these plant species because the increase in Al uptake under $N_2$ was highest in the Al-sensitive plant species. On the other hand, Al uptake by Al-tolerant wheat cultivars increased when the cultivars were treated with DNP, while Al uptake by Al-sensitive wheat cultivars was relatively unaffected (Zhang and Taylor 1989). Al transport across the PM in single cells of *Chara corallina* was measured using the rare $^{26}Al$ isotope, accelerator mass spectrometry (an emerging technology), and a surgical technique for isolating subcellular compartments (Taylor et al. 2000). Al transport across the PM was detectable within 30 min of the cells' exposure to Al. DNP and carbonylcyanide *m*-chlorophenylhydrazone (CCCP) increased Al transport through the PM by 12- to 13-fold. Al uptake in protoplasts isolated from 1-cm root-tip portions by DNP treatment was greater in an Al-sensitive maize cultivar (XL61) compared with in an Al-tolerant cultivar (DK789) (Ishikawa et al. 2001). Taylor et al. (2000) discussed the possible causes of the considerable increase in the Al transport across the PM under DNP or CCCP treatment; however, their data did not allow them to determine whether the increased Al uptake was a result of increased PM permeability or the disruption of a metabolism-dependent exclusion mechanism. The amounts of exuded citrate and malate in the Al media were greater in the Al-tolerant DK789 cultivar than in the Al-sensitive XL61 cultivar (Ishikawa et al. 2000), indicating at least one of the metabolism-dependent exclusion mechanisms may have contributed to the greater Al tolerance in the Al-tolerant maize cultivar. The effectiveness of DNP treatment on the enhancement of Al uptake by protoplasts isolated from the Al-sensitive maize cultivar (Ishikawa et al. 2001) suggested the increased Al uptake may be the result of increased PM permeability rather than the disruption of the OA exudation-dependent exclusion mechanism because the OA exudation was lower in the Al-sensitive cultivar than in the Al-tolerant cultivar. Protoplasts isolated from 1-cm root-tip portions of young seedlings of rice, maize, pea, and barley were treated with 100 μM $AlCl_3$ in the presence of isotonic 0.7 M mannitol for 10 min. Protoplast ghosts were prepared by treating protoplasts from the same sources with 0.2 mM $CaCl_2$ in the absence of 0.7 M mannitol and subsequently with 100 μM $AlCl_3$ for 10 min. Protoplast ghosts (burst protoplasts) took up more Al than the intact protoplast in all the plant species tested, but the ratios of the Al content of protoplast ghosts to protoplasts were in the order rice > maize > pea, barley, which is the same order as the Al tolerance in these plants. Greater Al exclusion barrier potentials of the PMs in the root tips of Al-tolerant plant species were suggested as a possible explanation for this observation (Table 1) (Ishikawa and Wagatsuma 1998).

In isotonic 0.7 M mannitol and in the absence of Al, 0.2 mM $CaCl_2$ induced no abnormal PM permeability in protoplasts isolated from the root-tip portions of both the Al-tolerant and the Al-sensitive pea cultivars. On the contrary, in moderately hypotonic 0.55 M mannitol in the presence of 100 μM $AlCl_3$, only the PMs of the protoplasts from the Al-sensitive pea cultivar were permeabilized considerably (Ishikawa et al. 2001). PM permeabilization in the protoplasts from the

**Table 1** Al content of protoplasts and protoplast ghosts (fmol Al/protoplast or ghost) and their ratios (Ishikawa and Wagatsuma 1998)

| | Rice | Maize | Pea | Barley |
|---|---|---|---|---|
| Protoplast (A) | 20.9 ± 4.3 | 58.7 ± 2.0 | 48.6 ± 5.9 | 62.3 ± 3.2 |
| Protoplast ghost (B) | 59.9 ± 4.2 | 104.3 ± 9.2 | 57.9 ± 1.7 | 74.4 ± 7.8 |
| (B)/(A) | 2.87 | 1.78 | 1.19 | 1.19 |

Al-sensitive pea cultivar occurred in moderately hypotonic conditions that may have enlarged the Al-bound PM. The investigations described above indicate that $N_2$, DNP, and other inhibitors can permeabilize the PM lipid bilayers, and low Al permeation through the PM lipid bilayers is the determinant for Al tolerance in plants.

## 3.2 *Immature Research Status on the Severely Toxic Cytosolic Al*

Although the amount of Al that permeates through the PM is considered to be the determinant of Al toxicity, accurate measures of Al concentrations in the cytosol together with accurate estimates of the volume of the cytosol of cells in the critical root-tip portions of plants are scarce. This lack of data makes it difficult to provide detailed cytosolic physiological insights into the mechanisms of Al toxicity, although in pea root cells, accumulated Al was found to trigger ROS production, inhibit respiration, and deplete ATP, all of which could be correlated with inhibition of root elongation (Yamamoto et al. 2002). To evaluate the function of the Al in the cytoplasm, several points need to be considered. Some of these points have been reviewed previously (Martin 1994; Harris et al. 1997); however, until recently, no useful information was available regarding: (1) the Al concentration in the cytosol; (2) the concentrations of physiologically essential ligands that can complex with Al ions resulting in the less vital cell status, for example, PPi, glucose-6-phosphate (G6P), fructose-6-phosphate (F6P), ATP, ADP, uridine diphosphate glucose (UDPG), NAD, protein phosphates, phosphorylated histones, and OAs, and (3) the form and concentration of each Al complex.

In *Chara corallina*, using the intracellular perfusion technique, inorganic pyrophosphate (PPi) was found to be present predominantly in the cytosol at a concentration of 193 μM; it was present in the vacuole at a concentration of only 2.20 μM. Inorganic phosphate (Pi) was distributed almost equally in the cytosol (12.0 mM) and the vacuole (6.70 mM) (Takeshige and Tazawa 1989). Using a metabolomics approach based on combined capillary electrophoresis and mass spectrometry (CE-MS), the metabolomics of a giant internodal cell of *Chara australis* was investigated (Oikawa et al. 2011), and the following concentrations were found in the cytoplasm: 138 μM ADP, 193 μM ATP, 360 μM F6P, 3099 μM G6P, 715 μM UDPG, about 200 μM citrate, and about 200 μM malate. However, in vitro $^{31}$P-NMR analyses showed that the Pi concentrations in the cytosol of sycamore (*Acer pseudoplatanus*) and Arabidopsis cells were much lower than the cytoplasmic Pi concentrations that were usually considered (60–80 μM rather than >1 mM) and that the levels dropped very rapidly following the onset of Pi starvation (Pratt et al. 2009). G6P, nucleotide triphosphates, and UDPG were also found in the cytoplasm; however, their specific positions in either the cytosol or the organelles could not be determined. Although low Al permeation through the PM lipid layer is considered to be a determinant for Al tolerance, several complex steps need to be elucidated to obtain a clear understanding of the symplastic Al functions.

## 3.3 Relationship Between Al Tolerance and Al Absorption to Cell Wall

Pfeffer et al. (1986) suggested that the enhanced Al uptake under a $N_2$ atmosphere (Wagatsuma 1983) may simply be a consequence of the more pronounced absorption of Al to the highly charged cell walls; thus, the excessive Al accumulation in the cell may reflect cell wall entrapment rather than cell penetration. However, the cation exchange capacity (CEC) values of cell walls isolated from the root tips (0–1 cm) of rice and barley were about 10 $cmol_c\,kg^{-1}$ dry weight (Ishikawa and Wagatsuma 1998). On the other hand, under $N_2$ conditions, Al accumulation in rice hardly increased while, in barley, Al accumulation increased considerably. The differences in Al accumulation in the roots of these two plants did not correspond to differences in the Al absorption to the charged cell walls, but rather to differences in Al permeation through the PM lipid bilayers. No differences were found between the CEC values (CECs measured at pH 5.0, similar to the pH of the Al treatment) of the purified cell walls isolated from the root tips (0–1 cm) of cultivars of five plant species (rice, maize, pea, wheat, and sorghum) (Fig. 2) with different Al tolerances (Fig. 1) (Ishikawa et al. 2001). Uptake, intoxication, and alleviation correlated well with ion concentrations at the PM surface computed assuming the PMs were bathed directly in the rooting medium with no effect from the cell walls (Kinraide 2004). The cell wall may have a small effect on ion uptake by the PM or on intoxication or

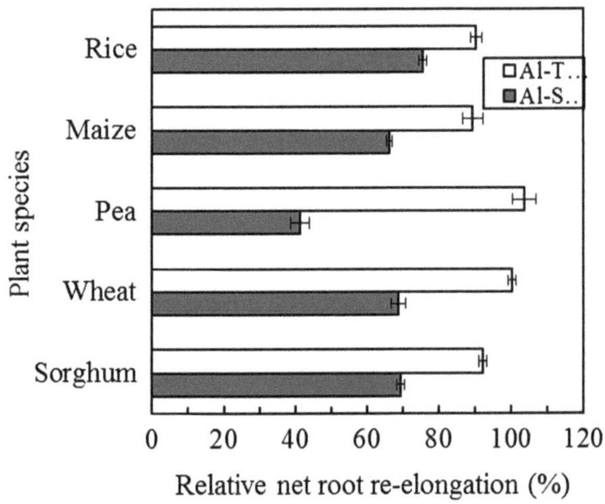


**Fig. 1** Difference in Al tolerances between each two cultivars of five plant species. Al tolerance was expressed as the relative net elongation of the longest root in an Al-free solution for 24 h following Al pretreatment for 1 h. Al-pretreatment conditions containing 0.2 mM Ca were as follows: 100 μM Al (pH 4.5) for rice, 50 μM Al (pH 4.7) for maize, 20 μM Al (pH 4.9) for wheat and pea, 10 μM Al (pH 5.0) for sorghum. *Al-T* Al-tolerant cultivar, *Al-S* Al-sensitive cultivar (Ishikawa et al. 2001)

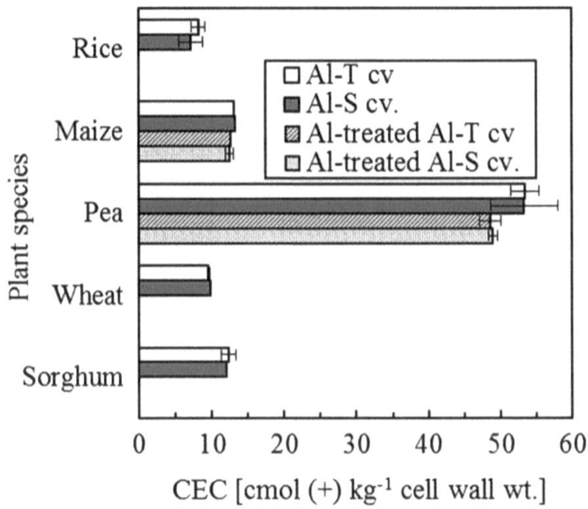


**Fig. 2** Cation exchange capacity (CEC) (at pH 5.0, which is similar to the pH in the Al treatment) of cell walls isolated from the root tips (0–1 cm) of five plants. Al-T cv and Al-S cv, CECs of the root tips of Al-tolerant and Al-sensitive cultivars, respectively, without Al treatment; Al-treated Al-T cv. and Al-treated Al-S cv., CECs of the root tips of Al-tolerant and the Al-sensitive cultivars, respectively, treated with 0.2 mM Al at pH 4.5 for 3 h (Ishikawa et al. 2001)

alleviation of intoxication originating at the PM surface. The investigations described here all suggest that the increase in Al uptake in Al-sensitive plant species under $N_2$ conditions can be ascribed not to entrapment of Al by the cell wall but to Al penetration through the PM; thus, the contribution of the cell wall to differences in Al tolerance is considered to be minor. However, some research groups have stressed the significant role of the cell wall in the Al tolerance (Yang et al. 2008; Horst et al. 2010).

## 3.4 *Callose Formation as Early Indicator of PM Permeabilization*

In the 1–2 mm root zone from the apex, high levels of Al-induced callose formation and Al accumulation were found (Sivaguru and Horst 1998). The distal part of the transition zone of the root apex, where the cells undergo a preparatory phase for rapid elongation (Baluška et al. 1996), has been accepted as the primary target of Al in an Al-sensitive maize cultivar. Callose concentrations in the root tips were correlated closely and positively with Al-induced inhibition of root elongation in 37 maize cultivars (Horst et al. 1997). Al-induced callose formation in the root tips appears to be an excellent indicator of Al injury, which can be used as a selection criterion for Al sensitivity. Callose deposition in the epidermal cell layers of maize roots was observed within minutes of exposure to Al (Jones et al. 2006). Chitosan and other polycations (poly-L-Lys and poly-L-Orn) can elicit callose synthesis, as can certain amphipathic compounds (polymyxin B, echinocandin B, acylcarnitine, and digitonin) that may be regarded as analogous to the nonspecific toxins produced from fungal hyphae in contact with the cells of an infected plant (Kauss 1987). Specific perturbation of membrane permeability is considered to be a prerequisite for callose formation; however, no quantitative correlation has been found between the degree of leakage and the extent of callose formation when different elicitors were compared (Köhle et al. 1985; Kauss and Jeblick 1986). This point should be kept in mind when elicitors are compared for their effectiveness of callose formation (Schmohl and Horst 2000; Schmohl et al. 2000). All of these findings suggest that Al permeability through the PM lipid bilayers is a determinant for Al toxicity.

## *3.5 New Technique for PM Isolation, and Direct Evidence of Significant Role of PM Permeability in Al Tolerance*

There are two aspects to the Al exclusion mechanism: one is the exclusion ascribed to the PM lipid bilayers, and the other is the exclusion ascribed to PM proteins such as the OA transporters. This point was the next step for further clarification of the exclusion mechanism. We developed a technique for PM isolation as an alternative to the laborious two-polymer phase partitioning method that was commonly applied, as follows: (1) separation of protoplasts from 1-cm root-tip portions by enzymatic digestion; (2) attachment of the purified protoplasts to glass plates coated with positively charged polylysine; and (3) preparation of PM ghosts by successive burst of the attached protoplasts using three separate solutions (25 mM PIPES, 5 mM EDTA, and 2 mM $MgCl_2$, at pH 7.0) with slow stirring for 60 s (Wagatsuma et al. 2005). The PMs were confirmed to be devoid of organelle membranes by fluorescence microscopy (DAPI for nucleus, $DiOC_6$ for membranes of the mitochondria and endoplasmic reticulum, FM4-64 for tonoplast), thin layer chromatography (DGDG for plastid membrane, cardiolipin for mitochondria inner membrane), and western blot (V-ATPase and V-PPase for tonoplast, ADP ribosylation factor for endoplasmic reticulum/Golgi membranes, $H^+$-ATPase for PM). PMs were solubilized with hot chloroform (50 °C), and the solubilized fraction was separated by shaking with 0.1 M KCl to remove proteins, concentrated to dryness under $N_2$ gas, and solubilized with dodecane. Following these procedures, about $1.25 \times 10^8$ protoplasts were obtained from 100 g fresh weight of root tips, and 6 mg PM lipids were obtained from $10^8$ protoplasts. We also established a system to examine lipid permeability using synthesized nylon-2,8 ultrathin and porous capsules (by mixing ethylenediamine, NaOH, chloroform, cyclohexane, and terephthaloyl dichloride) trapped previously with 0.1 % (W/V) methylene blue (MB) solution and coated thereafter with the PM lipid isolated from the root tips (Wagatsuma et al. 2005). Permeability of the PM lipid measured photometrically ($A_{680}$) over time in 0.2 mM Ca with or without 50 μM $AlCl_3$ was significantly

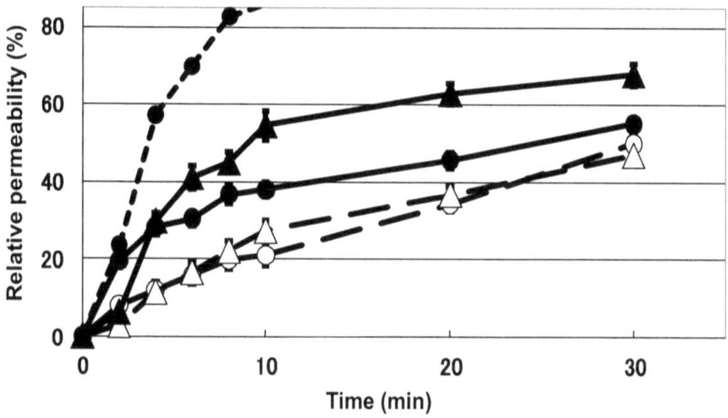


**Fig. 3** Relative permeability of the PM lipids isolated from the root tips of two maize cultivars in 0.2 mM Ca with or without 50 μM Al (pH 4.6). Permeability was measured photometrically ($A_{680}$). The numbers (*Y* axis) are the $A_{680}$ ratios at each measuring time (*X* axis) of the PM lipid-coated nylon capsules to the non-coated nylon capsules at 30 min. *Closed circles* on the *dotted line*, non-coated capsule; *closed triangle* on the *solid line*, Al-sensitive cv. Snowdent (with Al); *closed circle* on the *solid line*, Al-tolerant cv. Neopirika 90 (with Al); *open triangle* on the *broken line*, Al-sensitive cv. Snowdent (without Al); *open circle* on the *broken line*, Al-tolerant cv. Neopirika 90 (without Al) (unpublished data)

greater in the Al-sensitive triticale line ST22 (Wagatsuma et al. 2005) and the Al-sensitive maize cultivar Snowdent-125 (Fig. 3). This was the first direct evidence showing the primary and early role of the PM lipid in Al tolerance without involvement of OA or other protein-related exclusion mechanisms. The technique using nylon capsules is not simple to use, and some skill is required. An easier to use in vitro model of passive permeation, the parallel artificial membrane permeation assay (PAMPA), was established earlier by Kansy et al. (1998). PAMPA has been used widely to screen the permeation of medicinal substances through artificial lipid membranes (Di et al. 2003) and may be an alternative technique for research related to PM permeability in plant species.

## 4 PM Surface Negativity, Al Tolerance, and Related Molecular Background

### *4.1 Relationship Between the PM Surface Negativity and Al Tolerance*

Membrane potentials were calculated by McLaughlin and Murray (2005) using a modified version of Delphi (Gallagher and Sharp 1998) and visualized with GRASP (Nicholls et al. 1991). The 2:1 PC/PS bilayers showed an equipotential −25 mV profile located flatly about 1 nm from the surface of the bilayers. McLaughlin

(1989) applied the electrostatic Gouy–Chapman theory to an atomic model of the bilayers and found it to be surprisingly consistent with a wide range of experimental studies on membranes.

Five plant species with different Al tolerances, rice, oats, maize, pea, and barley were used, and the relationship between the surface negativity of root protoplasts and Al tolerance was investigated (Wagatsuma and Akiba 1989). Higher average zeta potentials of the protoplasts isolated from 0–0.5 cm tip portion of roots were observed in the Al-tolerant plant species compared with the Al-sensitive species. Lower zeta potentials were observed in the protoplasts isolated from 0 to 0.5 cm tip portions than in the protoplasts isolated from 0 to 2 cm tip portions. The basic MB dye was adsorbed strongly by the PMs of root cells in the tip portions of Al-sensitive plant species (Wagatsuma et al. 1991), and a simple and rapid technique to discriminate Al-tolerant protoplasts from an original protoplast population was developed. A technique for the collection of Al-tolerant plant cells was also developed (Wagatsuma et al. 1995), as follows. Equal volume of freshly prepared, positively charged silica microbeads (PCSMs) with a diameter 0.014 μm (0.05 % w/v) and purified protoplasts derived from the root tips ($2 \times 105\ mL^{-1}$) of rice, maize, or pea were mixed and then centrifuged on a discontinuous Ficoll gradient. Intact protoplasts from the Al-tolerant plant were recovered mostly in the bottom fraction, those from the Al-sensitive plant were recovered at the uppermost interface, and those from the intermediate Al-tolerant plant were collected at the middle interface. The mechanism for the isolation of the Al-tolerant protoplasts can be explained by the DLVO theory (Derjaguin and Landau, Verwey and Overbeek) as follows: the largest size of the aggregates of the protoplasts from the Al-tolerant plant PCSMs precipitated based on their relatively low surface negativity. The results described here suggest that the root-tip cells in Al-tolerant plant species have less surface negativity than the root-tip cells of Al-sensitive plants. Two types of molecules in PMs (polypeptides and PLs) have the ability to generate surface negativity. The amino acid residues in polypeptides, including glutamic and aspartic acid residues that can carry negative charges, were found to be very similar in the PMs of barley leaves and barley roots; however, the leaves contained 50 % more negatively charged PLs (Larsson et al. 1990), and the outer surface of the PMs from the leaves was more negatively charged than the PMs from the roots. Therefore, the observed differences in charge densities can probably be explained by differences in numbers of negatively charged PLs in the PMs.

Zhang et al. (1996) studied the effects of Al on the lipid composition of microsomal membranes isolated from 5-mm root tips of Al-resistant (PT741) and Al-sensitive (Katepwa) cultivars of *Triticum aestivum* L. The ratio of steryl lipids to PLs and the free sterol content tended to be higher in the Al-resistant cultivar. Zhang et al. (1996) also studied the relationship between the PM lipid composition and Al resistance in these two wheat cultivars; however, contrary to their expectations, the ratio of steryl lipids to PLs and the free sterol content in purified PMs from whole roots were higher in the Al-sensitive cultivar. In spite of the interspecific difference in the membrane electronegativity by Wagatsuma and Akiba (1989) and Wagatsuma et al. (1991, 1995), Zhang et al. (1997) could not identify the consistent physiological basis on the intraspecific differences in PM lipid composition in wheat.

The intrinsic surface charge density of PM vesicles isolated from 5-mm root tips of two wheat cultivars was 26 % more negative for PM vesicles from Al-sensitive cv Scout than for PM vesicles from Al-tolerant cv Atlas (−37.2 vs. −29.5 mC $m^{-2}$) (Yermiyahu et al. 1997). Sorption of Al by the PM vesicles from the root tips of cv Scout exceeded the sorption of Al by the vesicles isolated from the tipless roots of cv Scout and from the vesicles isolated from any parts of the roots of cv Atlas. The differences in PM surface negativity and Al sorptive capacity were evaluated to account for some of the difference in sensitivity to $Al^{3+}$, but the major part of the difference was speculated to arise from other tolerance mechanisms expressed in the cv Atlas root tips reduced the amount of $Al^{3+}$ that could reach the PM; for example, OA exudation. The Al sorptive capacity was compared based on the unit amount of protein in the roots of the two cultivars. It is possible that the differences in Al sorptive capacity would be greater if the capacity was estimated based on the unit amount of lipid instead of the unit amount of protein, because the protein content in PMs is greater in the root tips than in other proximal regions of roots, and the difference in protein content in the root-tip PMs between the two cultivars was unclear. Ahn et al. (2004) compared the zeta potential of PM vesicles, which lacked OA exudation ability, from 10 mm root-tip portion of two near-isogenic wheat lines, and found a slightly higher negative value in the Al-sensitive ES8 (−18 mV) than in the Al-tolerant ET8 (−15 mV). They evaluated the lower negativity of the zeta potential in ET8 as a partial contribution to Al tolerance because ET8 exudes 10 times greater malate than ES8. However, Ahn et al. (2001) found that the PM vesicles prepared from the root-tip portion (−23 mV at 0–5 mm from tip) had more negative zeta potential than the PM vesicles from the proximal portion (−18 mV at 5–20 mm from tip) of squash (*Cucurbita pepo* L.) roots. Wagatsuma and Akiba (1989) had already reported the more negative zeta potential of protoplasts isolated from the more root-tip portion of Al-sensitive plant species (0–5 mm vs. 0–20 mm). Considering all these findings, the surface negativity of the PMs in root-tip portions is considered to contribute to the different Al tolerance capacities among plants. The applicability of root cell PM surface electrical potential to the bioavailability and toxicity of various cations and anions has been characterized further by various researchers (Vulkan et al. 2004; Kinraide 2006; Wang et al. 2011, 2014). Short-term Al tolerance was screened, and the MB stainability of protoplasts isolated from root tips were measured using samples from 18 different plant species, cultivars, and lines (Wagatsuma et al. 2005). Microscopy observations showed that the MB dye was not only adsorbed on the surface of the protoplasts but also permeated into the protoplasts. MB stainability was negatively correlated with Al tolerance suggesting the importance of the permeation characteristics of the root PMs in addition to the PM negativity for Al tolerance in a wide range of plant species, cultivars, and lines.

## 4.2 Higher Sterols and Lower PL/Sterols Ratio in Root-Tip Cells of Al-Tolerant Plants

Research on the lipid composition of roots has also been carried out in connection with Al tolerance. Al tolerance among the *temperate japonica* rice ancestor cultivars of the Al-tolerant cv Sasanishiki, permeability of PM, Al uptake, OA release, and lipid composition of the PM of root-tip portions have been investigated (Khan et al. 2009). The Al-sensitive cultivar showed increased PM permeability and greater Al uptake in the root-tip portion by Al treatment; however, Al tolerance could not be explained by the OA release from the roots. The tolerant cultivar had a lower ratio of PLs to free sterols than the sensitive cultivar, suggesting that the PMs of the tolerant cultivar were less negatively charged and less permeabilized than the PMs of the sensitive cultivar. In addition, the tolerant cultivar showed a similar level of Al sensitivity when the ratio of PLs to free sterols was increased to match the ratio found in the Al-sensitive cultivar after treatment with uniconazole-P, a triazole-type fungicide that inhibits obtusifoliol-14α-demethylase (OBT 14DM), a key enzyme in the post-squalene sterol biosynthetic pathway (Benveniste 2004). Al tolerance was negatively correlated with the lipid ratio (PL/free sterol) in the root-tip portion (Fig. 4). The concentration of PL and galactolipids (GL) in the roots of the rice receiving −P pretreatment were lower in PL and higher in GL than those receiving +P pretreatment, and the seedling receiving −P pretreatment showed enhanced Al tolerance accompanied by the decrease in Al accumulation in the roots (Maejima et al. 2014). −P pretreatment slightly decreased the amount of free sterols in the roots together with a negligible decrease in uronic acid in the pectin and enhanced the low-Ca tolerance of the roots under low pH conditions at pH 4.2. Phosphatidate phosphohydrolase 1 (PAH1) and PAH2 were found to be responsible for the eukaryotic galactolipid synthesis pathway, and the membrane lipid remodeling mediated by these two enzymes was reported to be an essential

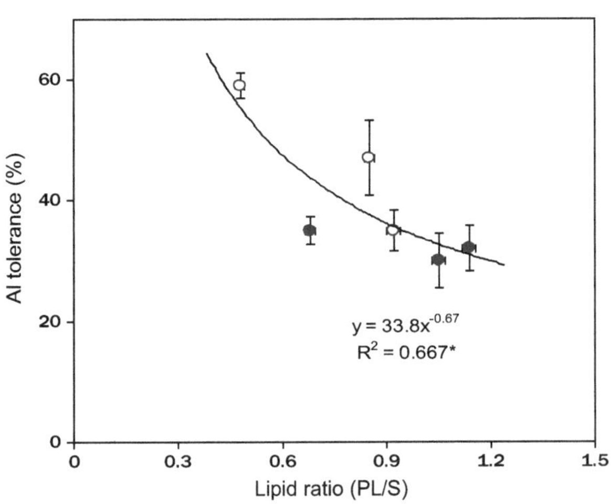


**Fig. 4** Relationship between the lipid ratio (PL/free sterol) and Al tolerance. *Open circles*, Al-tolerant rice cultivar Rikuu-132 after treatment with sterol inhibitors (uniconazole-P, fenpropimorph). *Closed circles*, Al-sensitive rice cultivar Rikuu-20 after treatment with the same sterol inhibitors (Khan et al. 2009)

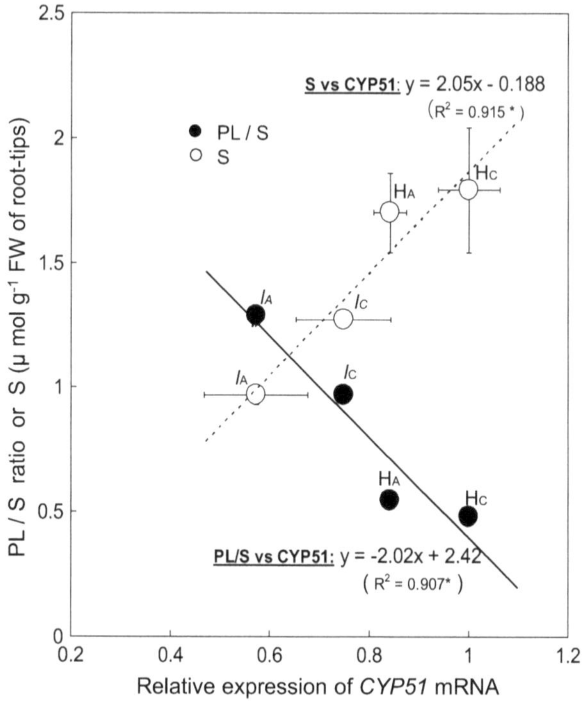


**Fig. 5** Relationship between the relative transcript levels of *CYP51* and the PL/free sterol ratio or free sterol. *Open circles*, free sterols and the transcript levels of the two pea genotypes (Al-tolerant cv, Al-sensitive mutant) with or without Al treatment. *Closed circles*, the PL/free sterol ratios and the transcript levels of the two pea genotypes with or without Al treatment (Wagatsuma et al. 2015)

adaptation mechanism to cope with P starvation (Nakamura et al. 2009). The *pah1pah2* double mutant of Arabidopsis (*Arabidopsis thaliana*) showed enhanced Al sensitivity under low-P conditions where greater levels of negatively charged PL occurred in the PM. The resultant increased PM surface negativity compared with wild-type plants increased $\{Al^{3+}\}_{PM}$ and Al uptake in the roots (Kobayashi et al. 2013). Compared with the Al-tolerant pea genotypes, the Al-sensitive genotype accumulated more Al in the root tips, had less intact PM, and showed a lower expression level of *PsCYP51*, which encodes OBT 14DM (Wagatsuma et al. 2015). The ratio of PL to free sterols was higher in the sensitive genotype than in the tolerant genotype, suggesting that the free sterol biosynthetic pathway play an important role in Al tolerance (Fig. 5).

A transgenic Arabidopsis line with knocked-down *AtCYP51* expression showed an Al-sensitive phenotype with greater reduction of root elongation (Fig. 6a) and PM permeability, greater accumulation of Al in the root-tip portion, and lower free sterols and higher PL/free sterol ratios (Fig. 6b) than the wild-type. Uniconazole-P, an inhibitor of OBT 14DM, suppressed the Al tolerance of the Al-tolerant genotypes of maize, sorghum (*Sorghum bicolor*), rice, wheat, and triticale (×*Triticosecale* Wittmark cv. Currency). These results suggested that the higher free sterol content, regulated by *CYP51*, with concomitant lower PL content in the root tips results in the lower negativity of the PM. This mechanism appears to be a common strategy for Al tolerance among several plant species.

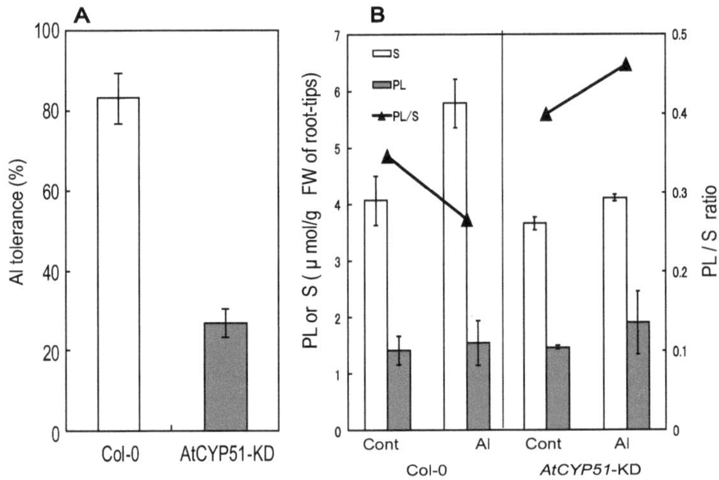


**Fig. 6** (**a**) Al tolerance of Arabidopsis Col-0 and *CYP51* knocked-down (KD) line. Three-day-old synchronously germinated Arabidopsis seedlings of Col-0 and *CYP51*-KD line were treated with or without 4 μM $AlCl_3$ (pH 5.0) for 7 days. Al tolerance was calculated as the ratio of root length in Al treatment to that in the control. (**b**) PL or free sterol content (*left Y* axis) and PL/free sterol ratio (*right Y* axis) in Arabidopsis seedlings of Col-0 and *CYP51*-KD line. After 7-day treatment with or without 4 μM $AlCl_3$ (pH 5.0), PL and free sterols were extracted from whole seedlings and then quantified. Free sterol is expressed as β-sitosterol equivalents (Wagatsuma et al. 2015)

## *4.3 Significant Role of* HMG *in Al Tolerance*

These results suggested that it may be possible to obtain plants with greater Al tolerance if the amount of free sterol in the root tips could be increased. Transgenic Arabidopsis plants that overexpressed *CYP51*, however, showed no change in sterol profiles and morphology (Kim et al. 2005). However, Schaller et al. (1995) showed that transgenic tobacco lines with the increased expression levels of *HMG* mRNA had sixfold increased levels of total sterols compared with wild-type lines. *HMG* encodes 3-hydroxy-3-methylglutaryl CoA reductase (HMGR), which is considered to be a key limiting enzyme in the upstream biosynthesis of phytosterols passing through the mevalonate pathway (Schaller et al. 1995). *HMG1* expression in Arabidopsis seedlings was found to increase under dark treatment (Enjuto et al. 1994; Learned 1996); however, its expression in roots was unresponsive to the illumination conditions and was confined to the elongation zone (Learned and Connolly 1997). In the roots of 4-day-old rice seedling (*japonica*-type cultivar Ilpumbyeo), the expression of *HMG2* was found to be slightly higher under dark conditions, while the expression of *HMG3* was similar in both light and dark conditions (Ha et al. 2001). Except for the two rice cultivars with extreme difference in Al tolerance, Kasalath, (Ka) the most Al-sensitive, and Rikuu-132, ($R_{132}$) the

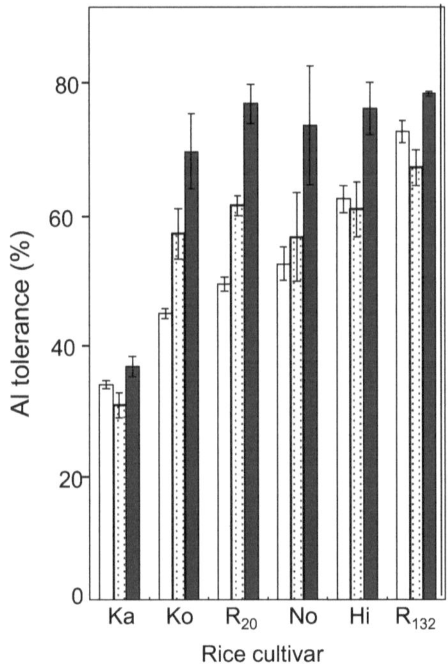


**Fig. 7** Al tolerance of six rice cultivars under different light conditions. Five-day-old rice seedlings were treated for 24 h with 10 μM $AlCl_3$ in the presence of 0.2 mM $CaCl_2$ (pH 4.9). The bars from *left* to *right* indicate the illumination conditions light, dark, and dark in the presence of 1 mM mevalonate and 1 mM glucose, respectively. *Ka* Kasalath, *Ko* Koshihikari, *R20* Rikuu-20, *No* Norin-21, *Hi* Hidekomochi, *R132* Rikuu-132 (unpublished data)

most Al tolerant, the Al tolerance of other *japonica*-type rice cultivars with intermediate Al tolerances increased under 24 h dark conditions, especially in the presence of 1 mM mevalonate and 1 mM glucose (DMG) compared with their Al tolerance under light conditions (Fig. 7). Under DMG, Al accumulation in the root-tip portion was unchanged in the two cultivars with extreme difference in Al tolerances (Ka and $R_{132}$); however, Al accumulation decreased significantly in the Al-sensitive cultivar Koshihikari (Ko) compared with its accumulation under light conditions (data not shown). Relative transcript levels of *HMG2* and *HMG3* compared with 18S rRNA in Ko and $R_{132}$ were higher under DMG than under light conditions (data not shown). Under DMG, the relative value of the total sterols in the root tips (0–1 cm) in the Al treated cultivars compared with the control was considerably higher in the Al-sensitive Ko than that in the Al-tolerant $R_{132}$; however, this was reversed under light (Fig. 8). (Sterols contribute to Al tolerance by reducing PM permeability.) Al tolerance of the Al-sensitive *japonica*-type rice cultivar was enhanced under dark conditions especially in the presence of the HMGR-related intermediates (mevalonate and glucose), which have been associated with increased root-tip sterols and *HMG* expression. These results may provide promising insights for breeding new Al-tolerant plants in the future.

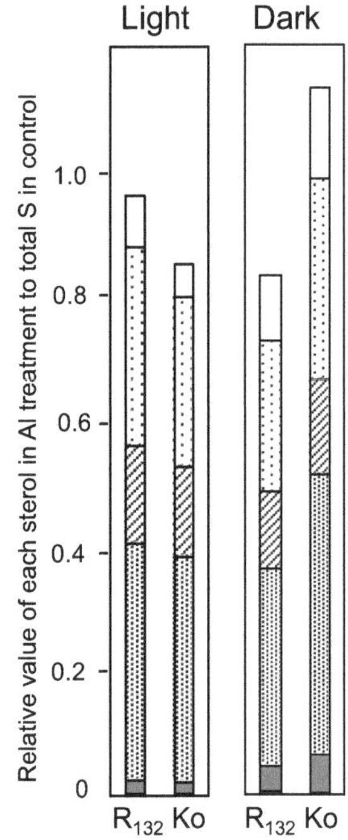


**Fig. 8** Relative value of each sterol in the Al treated cultivars to total free sterols in the control. Within each bar, from the *top* to the *bottom*; isofucosterol, sitosterol, stigmasterol, campesterol, and 24-methylene cholesterol. Light, treatment under light conditions; dark, treatment under dark with mevalonate and glucose (unpublished data)

# 5 Contribution of Phenolics to PM Lipid Bilayers Permeability in Highly Al-Tolerant Plants

## *5.1 Higher Amounts of Root Phenolics in Highly Al-Tolerant Plants*

Plants contain a high proportion of phenolics with diverse structures and functions. The contribution of phenolics to greater Al tolerance has been reported in many plants species. Common woody plants, namely *Pinus thunbergii* Parl. (Sanshu black pine), *Camellia sinensis* L. (tea), *Gleditsia triacanthos* L. (honey locust), *Robinia pseudoacacia* L. (black locust), *Picea abies* Karst. (Norway spruce), and *Cryptomeria japonica* (Japanese cedar), were found to be more tolerant to combined stress conditions, including high Al, low P, and low pH, than rice, which is known to be one of the most tolerant crop plants (Ofei-Manu et al. 2001). The Al tolerance of woody plants was positively correlated with the concentration of soluble phenolics in their roots. In vitro binding affinity to Al ions at pH 7.0, which mimics cytosolic pH, was significantly higher at equimolar concentrations of quercetin, catechin, and chlorogenic acid, but lower with citric, oxalic, and malic acids, suggesting the contribution of higher amounts of root phenolics to the detoxification of Al ions in the cytosol. Based on the water content of the fresh roots for younger woody plants, the actual concentration of the phenolics in the cytoplasm was estimated to be high; for example, 9.9 mM for *G. triacanthos*, 40.5 mM for *P. abies*, and 57.2 mM for *C. sinensis*. Thus, phenolics were found to be effective detoxifiers of Al ions in the cytoplasm before the formation of $[Al(H_{-1}Cit)]^{-1}$, $[Al(OH)(H_{-1}Cit]^{2-}$, or $[Al_3(H_{-1}Cit)_3(OH)]^{4-}$, which has higher stability constants than the Al–ATP complex at near neutral pH (Harris et al. 1997). High concentrations of phenolics in the roots may be connected with the hyper Al-tolerance of these woody plants. Tropical forage grass *Brachiaria decumbens* (signal grass) contains high concentration of the phenolics in the root-tip portions and is a higher Al-tolerant plant than the most Al-tolerant rice cultivar Rikuu-132 (Watanabe et al. 2011). Mangrove trees grow under reductive and sometimes acidic conditions, both of which are injurious to plant growth; however, mangrove roots contain large amounts of tannins that combine with ferric ions existing at toxic levels in the soil environment (Kimura and Wada 1989). This mechanism is similar to the alleviating effect of the root phenolics, which can form complexes with metal ions. Phenolics found in the roots of the hyper Al-tolerant woody plants can combine with Al and contribute to Al tolerance. They include the proanthocyanins (polymers of catechin and epicatechin) for *Cinnamomum camphora* (camphor tree) (Osawa et al. 2011) and oenothein B (a dimeric hydrolysable tannin with many adjacent phenolic hydroxyl groups) for *Eucalyptus camaldulensis* (Tahara et al. 2014). Proanthocyanins were predominant in the cytosol (but may also be located near cell wall area based on our assumption), and oenothein B was thought to be located mostly in the symplast, especially in the vacuole.

## 5.2 Interaction of Phenolics with Membrane Lipid Bilayers

Biopolymer lignin is built from reactive monolignols and is essential to terrestrial plants (Boija et al. 2007). Monolignols glucosides are aromatic monomers (*p*-coumaryl alcohol, coniferyl alcohol, and sinapyl alcohol) that are synthesized in the cell, glycosylated, and transported from the vacuole to the cell wall (Whetten and Sederoff 1995). The monomers are transported through the PM by a mechanism which is not fully understood; i.e., exocytosis, specific transport systems, or partition-dependent diffusion (Boija and Johansson 2006). The partitioning of phenolic compounds into the lipid bilayer discs and liposomes was used to create lignin precursor models (Boija et al. 2007). The lignin precursor models could be partitioned well into all types of model membrane, indicating that passive diffusion is a possible mechanism in the transport of monolignol and dilignol through the PM as demonstrated by the interaction involving both hydrophobic effects and polar interactions. This possibility was supported by experiments using a bilayer disk technique with lipid bilayers with a similar lipid composition as the PM of maize roots.

The effects of representative flavonoids, isoflavonoids, and their metabolites on membrane fluidity and their preferential localization in the membrane were investigated using large unilamellar vesicles as membrane models. The flavonoids and isoflavonoids partitioned into the hydro-phobic core of the membrane and caused a dramatic decrease in the lipid fluidity in this region of the membrane (Arora et al. 2000). Sophorafla-vanone, which has intensive antibacterial activity, reduced the fluidity of outer and inner layers of the membranes after partitioning into the membranes (Tsuchiya and Iinuma 2000). The galloyl moiety of epicatechin gallate, a green tea polyphenol, contributed to an increase in the hydrophobicity of catechin molecules, and consequently to the high affinity of galloyl-type catechins for PL membranes, as well as to stabilization by cation-$\pi$ interaction between the galloyl ring and quaternary amine of the PL head-groups (Uekusa et al. 2011). The partition coefficients of catechins in an *n*-octanol/PBS system were the same as partition coefficients of catechins incorporated into liposomes (Kajiya et al. 2001). These results indicate that in addition to the number of hydroxyl groups on the B-ring and the presence of a galloyl moiety, the stereochemical structure of the C-ring also governs the hydrophobicity and affinity of catechins for the lipid bilayers. Together these findings indicate the reducing properties of phenolics on membrane lipid layer fluidity by their solubilization into the lipid layer. Some of the physiological effects of tea consumption (antioxidant activity, antibacterial effect, anticarcinogenic effects, antihypercholesterolemia, improvement of hyperglycemia, and cutaneous photoprotection from UV radiation) may be attributable to a decrease in membrane fluidity as a result of the incorporation of these catechins into the membranes.

If phenolics synthesized in the cytoplasm are transported through the PM by exocytosis and/or partition-dependent diffusion (Boija and Johansson 2006), these phenolics may be solubilized at different depths within the lipid bilayers and reduce the fluidity of the membrane lipid bilayers. The reduction of membrane fluidity and permeability are major contributory factors for the enhancement of Al tolerance. Identification of the phenolics within the PM lipid bilayers is considered to be an important future task for the clarification of the hyper Al tolerance mechanism, especially in woody plants.

# 6 Combined Studies on Ionome and Lipidome for Comprehensive Recognition of Ionic Stress Response in Plants

The interactions between membrane lipids and the responses of plants to diverse ionic environments have been investigated widely. Not only Al but also heavy metal ions have been reported to decrease sterols in the roots and increase the permeability of root PM. Seven days exposure to 50 μM cadmium ($CdCl_2$) or copper ($CuSO_4$) decreased the contents of glycolipids, PL, and neutral lipids in the roots of tomato plants (*Solanum esculenttum* L.) (Ouariti et al. 1997), and 50 μM $CdSO_4$ decreased the free sterols in the root PM of pea (Hernández and Cooke 1997). On the contrary, P starvation decreased PL and increased free sterols inducing less negatively charged PM of the roots. P starvation decreased the PM phosphoglycerolipids and increased DGDG in the roots of oat (*Avena sativa* L.) (Andersson et al. 2003). DGDG was not the only non-phosphorus-containing lipid that replaced PL; the glucosylceramides and sterylglycosides content also increased in the PM of P-starved oat roots (Andersson et al. 2005). P starvation triggered membrane lipid remodeling, a process that converts significant portions of the PL to the non-phosphorus-containing galactolipids, indicating that PAH1 and PAH2 are the essential enzymes for adaptation to P starvation (Nakamura et al. 2009). The increase in the (Z)-isomer of sphingolipids makes the PM of the transgenic Arabidopsis more rigid, thereby conferring greater Al tolerance (Ryan et al. 2007). Perturbation of sphingolipid biosynthesis in the Arabidopsis *tsc10a* mutant led to an altered leaf ionome, including increases in Na, K, and Rb and decreases in Mg, Ca, Fe, and Mo (Chao et al. 2011). The Purdue Ionomics Information Management System (PiiMS) provides integrated work flow control, data storage, and analysis to support high-throughput data acquisition, along with integrated tools for data search, retrieval, and visualization for hypothesis development (Baxter et al. 2007). It has been speculated that Ca requirement will be low in the PM of the P-deficient rice roots (Maejima et al. 2014). The contents of the roots of the Arabidopsis *pah1pah2* mutant were higher in cationic nutrients (Ca, Cu, Mn, and Zn), but similar in the neutral B and anionic Mo nutrients under P-starved conditions compared with these nutrients in the wild-type. These ionomic characteristics can be explained by the greater negativity of the root PM in the mutant plants (data not shown). These data indicate the significant role of PM lipids in multiple tolerances to complex ionic stresses and also provide a promising future research direction. Combined studies on the ionome and lipidome should reveal new insights into the role of PM in plants' responses to ionic stresses.

## References

Ahn SJ, Sivaguru M, Osawa H, Chung GC, Matsumoto H (2001) Aluminum inhibits the $H^+$-ATPase activity by permanently altering the plasma membrane surface potentials in squash roots. Plant Physiol 126:1381–1390

Ahn SJ, Rengel Z, Matsumoto H (2004) Aluminum-induced plasma membrane surface potential and $H^+$-ATPase activity in near-isogenic wheat lines differing in tolerance to aluminum. New Phytol 162:71–79

Andersson MX, Stridh MH, Larsson KE, Liljenberg C, Sandelius AS (2003) Phosphate-deficient oat replaces a major portion of the plasma membrane phospholipids with the galactolipid digalactosyldiacylglycerol. FEBS Lett 537:128–132

Andersson MX, Larsson KE, Tjellström H, Liljenberg C, Sandelius AS (2005) Phosphate-limited oat. The plasma membrane and tonoplast as major targets for phospholipid-to-glycolipid replacement and stimulation of phospholipases in the plasma membrane. J Biol Chem 280:27578–27586

Arora A, Byrem TM, Nair MG, Strasburg GM (2000) Modulation of liposomal membrane fluidity by flavonoids and isoflavonoids. Arch Biochem Biophys 373:102–109

Baluška F, Volkmann D, Barlow PW (1996) Specialized zones of development in roots: view from the cellular level. Plant Physiol 112:3–4

Baxter I, Ouzzani M, Orcun S, Kennedy B, Jandhyala SS, Salt DE (2007) Purdue ionomics information management system. An integrated functional genomics platform. Plant Physiol 143:600–611

Benveniste P (2004) Biosynthesis and accumulation of sterols. Annu Rev Plant Biol 55:429–457

Boggs JM (1987) Lipid intermolecular hydrogen bonding: influence on structural organization and membrane function. Biochim Biophys Acta 906:353–404

Boija E, Johansson G (2006) Interactions between model membranes and lignin-related compounds studied by immobilized liposome chromatography. Biochim Biophys Acta 1758:620–626

Boija E, Lundquist A, Edwards K, Johansson G (2007) Evaluation of bilayer disks as plant cell membrane models in partition studies. Anal Biochem 364:145–152

Brown DJ, DuPont FM (1989) Lipid composition of plasma membranes and endomembranes prepared from roots of barley (*Hordeum vulgare* L.): effects of salt. Plant Physiol 90:955–961

Cevc G (1982) Water and membranes: the interdependence of their physico-chemical properties in the case of phospholipid bilayers. Studia Biophysica 91:45–52

Chao D-Y, Gable K, Chen M, Baxter I, Dietrich CR, Cahoon EB, Guerinot ML, Lamer B, Lü S, Markham JE, Morrissey J, Han G, Gupta SD, Harmon JM, Jaworski JG, Dunn TM, Salt DE (2011) Sphingolipids in the root play an important role in regulating the leaf ionome in *Arabidopsis thaliana*. Plant Cell 23:1061–1081

Deleers M, Servais J-P, Wülfert E (1985) Micromolar concentrations of $Al^{3+}$ induce phase separation, aggregation and dye release in phosphatidylserine-containing lipid vesicles. Biochim Biophys Acta 813:195–200

Deleers M, Servais J-P, Wülfert E (1987) Aluminum-induced lipid phase separation and membrane fusion does not require the presence of negatively charged phospholipids. Biochim Int 14:1023–1034

Di L, Kerns EH, Fan K, McConnell OJ, Carter GT (2003) High throughput artificial membrane permeability assay for blood-brain barrier. Eur J Med Chem 38:223–232

Enjuto M, Balcells L, Campos N, Caelles C, Arró M, Boronat A (1994) *Arabidopsis thaliana* contains two differentially expressed 3-hydroxy-3-methylglutaryl-CoA reductase genes, which encode microsomal forms of the enzyme. Proc Natl Acad Sci U S A 91:927–931

Gallagher K, Sharp K (1998) Electrostatic contributions to heat capacity changes of DNA-ligand bindings. Biophys J 75:769–776

Ha S-H, Lee S-W, Kim Y-M, Hwang Y-S (2001) Molecular characterization of *Hmg2* gene encoding a 3-hydroxy-3-methylglutaryl-CoA reductase in rice. Mol Cells 11:295–302

Haller I, Freiser MJ (1976) Structural changes in bilayer membranes by multivalent ions. Biochim Biophys Acta 455:739–748

Handley R, Overstreet R (1963) Uptake of strontium by roots of *Zea mays*. Plant Physiol 38:180–184

Harris WR, Berthon G, Day JP, Exley C, Flaten TP, Forbes WF, Kiss T, Orvig C, Zatta PF (1997) Speciation of aluminum in biological systems. In: Yokel RA, Golub MS (eds) Research issues in aluminum toxicity. Taylor & Francis, Washington, DC, pp 91–116

Hauser H, Phillips MC (1979) Interactions of the polar groups of phospholipid bilayer membranes. Prog Surf Membr Sci 13:297–413

Hernández LE, Cooke DT (1997) Modification of the root plasma membrane lipid composition of cadmium-treated *Pisum sativum*. J Exp Bot 48:1375–1381

Hiatt AJ, Lowe RH (1967) Loss of organic acids, amino acids, K, and Cl from barley roots treated anaerobically and with metabolic inhibitors. Plant Physiol 42:1731–1736

Horst WJ, Püschel A-K, Schmohl N (1997) Induction of callose formation is a sensitive marker for genotypic aluminium sensitivity in maize. Plant Soil 192:23–30

Horst WJ, Wang Y, Eticha D (2010) The role of the root apoplast in aluminium-induced inhibition of root elongation and in aluminium resistance of plants: review. Ann Bot 106:185–197

Huett DO, Menary RC (1979) Aluminium uptake by excised roots of cabbage, lettuce and kikuyu grass. Aust J Plant Physiol 6:643–653

Ishikawa S, Wagatsuma T (1998) Plasma membrane permeability of root-tip cells following temporary exposure to Al ions is a rapid measure of Al tolerance among plant species. Plant Cell Physiol 39:516–525

Ishikawa S, Wagatsuma T, Ikarashi T (1996) Comparative toxicity of $Al^{3+}$, $Yb^{3+}$, and $La^{3+}$ to root-tip cells differing in tolerance to high $Al^{3+}$ in terms of ionic potentials of dehydrated trivalent cations. Soil Sci Plant Nutr 42:613–625

Ishikawa S, Wagatsuma T, Sasaki R, Ofei-Manu P (2000) Comparison of the amount of citric and malic acids in Al media of seven plant species and two cultivars each in five plant species. Soil Sci Plant Nutr 46:751–758

Ishikawa S, Wagatsuma T, Takano T, Tawaraya K, Oomata K (2001) The plasma membrane intactness of root-tip cells is a primary factor for Al-tolerance in cultivars of five plant species. Soil Sci Plant Nutr 47:489–501

Jones DL, Kochian LV (1997) Aluminum interaction with plasma membrane lipids and enzyme metal binding sites and its potential role in Al cytotoxicity. FEBS Lett 400:51–57

Jones DL, Blancaflor EB, Kochian LV, Gilroy S (2006) Spatial coordination of aluminium uptake, production of reactive oxygen species, callose production and wall rigidification in maize roots. Plant Cell Environ 29:1309–1318

Kajiya K, Kumazawa S, Nakayama T (2001) Steric effects on interaction of tea catechins with lipid bilayers. Biosci Biotechnol Biochem 65:2638–2643

Kansy M, Senner F, Gubernator K (1998) Physicochemical high throughput permeation assay in the description of passive absorption processes. J Med Chem 41:1007–1010

Kauss H (1987) Some aspects of calcium-dependent regulation in plant metabolism. Annu Rev Plant Physiol 38:47–72

Kauss H, Jeblick W (1986) Influence of free fatty acids, lysophosphatidylcholine, platelet activating factor, acylcarnitine, and Echinocandin B on 1,3-β-D-glucan synthase and callose synthesis. Plant Physiol 80:7–13

Khan MSH, Tawaraya K, Sekimoto H, Koyama H, Kobayashi Y, Murayama T, Chuba M, Kambayashi M, Shiono Y, Uemura M, Ishikawa S, Wagatsuma T (2009) Relative abundance of $\Delta^5$-sterols in plasma membrane lipids of root-tip cells correlates with aluminum tolerance of rice. Physiol Plant 135:73–83

Kim HB, Schaller H, Goh C-H, Kwon M, Choe S, An CS, Durst F, Feldman KA, Feyereisen R (2005) *Arabidopsis cyp51* mutant shows postembryonic seedling lethality associated with lack of membrane integrity. Plant Physiol 138:2033–2047

Kimura M, Wada H (1989) Tannins in mangrove tree roots and their role in the root environment. Soil Sci Plant Nutr 35:101–108

Kinraide TB (2004) Possible influence of cell walls upon ion concentrations at plasma membrane surfaces. Toward a comprehensive view of cell-surface electrical effects upon ion uptake, intoxication, and amelioration. Plant Physiol 136:3804–3813

Kinraide TB (2006) Plasma membrane surface potential ($\Psi_{PM}$) as a determinant of ion bioavailability: a critical analysis of new and published toxicological studies and a simplified method for the computation of plant $\Psi_{PM}$. Environ Toxicol Chem 25:3188–3198

Kobayashi Y, Kobayashi Y, Watanabe T, Shaff JE, Ohata H, Kochian LV, Wagatsuma T, Kinraide TB, Koyama H (2013) Molecular and physiological analysis of $Al^{3+}$ and $H^+$ rhizotoxicities at moderately acidic conditions. Plant Physiol 163:180–192

Kochian LV, Jones DL (1997) Aluminum toxicity and resistance in plants. In: Yokel RA, Golub MS (eds) Research issues in aluminum toxicity. Taylor & Francis, Washington, DC, pp 69–89

Kochian LV, Piñeros MA, Hoekenga OA (2005) The physiology, genetics and molecular biology of plant aluminum resistance and toxicity. Plant Soil 274:175–195

Köhle H, Jeblick W, Poten F, Blaschek W, Kauss H (1985) Chitosan-elicited callose synthesis in soybean cells as a $Ca^{2+}$-dependent process. Plant Physiol 77:544–551

Larsson C, Mφller IM, Widell S (1990) Introduction to the plant plasma membrane – its molecular composition and organization. In: Larsson C, Mφller IM (eds) The plasma membrane. Springer, Berlin, pp 1–15

Learned RM (1996) Light suppresses 3-hydroxy-3-methylglutaryl coenzyme A reductase gene expression in *Arabidopsis thaliana*. Plant Physiol 110:645–655

Learned RM, Connolly EL (1997) Light modulates the spatial patterns of 3-hydroxy-3-methylglutaryl coenzyme A reductase gene expression in *Arabidopsis thaliana*. Plant J 11:499–511

Ma JF (2007) Syndrome of aluminum toxicity and diversity of aluminum resistance in higher plants. Int Rev Cytol 264:225–252

Maejima E, Watanabe T, Osaki M, Wagatsuma T (2014) Phosphorus deficiency enhances aluminum tolerance of rice (*Oryza sativa*) by changing the physicochemical characteristics of root plasma membranes and cell walls. J Plant Physiol 171:9–15

Martin RB (1994) Aluminum: a neurotoxic product of acid rain. Acc Chem Res 27:204–210

Matsumoto H (2003) Acid soils and aluminum stress in plant roots. Root Res 12:149–162 (in Japanese with English abstract)

McLaughlin S (1989) The electrostatic properties of membranes. Annu Rev Biophys Biophys Chem 18:113–136

McLaughlin S, Murray D (2005) Plasma membrane phosphoinositide organization by protein electrostatics. Nature 438:605–611

Meier EM, Shummer D, Sandhoff K (1990) Evidence for the presence of water within the hydrophobic core of membranes. Chem Phys Lipids 55:103–113

Miller IR (1987) Zwitterionic water chains $H^+/OH^-$ transporters. Biophys J 52:497–500

Nakamura Y, Koizumi R, Shui G, Shimojima M, Wenk MR, Ito T, Ohta H (2009) *Arabidopsis* lipins mediate eukaryotic pathway of lipid metabolism and cope critically with phosphate starvation. Proc Natl Acad Sci U S A 106:20978–20983

Nicholls A, Sharp KA, Honig B (1991) Protein folding and association: insights from the interfacial and thermodynamic properties of hydrocarbons. Proteins 11:281–296

Ofei-Manu P, Wagatsuma T, Ishikawa S, Tawaraya K (2001) The plasma membrane strength of the root-tip cells and root phenolic compounds are correlated with Al tolerance in several common woody plants. Soil Sci Plant Nutr 47:359–375

Oikawa A, Matsuda F, Kikuyama M, Mimura T, Saito K (2011) Metabolomics of a single vacuole reveals metabolic dynamism in an alga *Chara australis*. Plant Physiol 157:544–551

Osawa H, Endo I, Hara Y, Matsushima Y, Tange T (2011) Transient proliferation of proanthocyanidin-accumulating cells on the epidermal apex contributes to highly aluminum-resistant root elongation in camphor tree. Plant Physiol 155:433–446

Ouariti O, Boussama N, Zarrouk M, Cherif A, Ghorbal MH (1997) Cadmium- and copper-induced changes in tomato membrane lipids. Phytochemistry 45:1343–1350

Pfeffer PE, Tu S-I, Gerasimowicz WV, Cavanaugh JR (1986) *In vivo* $^{31}P$ NMR studies of corn root tissue and its uptake of toxic metals. Plant Physiol 80:77–84

Pratt J, Boisson A-M, Gout E, Bligny R, Douce R, Aubert S (2009) Phosphate (Pi) starvation effect on the cytosolic Pi concentration and Pi exchanges across the tonoplast in plant cells: an *in vivo* $^{31}P$-nuclear magnetic resonance study using methylphosphonate as a Pi analog. Plant Physiol 151:1646–1657

Ryan PR, Liu Q, Sperling P, Dong B, Franke S, Delhaize E (2007) A higher plant Δ8 sphingolipid desaturase with a preference for (Z)-isomer formation confers aluminum tolerance to yeast and plants. Plant Physiol 144:1968–1977

Schaller H, Grausem B, Benveniste P, Chye M-L, Tan C-T, Song Y-H, Chua N-H (1995) Expression of the *Hevea brasiliensis* (H. B. K.) Müll. Arg. 3-hydroxy-3-methylglutaryl-coenzyme A reductase 1 in tobacco results in sterol overproduction. Plant Physiol 109:761–770

Schmohl N, Horst WJ (2000) Cell wall pectin content modulates aluminium sensitivity of *Zea mays* (L.) cells grown in suspension culture. Plant Cell Environ 23:735–742

Schmohl N, Pilling J, Fisahn J, Horst WJ (2000) Pectin methylesterase modulates aluminium sensitivity in *Zea mays* and *Solanum tuberosum*. Physiol Plant 109:419–427

Sivaguru M, Horst WJ (1998) The distal part of the transition zone is the most aluminum-sensitive apical root zone of maize. Plant Physiol 116:155–163

Tahara K, Hashida K, Otsuka Y, Ohara S, Kojima K, Shinohara K (2014) Identification of a hydrolyzable tannin, oenothein B, as an aluminum-detoxifying ligand in a highly aluminum-resistant tree, *Eucalyptus camaldulensis*. Plant Physiol 164:683–693

Takeshige K, Tazawa M (1989) Determination of the inorganic pyrophosphate level and its subcellular localization in *Chara corallina*. J Biol Chem 25:3262–3266

Taylor GJ (1991) Current views on the aluminum stress response; the physiological basis of tolerance. Curr Topics Plant Biochem Physiol 10:57–93

Taylor GJ, McDonald-Stephens JL, Hunter DB, Bertsch PM, Elmore D, Rengel Z, Reid RJ (2000) Direct measurement of aluminum uptake and distribution in single cells of *Chara corallina*. Plant Physiol 123:987–996

Tsuchiya H, Iinuma M (2000) Reduction of membrane fluidity by antibacterial sophoraflavone G isolated from *Sophora exigua*. Phytomedicine 7:161–165

Uekusa Y, Kamiya-Ishijima M, Sugimoto O, Ishii T, Kumazawa S, Nakamura K, Tanji K, Naito A, Nakayama T (2011) Interaction of epicatechin gallate with phospholipid membranes as revealed by solid-state NMR spectroscopy. Biochim Biophys Acta 1808:1654–1660

Vulkan R, Yermiyahu U, Mingelgrin U, Rytwo G, Kinraide TB (2004) Sorption of copper and zinc to the plasma membrane of wheat root. J Membr Biol 202:97–104

Wagatsuma T (1983) Effect of non-metabolic conditions on the uptake of aluminum by plant roots. Soil Sci Plant Nutr 29:323–333

Wagatsuma T, Akiba R (1989) Low surface negativity of root protoplasts from aluminum-tolerant plant species. Soil Sci Plant Nutr 35:443–452

Wagatsuma T, Nakashima T, Tawaraya K (1991) Identification of aluminum-tolerant protoplasts in the original root protoplast population from several plant species differing in aluminum tolerance. In: Wright RJ et al (eds) Plant-soil interactions at low pH. Kluwer, Dordrecht, pp 789–793

Wagatsuma T, Jujo K, Ishikawa S, Nakashima T (1995) Aluminum-tolerant protoplasts from roots can be collected with positively charged silica microbeads: a method based on differences in surface negativity. Plant Cell Physiol 36:1493–1502

Wagatsuma T, Khan MSH, Rao IM, Wenzl P, Tawaraya K, Yamamoto T, Kawamura T, Hosogoe K, Ishikawa S (2005) Methylene blue stainability of root-tip protoplasts as an indicator of aluminum tolerance in a wide range of plant species, cultivars and lines. Soil Sci Plant Nutr 51:991–998

Wagatsuma T, Khan MSH, Watanabe T, Maejima E, SekimotoH YT, Nakano T, Toyomasu T, Tawaraya K, Koyama H, Uemura M, Ishikawa S, Ikka T, Ishikawa A, Kawamura T, Murakami S, Ueki N, Umetsu A, Kannari T (2015) Higher sterol content regulated by *CYP51* with concomitant lower phospholipid content in membranes is a common strategy for aluminum tolerance in several plant species. J Exp Bot 66:907–918

Wang P, Kinraide TB, Zhou D, Kopittke PM, Peijnenburg WJGM (2011) Plasma membrane surface potential: Dual effects upon ion uptake and toxicity. Plant Physiol 155:808–820

Wang Y-M, Kinraide TB, Wang P, Hao X-Z, Zhou D-M (2014) Surface electrical potentials of root cell plasma membranes: implications for ion interactions, rhizotoxicity, and uptake. Int J Mol Sci 15:22661–22677

Watanabe T, Khan MSH, Rao IM, Wasaki J, Shinano T, Ishitani M, Koyama H, Ishikawa S, Tawaraya K, Nanamori M, Ueki N, Wagatsuma T (2011) Physiological and biochemical mechanisms of plant adaptation to low-fertility acid soils of the tropics: the case of brachiaria-grasses. In: Güngör E (ed) Principles, application and assessment in soil science. InTech Open Access, Rijeka, Croatia, pp 87–116

Whetten R, Sederoff R (1995) Lignin biosynthesis. Plant Cell 7:1001–1013

Yamamoto Y, Kobayashi Y, Devi R, Rikiishi S, Matsumoto H (2002) Aluminum toxicity is associated with mitochondrial dysfunction and the production of reactive oxygen species in plant cells. Plant Physiol 128:63–72

Yang JL, Li YY, Zhang YJ, Zhang SS, Wu YR, Wu P, Zhang SJ (2008) Cell wall polysaccharides are specifically involved in the exclusion of aluminum from the rice root apex. Plant Physiol 146:602–611

Yermiyahu U, Brauer DK, Kinraide TB (1997) Sorption of aluminum to plasma membrane vesicles isolated from roots of Scout 66 and Atlas 66 cultivars of wheat. Plant Physiol 115:1119–1125

Yoshida S, Uemura M (1986) Lipid composition of plasma membranes and tonoplasts isolated from etiolated seedlings of mung bean (*Vigna radiata* L.). Plant Physiol 82:807–812

Zhang G, Taylor GJ (1989) Kinetics of aluminum uptake by excised roots of aluminum-tolerant and aluminum-sensitive cultivars of *Triticum aestivum* L. Plant Physiol 91:1094–1099

Zhang G, Ślaski JJ, Archambault DJ, Taylor GJ (1996) Aluminum-induced alterations in lipid composition of microsomal membranes from an aluminum-resistant and aluminum-sensitive cultivar of *Triticum aestivum*. Physiol Plant 96:683–691

Zhang G, Ślaski JJ, Archambault DJ, Taylor GJ (1997) Alteration of plasma membrane lipids in aluminum-resistant and aluminum-sensitive wheat genotypes in response to aluminum stress. Physiol Plant 99:302–308

# Breeding for Al Tolerance by Unravelling Genetic Diversity in Bread Wheat

**Ana Luisa Garcia-Oliveira, Charlotte Poschenrieder, Juan Barceló, and Paula Martins-Lopes**

**Abstract** Globally, Aluminium (Al) toxicity not only restricts cultivation of crop plants but also causes substantial losses in their production in areas where acidic soils are more prevalent. As plants are sessile, their roots are continuously exposed to Al when growing in acid mineral soils. Thus, the evolution of Al tolerance mechanisms is a prerequisite for plants to perform in these soils. Wheat is a major crop consumed by most of the human population around the world, and its demand is ever increasing. However, wheat is rather sensitive to Al toxicity, more than other major cereal crops, especially rice and maize. In this context, it has become imperative to develop Al-tolerant wheat cultivars which will help ameliorate this problem in a sustainable manner. Therefore, in order to develop improved cultivars for Al tolerance, information on both the manifestation of Al toxicity and the existence of natural variation is a prerequisite which facilitates the further elucidation of different mechanisms on the physiological, genetic and molecular levels. The improvement of any trait by plant breeding mainly relies on the availability of efficient screening techniques, but the pace of improvement depends on easy and reliable phenotyping techniques. In this chapter, we presented the advances made so far on Al tolerance in wheat with special focus on future perspectives, aiming to help for further improvement of Al tolerance in wheat in a sustainable way.

## 1 Introduction

Adverse environmental conditions cause more than 50 % crop loss worldwide (Bray et al. 2000). However, the degree of losses due to abiotic stresses varies and depends upon the intensity and duration of the stress. For instance, one-third of the world's soils is alkaline (Guerinot 2007) and nearly one-half is predicted to be

A.L. Garcia-Oliveira (✉)
International Institute of Tropical Agriculture (IITA), Nairobi, Kenya
e-mail: A.Oliveira@cgiar.org

C. Poschenrieder • J. Barceló
Lab. Fisiología Vegetal, Facultad de Biociencias, Universidad Autónoma de Barcelona, 08193 Bellaterra, Spain

P. Martins-Lopes
Dept Genetica e Biotecnologia, Universidade Tras-os-Montes e Alto Douro, Vila Real



S.K. Panda, F. Baluška (eds.), *Aluminum Stress Adaptation in Plants*, Signaling and Communication in Plants 24, DOI 10.1007/978-3-319-19968-9_7

acidic (Granados et al. 1993). In all those zones where acidic soils are widespread, the suboptimal plant performance can usually be attributed to a combination of abiotic stress factors, such as aluminium (Al), manganese (Mn) and hydrogen (H) toxicity coupled with nutrient deficiency, particularly phosphorus (P), iron (Fe), molybdenum (Mb), magnesium (Mg) and boron (B).

Wheat is one of the major food crops on which more than half of the world population is directly or indirectly dependent for their food requirements. Globally, it is grown on more land area than any other commercial crop due to its ability to grow in a wide range of climatic environments and geographic regions (Dixon et al. 2009). In comparison to other major cereal crops like maize or rice, wheat is considered as rather sensitive to Al toxicity. This results in low crop productivity and reduced yield potential in regions where Al toxicity is more prevalent. Thus, Al toxicity has attracted the attention of wheat researchers to unprecedented levels. The present chapter overviews Al toxicity and tolerance in wheat with special emphasis on updated knowledge of the genetic variability, the different underlying mechanisms at the physiological, genetic and molecular level, as well as the screening techniques for Al tolerance in wheat. Finally, we describe the present status of wheat improvement for Al tolerance using conventional and molecular approaches. Future perspectives are discussed.

## 2 Relevance of Al Toxicity

In the earth crust, Al is the third most abundant element after oxygen and silicon. Under low pH conditions (acidic soils), Al dissolves in various ionic forms [$Al^{+3}$, $Al(OH)^{+2}$ and $Al(OH)_2{}^{+}$]. Among these ionic forms, $Al^{3+}$ is the most phytotoxic form for rhizosphere of wheat. Additionally, $Al(OH)^{2+}$ and $Al(OH)_2{}^{+}$ appear to be toxic for dicots (Delhaize and Ryan 1995; Kochian 1995). In order to remediate Al phytotoxicity in acidic soils, the application of lime is a common agronomic practice. This enhances the soils' pH which is the best way to correct soil acidity to some extent. Unfortunately, this is not always economically or physically feasible because of either or both the huge amount of lime required greatly depending upon the soil's pH and texture and subsoil acidity. Liming is poorly effective in correcting subsoil acidity. Moreover, heavy application of lime may have adverse effects on crop plants or even cause deficiencies of certain nutrients (Whitten 1997). Thus, an alternative effective and more eco-friendly solution is to develop and use Al-tolerant cultivars.

### *2.1 Al Phytotoxicity in Wheat*

Since the identification of poor plant growth and 'crestamento', a burning effect in wheat on acid soil in Brazil, the symptoms and effects of Al phytotoxicity in wheat have been extensively investigated. The primary symptoms and effects of Al phytotoxicity particularly the inhibition of root growth is well known and can be noticed on wheat roots as earliest within few seconds to minutes (Ownby and Popham 1990; Ryan et al. 1992). Wheat roots upon expose to Al stress respond by an initial acute inhibition of their elongation followed by later chronic effects (Parker 1995), resulting in a reduced and damaged root system. This impairs water and mineral nutrient uptake at later stages. In wheat, Al also causes extensive damage in other cellular components and processes such as inhibition of DNA synthesis (Wallace and Anderson 1984), alteration of cell membrane potential (Kinraide 1988) and reduction of root apex $H^+$ efflux (Ryan et al. 1992).

Kinetic studies indicate that the Al uptake by roots in wheat is biphasic, consisting of an initial rapid uptake followed by a linear uptake phase (Zhang and Taylor 1989, 1990). Comprehensive analysis of Al uptake during the linear phase also suggested two mechanisms: metabolism-dependent Al binding in the apoplasm and permeation of the cell membrane (Zhang and Taylor 1989, 1990). Toxic levels of Al accumulate within 6 h in the root's apical region (2 mm) of sensitive cultivars reaching higher concentration (about seven- to eightfold) than tolerant cultivars (Rincón and Gonzales 1992). However, the remarkable differences in the Al uptake by the roots of Al-tolerant and sensitive genotypes could already be observed within 4 h of exposure to Al stress (Delhaize et al. 1993a). The degree of Al sensitivity was clearly related to differences in Al accumulation in the growing root tissues of wheat plants (Samuels et al. 1997).

## 2.2 *Genetic Diversity for Al Tolerance*

Probably wheat is the first crop in which evidence of intraspecific natural variations for tolerance to Al phytotoxicity was reported by Brazilian scientists already during the second decade of the last century. Several, well-documented studies clearly demonstrate that substantial genetic variation for Al tolerance exists in wheat (Carver and Ownby 1995; Hu et al. 2008; Raman et al. 2010). By far, the most appropriate donor lines for increasing Al tolerance in bread wheat have been originated from Brazil and Portugal where vast extensions of land have acid soils (Rajaram et al. 1988; Silva et al. 1991). In addition, Chinese collection of wheat landraces distributed mainly within regions of China where acidic soils are predominant could also be an alternative source of Al tolerance for the regions where soil acidity is not so strong, because Chinese accessions exhibit moderate level of tolerance to Al phytotoxicity in comparison to the genotypes originated from Brazil and Portugal (Han et al. 2013).

Most of the modern wheat cultivars have source of Al tolerance from Brazilian landraces which were utilised in the wheat breeding programme at CIMMYT. Thus, there is an urgent need to further identify the novel source of Al tolerance in wheat, for which a systematic emphasis on the primary gene pool of wheat that includes hexaploid, tetraploid and diploid species, from regions where Al toxicity is a major problem, should be explored.

Generally, cultivated members of the tribe Triticeae such as rye and triticale could also provide an opportunity to introgress the novel alleles for Al tolerance in wheat, as the level of tolerance to Al phytotoxicity in rye and triticale is higher than in wheat. It is noteworthy that the high degree of tolerance to Al phytotoxicity of Portuguese bread wheat genotypes such as Barbela-derived lines has been attributed to the possible introgression from rye (Silva et al. 1991; Ribeiro-Carvalho et al. 2001) as both crops were simultaneously and nearby cultivated by farmers in Trás-os-Montes region of Portugal. Portuguese rye genotypes have also been reported to exhibit outstanding performance under Al stress conditions (Pinto-Carnide and Guedes-Pinto 1999).

Besides the primary and secondary gene pools, the tertiary gene pool which includes wild relatives such as *Aegilops* species and *Leymus racemosus* (an allopolyploid perennial grass) can enrich the bread wheat particularly at the point of identification of Al tolerance genes and associated regulatory regions which could accelerate the improvement of Al tolerance in bread wheat. In the past, high levels of tolerance have been identified in the *Aegilops uniaristata* (Berzonsky and Kimber 1986) and introgressed successfully into wheat (Miller et al. 1997). Recently, Mohammed et al. (2013) investigated the wheat—*Leymus racemosus* addition and wheat—*Leymus racemosus* substitution lines for Al tolerance and found that *Leymus racemosus* chromosomes E had the greatest tolerance even at a very high level of Al stress.

# 3 Mechanisms of Al Tolerance in Wheat

## *3.1 Physiological Mechanisms*

The physiological basis of Al tolerance has been extensively investigated in wheat. Various studies conducted on a collection of different wheat cultivars so far seem to suggest that differences in tolerance to toxic concentrations of Al are based on several mechanisms. Broadly, these mechanisms can be divided into two categories: (A) external detoxification and (B) internal detoxification of Al.

### 3.1.1 External Detoxification

The mechanisms of external detoxification of Al enhance plant tolerance to Al stress by restricting the Al uptake. These mechanisms reduce the harmful interactions of toxic forms of Al with sensitive sites in the apoplast, such as the plasma membrane and the cell wall and decrease the quantity of Al reaching sensitive sites in the symplasm. Mechanisms for Al exclusion can implicate several mechanisms such as efflux of organic acid anions that chelate the Al in the rhizosphere in nontoxic form, root-induced changes in rhizosphere pH, production of root-cap mucilage, secretion of secondary metabolites from root apices, changes in cell wall chemistry and the active export of Al from cells (Delhaize et al. 2012a).

### 3.1.2 Exudation of Al-Chelating Organic Acid Anions

So far, the efflux of Al-chelating organic acid (OA) anions from root apex has been shown to be one of the most effective and widespread Al exclusion mechanisms in wheat which is demonstrated by strong genetic and molecular studies (Delhaize et al. 1993b; Ryan et al. 1995a, b, 2009; Tovkach et al. 2013; Garcia-Oliveira et al. 2014). It is hypothesised that OAs form harmless complexes with Al in the apoplast that protect the sensitive root apex and reduce uptake of Al into the roots by preventing Al from binding to the fixed negative sites of the cell wall and plasma membrane. Several OAs of low molecular weight such as malate, citrate, oxalate, succinate, tartarate and fumarate have been found to be excreted from wheat roots (Foy et al. 1990; Christiansen-Weniger et al. 1992; Delhaize et al. 1993b; Basu et al. 1994; Ryan et al. 2001). Higher rates of Al-induced exudation of malate in Al-tolerant than in Al-sensitive varieties have been observed in most studies comparing wheat genotypes (Delhaize et al. 1993b). More recently, however, it has been observed that not only the exudation rates but also the kind of organic acids differ among the wheat genotypes. For instance, some Brazilian and Portuguese Al-tolerant bread wheat genotypes constitutively secret high level of citrate both in the presence or absence of Al (Ryan et al. 2009; Garcia-Oliveira et al. 2014).

The efflux of malate has been shown to be triggered by external Al from the roots of different Al-tolerant wheat genotypes (Delhaize et al. 1993b). Recent pharmacological evidence indicates that direct protein phosphorylation by protein kinase C (PKC) is a prerequisite for the activation of the TaALMT1 transporter and subsequent enhancement of malate efflux transport activity by extracellular Al (Ligaba et al. 2009). Generally, in response to Al stress, OAs exudation from roots of different plants follows two types of pattern. In Pattern I type, plant roots secrete the OAs immediately after contact with Al stress whereas in Pattern II type plants roots need a lag time of several hours before the start of Al-induced secretion of OAs. Wheat follows Pattern I for malate exudation (Delhaize et al. 1993b). Although in case of citrate exudation, it was noticed that genotype Carazinho releases it constitutively under control and Al stress, whereas Barbela 7/72/92 starts to release citrate only after 2 h of Al exposure. However, it also showed constitutive release under control condition (Ryan et al. 2009; Garcia-Oliveira et al. 2014). Thus, it seems that the kind, timing and quantity of OAs secreted by roots of wheat are genotype dependent.

### 3.1.3 Production of Root-Cap Mucilage and Other Metabolites

The outer layer of root cap of several plant species secretes mucilage, a gelatinous polysaccharide substance which has high capacity of Al binding and acts as a diffusion barrier to Al. Pioneering work by Horst et al. (1982) showed that tolerance to Al phytotoxicity in cowpea correlated with mucilage on the root. Similarly, there is also evidence that mucilage contribute to the reduction of Al toxicity in wheat. Puthota et al. (1991) observed higher mucilage production in the Al-tolerant wheat cultivar Atlas 66 than in a sensitive cultivar. Henderson and Ownby (1991) also found a positive association between the amounts of mucilage produced by wheat roots and the levels of Al tolerance and suggested that the organic acids present in mucilage might chelate Al before it comes in contact with the cell surface. Archambault et al. (1996) found that Al bound to the mucilage of wheat root accounted for approximately 25-35 % of Al remaining after desorption with citric acid. Interestingly, it was noticed that in some wheat genotypes, mucilage only occurs in some roots, but not in all roots of the same plant (Garcia-Oliveira, personal communication).

Other mechanisms include the exudation of phosphate and polypeptide from roots (Pellet et al. 1996; Basu et al. 1997). In root exudation experiments, Al-resistant wheat cultivar Atlas exhibited phosphate as well as malate release in response to Al exposure, and it was suggested that both exudation processes act in concert to enhance Al exclusion and Al resistance in Atlas (Pellet et al. 1996). However, further studies were unable to support the report that phosphate efflux contributed to the Al resistance of Atlas 66 (Ryan et al. 2009). Similarly, Basu et al. (1994) reported that treatment of wheat seedlings with Al leads to the accumulation of suite of polypeptides in root exudates. Al-resistant wheat cultivar Maringa showed enhanced accumulation of polypeptides (12-, 23- and 43.5-kDa) in root exudates, and the 23-kDa polypeptide co-segregated with the Al-resistant phenotype in $F_2$ populations (Basu et al. 1999). Exudation of phosphate and polypeptide seems to be an important Al exclusion mechanism, but, so far, no conclusive data have been presented which suggest the minor role of these mechanisms in differential Al tolerance in wheat (Tang et al. 2002).

### 3.1.4 Binding Al in the Cell Wall

Another hypothesis for differential tolerance to Al in wheat relates to the negative electrical charges present at the cell surface. In general, it is usually believed that binding of Al to charged sites on the cell surface is a prerequisite for its uptake and toxicity. Earlier reports exhibited the negative correlation between root cation exchange capacity (CEC) and Al tolerance in plant species, including some wheat cultivars (Foy et al. 1967; Blamey et al. 1990, 1993). However, this assumption is still controversial, and it has been suggested that such mechanisms do not play a significant role in the differential Al tolerance in wheat (Kinraide et al. 1992).

#### 3.1.5 Internal Detoxification

Some plant species like tea (*Camellia sinensis*), buckwheat (*Fagopyru esculentum*) and Hydrangea species have the remarkable ability to accommodate high amounts of Al in their root and shoot tissue without exhibiting any adverse effect on their normal growth and development. Usually, these species are endemic to regions with acid soils which seem to evolve internal detoxification mechanisms to cope with Al phytotoxicity either by chelating the toxic form of Al in their cytosol or by sequestering it to organelles (see Chap. 5). Although, a substantial body of literatures wind up with a common conclusion of external detoxification of Al in wheat, there is also evidence suggesting that internal mechanisms for Al detoxification also exist in wheat. For instance, Zhang and Taylor (1988) reported that Al-tolerant wheat cultivars accumulate more Al in their root than Al sensitive one. Recently, Silva et al. (2010) also found that the Al-tolerant wheat genotype Barbela 7/72/92 has the ability to accumulate more Al in the shoots than the Al-sensitive wheat genotype, Anahuac. More recently, we also observed the high transcript level of TaALMT1 (*Triticumaestivum*aluminium-activated malate transporter) and TaMATE1 (*Triticumaestivum*multidrug and toxic compound extrusion) in the shoots of Barbela 7/72/92 compared to the roots of Al-sensitive genotype Anahuac (Garcia-Oliveira et al. 2014). Furthermore, we also noticed a significant induction, at the transcript level, of TaALMT1 in the shoots of Barbela 7/72/92 after 6 h of Al exposure; the expression remained at the same high level for more than 24 h which indicates the important role of TaALMT1 for internal detoxification of Al in the aerial part of the Barbela 7/72/92. Until recently, to our knowledge, there is no comparative report on the amount of organic acids present in the wheat tissues particularly roots and aerial parts of the Al-tolerant and -sensitive genotypes of wheat genotypes under Al stress.

More recently, Mohammed et al. (2013) also reported that addition of *Leymus racemosus* (a wild relative Allopolyploid *Triticeae*) chromosomes A and E to wheat not only significantly enhanced the Al tolerance of alien addition lines A and E but also maintained the same amount of Al in the roots of these lines as Chinese Spring. Specifically, line E also translocates a high amount of Al to their shoots. Considering these recent results, it gets evident that at least in wild relatives of wheat, internal detoxification of Al plays an important role in Al tolerance mechanisms.

### 3.2 *Genetic Mechanism*

#### 3.2.1 Inheritance of Al Tolerance

On the basis of segregation pattern observed in $F_2$ generation derived from a cross between Al-tolerant and -sensitive genotypes, Beckman (1954) suggested the presence of a single dominant gene for Al tolerance in bread wheat. Similarly, Kerridge and Kronstad (1968) also reported that a moderately Al-tolerant cultivar,

Druchamp, differed from a sensitive cultivar, Brevor, by a single gene governing seedling root growth under Al stress. Subsequently, in the last quarter of the twentieth century, the wider recognition of Al phytotoxicity as the predominant growth-limiting factor in acid soils has swung the pendulum of attention to Al tolerance and its genetic control (Carver and Ownby 1995). Contrarily to earlier speculations, wide genetic range of Al tolerance observed in the wheat germplasm indicated that Al tolerance inheritance was much more complex (Lafever et al. 1977). However, most of studies suggested that Al tolerance in bread wheat is mainly governed by one to two dominant genes (Choudhry 1978; Camargo 1981, 1984; Campbell and Lafever 1981). The result of these classical bi-parental crosses performed during late 1970s and early 1980s were further supported by QTL mapping studies in the era of molecular markers during the first decade of the present century (Table 1) which is described in the following section.

### 3.2.2 Chromosomal Distribution of Loci Conferring Al Tolerance

Among cereal crops, wheat offers high buffering capacity and can tolerate a high degree of aneuploidy because of its polyploidy nature. Historically, a series of unique and valuable cytogenetic stocks have been developed by using traditional cytogenetic techniques (Sears 1954), which are still widely used to locate genes and DNA markers to individual chromosomes in wheat (Garcia-Oliveira et al. 2013). Aniol and Gustafson (1984) and Papernik et al. (2001) identified a number of genetic loci on chromosome arms 2DL, 3DL, 4BL, 4DL, 6AL, 7AS and chromosome 7D, using ditelosomic and nullisomic-tetrasomic lines of Chinese Spring wheat stocks that are important for conferring Al tolerance. Nevertheless, it was unclear whether all these loci identified from the deletion lines contribute to the natural variation for Al tolerance found in wheat germplasm.

In the last decade of twentieth century, with the advent of molecular markers, availability of different mapping populations and development of robust statistical analysis methods not only facilitated the identification of number of genetic loci that involved in complex traits but also helped to understand the gene action and their relative contributions in governing such traits. Further refinement in DNA technology also enabled plant scientists to elucidate the genes that underlie the QTL detected in the studied mapping population. At the molecular level, Riede and Anderson (1996) tagged the first major QTL influencing Al tolerance in bread wheat with DNA marker Xbcd1230 on chromosome 4DL using RFLP analysis of RIL populations derived from a cross between an Al-tolerant genotype BH1146 and an Al-sensitive genotype Anahuac which exhibited 85 % of the phenotypic variation for Al tolerance, showing the greatest effect on root growth of wheat plants in Al containing nutrient solutions. Subsequently, with the availability of TaALMT1 sequence in bread wheat, Raman et al. (2005) mapped the TaALMT1 in two different mapping population of wheat and was co-localised with the QTL for Al tolerance on chromosome 4DL which was previously reported in aneuploid stocks. Ryan et al. (2009) identified that Al tolerance mechanism in a Brazilian bread wheat

**Table 1** Details of QTL for Al tolerance in wheat mapped using molecular markers

| Marker used for mapping (#) | Size and type of mapping population (cross) | Chromosome (# QTL) | Locus | Nearest marker or flanking interval | Donor parent | $R^2$ (%) | Reference(s) |
|---|---|---|---|---|---|---|---|
| *Bread wheat* | | | | | | | |
| RFLP (83) | RIL (BH1146 × Anahuac) | 4DL (1) | $Alt_{BH}$ | *Xbcd*1230 | BH1146 | RGS (85) | Riede and Anderson (1996) |
| SSR[a] | DH (Sunco × Tasman) | 4D (1) | *ALMT1* | Xalmt1 | Tasman | NRG (80) | Raman et al. (2005) |
| SSR[a] | DH (Cranbrook × Halberd) | 4D (1) | *ALMT1* | Xalmt1 | Halberd | NRG (93) | Raman et al. (2005) |
| SSR (131) | 118, RIL (Atlas 66 × Century) | 4DL (1) | $Alt_{BH}$ | Xgdm125-Xwmc331 | Atlas 66 | SRG (50), RTI (50), HSS (50) | Ma et al. (2005) |
| AFLP (381), SSR (168) | 90, RIL (Annong 8455 × CS-Sumai3 3 7A) | 4DL (1)<br>5AS (1)<br>2DL (1) | – | Xcfd23-Xwmc331<br>XmCTGA.pACT233-XmCACG.pGTG138<br>XmACGC.pAG231-XmCTCG.pAGG142 | CS | SRG(18), RRL (13)<br>SRG(11), RRL (9)<br>SRG(13), RRL (11) | Ma et al. (2006) |
| SSR (50) | 192, RIL (Atlas 66 × Chisholm) | 4DL (1)<br>3BL (1) | $Alt_{BH}$ | *ALMT1* | Atlas 66 | RRG (43), HSS (49)<br>RRG (9), HSS (11) | Zhou et al. (2007) |
| SSR (116) | 199, RIL (FSW × ND35) | 4DL (1)<br>3BL (1)<br>2A (1) | $Alt_{BH}$ | Xgdm125–Xups4<br>Xbarc164-Xbarc344<br>Xgwm515-Xgwm296 | FSW | NRG (46 ), HSS (56)<br>NRG (42 ), HSS (47)<br>NRG (6), HSS (9) | Cai et al. (2008) |
| SSR (7) | 84, IL CS × Ae. tauschii | 4D (1) | $Alt_{BH}$ | Xgdm125-Xgwm976 | CS | RTI (31) | Navakode et al. (2009) |

(continued)

**Table 1** (continued)

| Marker used for mapping (#) | Size and type of mapping population (cross) | Chromosome (# QTL) | Locus | Nearest marker or flanking interval | Donor parent | $R^2$ (%) | Reference(s) |
|---|---|---|---|---|---|---|---|
| SSR (14) | 57, DH CS × CS′ ('Synthetic' 3B) | 3B (1) | – | Xgwm1029-Xgwm1005 | CS | RTI (49) | Navakode et al. (2009) |
| DArT, SSR (676) | 67, F2 (Carazinho × EGA-Burke) | 4BL (1) | *Xcec*/*MATE1* | Xgwm495 | Carazinho | Citrate efflux (51)<br>NRG (56) | Ryan et al. (2009) |
| DArT, SSR | 45, F2:3 (Carazinho × Egret) | 4BL (1) | *MATE1* | Xgwm495 | Carazinho | Citrate efflux (96) | Ryan et al. (2009) |
| SSR (132) | 217, RIL (FSW × Wheaton) | 4DL (1)<br>3BL (1) | $Alt_{BH}$ | Xgdm125-Xwmc331<br>Xbarc344-Xbarc164 | FSW | NRG (66), HSS (70)<br>NRG (4), HSS (3) | Dai et al. (2013) |

$R^2$: Phenotypic variation explained (%)

*RFLP* restricted fragment length polymorphism, *SSR* simple sequence repeat, *AFLP* amplified fragment length polymorphism, *DArT* diversity array technology, *CS* Chinese spring, *RIL* recombinant inbred line, *IL* introgression line, *DH* double haploid, *RGS* root growth under stress, *SRG* stress root growth, *RTI* root-tolerance index (%), *RRG* relative root growth, *HSS* hematoxylin staining score, *NRG* net root growth

[a]Linkage map originally developed by Chalmers et al. (2001) was used

genotype Carazinho relies on constitutive citrate efflux. The authors developed two segregating population by crossing Carazinho as a common parent with EGA-Burke and Egret. Using DArT and SSR analysis of both segregating populations, they identified QTL, Xcec, at chromosome 4BL that accounts for more than 50 % of the phenotypic variation in citrate efflux from roots of Carazinho.

Numerous QTL mapping studies were performed using different types of mapping population derived from diverse parents, and most of these studies demonstrated that one to three QTL might be involved in Al tolerance in bread wheat (Table 1). It is noteworthy that in spite of large number of loci identified for Al tolerance in wheat through aneuploid stocks, only few QTL were reported in different mapping population. The outcome of these studies suggests that it might be possible that the same casual allele exists in the population as traditional linkage mapping population is mainly based on bi-parental crosses. Thus, in such mapping populations, only two alleles at a given locus are analysed simultaneously, and low recombination frequencies can lead to poor resolution of QTL at a time.

Recently, QTL identification without the analysis of a mapping population has become possible with genome-wide association mapping which represents a complementary approach to the linkage mapping studies and is based on linkage disequilibrium in genetically diverse germplasms. Association mapping emerged as a potential genomic tool to identify alleles and loci that show significant effects on the target trait particularly complex traits. A genome-wide association analyses of wheat Al tolerance identified at least 16 genetic loci on chromosomes 1A, 1B, 2A, 2B, 2D, 3A, 3B, 4A, 4B, 4D, 5B, 6A, 6B, 7A and 7B associated with this trait in wheat (Raman et al. 2010). Some of these loci on chromosome(s) 2A, 2DL, 3BL, 3DL, 4BL, 4DL, 5AS, 6AL, 6D and 7AS correspond to previous genetic and cytogenetic studies that identified loci for Al tolerance in wheat (Aniol and Gustafson 1984; Papernik et al. 2001; references in Table 1), whereas others appeared to be novel. Interestingly, both genome-wide mapping and gene-specific functional markers identified the major locus *TaALMT1* on chromosome 4D as being significantly associated with Al tolerance which has previously been established in most of the bi-parental populations (Table 1) and among accessions of modern cultivars, landraces and subspecies (Raman et al. 2008, 2010). Noticeably, among the markers targeting *TaALMT1*, those that detected alleles in the promoter predicted most of the phenotypic variation for Al tolerance in diverse wheat germplasm comprising cultivars, landraces, subspecies of *T. aestivum*, and the wild ancestor of wheat, *Aegilops tauschii* (Sasaki et al. 2006; Raman et al. 2008, 2010).

Although association mapping overcomes some barriers of QTL mapping application in plant breeding in direct way, but, it is necessary to consider some important aspects of this methodology for avoiding the false associations between trait-markers. Thus, considering the limitation of association mapping, in order to establish whether these loci are responsible for the natural variation in Al tolerance in wheat, it would need to be verified by other methods, such as linkage mapping with bi-parental crosses.

## 3.3 *Molecular Mechanism*

### 3.3.1 Al Tolerance Genes in Wheat

Among cereals, wheat has been employed extensively in exploring many aspects related to Al phytotoxicity. However, fewer studies have been performed on the identification of genes conferring tolerance to Al phytotoxicity in wheat than in other cereals, particularly rice. As described in earlier section of this chapter, several loci for Al tolerance have been identified in wheat using genetic and cytogenetic stocks, but, so far only few genes have been cloned in wheat.

### 3.3.2 Al Chelating Organic Acid Anions Transporters

Physiological evidences in favour of organic acids efflux helped to identify two genes namely *TaALMT1* and *TaMATE1* underlying loci at 4DL and 4BL, respectively, and are considered as major genes for Al tolerance in wheat, because the loci harbouring these genes have been described as responsible for most of the genotypic variation for Al tolerance in bread wheat.

### 3.3.3 TaALMT1

A solid understanding of the genetics and physiology of Al tolerance in wheat facilitated the identification of first Al tolerance gene *TaALMT1* in plants which drives natural variation in wheat for tolerance to Al phytotoxicity (Sasaki et al. 2004). *TaALMT1* encodes an Al-activated anion channel on the plasma membrane, and the absence or loss of this gene coincided with the loss of both Al tolerance and malate efflux from root apices (Raman et al. 2005; Yamaguchi et al. 2005). Henceforth, *ALMT1* was extensively studied for Al tolerance and characterised from several plant species. In bread wheat, *ALMT1* underlies a major Al tolerance locus called $Alt_{BH}$, which was previously mapped to the long arm of chromosome 4D. Several alleles have been identified on the basis of differences in coding sequences of ALMT1, although none of them associated with Al tolerance in wheat. Interestingly, the transcript level of *TaALMT1* expression exhibits positive correlation with Al tolerance. Promoter analysis of *TaALMT1* showed a specific pattern of variations which can be classified into eight categories viz. types I to type VIII. Type I pattern revealed to have the simplest structure, while the other patterns presented blocks of sequence in duplicates and triplicates. In addition, types VII and VIII are a variation of promoter type VI (Sasaki et al. 2006; Garcia-Oliveira et al. 2014).

The blocks of sequences in the *ALMT1* upstream region contains different *cis*-acting regulatory motifs, such as block A which contains several recognition sequences for MYB and MYC transcription factors and ABA-responsive elements

(Sasaki et al. 2006). Block C contains the ATATT motif previously found in *rolD* promoter in *Agrobacterium* and correlated to strong root expression (Elmayan and Tepfer 1995). In addition, blocks A and C share the ACGT and CACT motifs which had been implicated in the plant responses to dehydration stress and carbon metabolism whereas blocks B and C share the GATA box motif which has been implicated in the high tissue-specific expression of some genes (Sasaki et al. 2006). More recently, Tokizawa et al. (2015) performed the promoter scanning analysis using an *Arabidopsis* dataset (i.e. overrepresented octamers in the promoter of suppressed genes in the stop1 mutant) and showed that the TaALMT1 promoter of wheat contains a set of STOP1-binding motifs and *cis*-acting elements for CAMTAs and was associated with *cis*-acting elements for TCP-domain transcription factor(s)/ASR5. Interestingly, Al-tolerant wheat near-isogenic line (ET8) contained three sets of STOP1/CAMTA binding sites, whereas an Al-sensitive near-isogenic line carried a single set (Tokizawa et al. 2015).

These *cis*-acting regulatory motifs located within different blocks observed in the promoter sequences might modulate the relative expression of *ALMT1* gene. To further promote typology, putative protein phosphorylation was also indicated to be involved in the Al responsive malate exudation from roots (Osawa and Matsumoto 2001) where the organic anion-specific channel could be a terminal target that responds to Al signalling. This was further confirmed in *Xenopus* heterologous system (Ligaba et al. 2009) where N- and C-terminal domains were shown to be involved in the Al response of plants through *TaALMT1* gene (Furuichi et al. 2010; Ligaba et al. 2013).

In wheat, ALMT1 transcript is highly expressed in roots (Sasaki et al. 2006; Ryan et al. 2009); however, a relatively higher level was also noticed in the shoots of Al-tolerant genotype when compared to the roots of sensitive genotype (Garcia-Oliveira et al. 2014). For Al-tolerant genotypes such as Barbela 7/72/92, transcript of *TaALMT1* gene was found highly expressed under control conditions.

### 3.3.4 TaMATE1

In plants, the first MATE (Multidrug And Toxic Compound Extrusion)/AACT (Al-activated citrate transporter) gene involved in Al tolerance was identified by fine mapping the major Al tolerance loci $Alt_{SB}$ and *Alp* in sorghum (Magalhães et al. 2007) and barley (Furukawa et al. 2007), respectively. In both crops, the relative tolerance to Al phytotoxicity was observed to be highly correlated with the level of MATE1 transcript. In bread wheat, Ryan et al. (2009) identified $Xce_c$ locus on the chromosome 4BL accounting for more than 50 % of the phenotypic variation in citrate efflux from roots of Brazilian cultivar Carazinho. A subsequent study has demonstrated that $Xce_c$ locus encodes a citrate transporter (TaMATE1), and it was concluded that upstream variation [Sukkula-like transposable element (TE) insert] in TaMATE1B homoeologue extends its transcript expression to the root apex in Carazinho where it confers citrate efflux and enhances Al tolerance (Tovkach et al. 2013). Similarly, constitutive citrate efflux in Portuguese bread wheat genotype Barbela 7/72/92 was also observed and seems to be correlated with the high level of TaMATE1B transcript. In addition, a similar TE, inserted 25 bp upstream of the ATG start site of *TaMATE1B* homoeologue, was identified in Al-tolerant genotype Barbela 7/72/92 (Garcia-Oliveira et al. 2014).

Earlier it was considered that upstream variation in the TaMAT1B is a rare event, but our group showed that the presence of Sukkula-like transposable element in the TaMATE1B promoter is common in the Portuguese Al-tolerant bread wheat genotypes. It is interesting to note that Al-tolerant genotypes possessing this TE also differed in their root regrowth and citrate exudation levels which indicate the presence of different *cis* elements in the TE insert upstream of *TaMATE1B* gene that might be associated with varied level of citrate efflux. In addition, significant differences in the transcript level of TaMATE1D homoeologue were also detected between the Al-tolerant and susceptible genotypes whereas TaMATE1A seemed to be silent (Garcia-Oliveira et al. 2014).

# 4 Other Candidate Gene(s) for Al Tolerance

In wheat, a number of Al responsive genes such as *wali* (wheat aluminium induced), *war* (wheat aluminium regulated) and *MnSOD* (manganese superoxide dismutases) that belong to general and/or specific stress related proteins have been cloned and considered to be potential candidates for Al tolerance in wheat (Snowden and Gardner 1993; Hamel et al. 1998; Basu et al. 2001). Recently, Navakode et al. (2014) identified a new locus for Al tolerance on 1DL chromosome in wheat and was assigned to the chromosomal bin 1DL2-0.41-1.00 where an Al responsive candidate gene wali5 was previously identified in wheat (Snowden and Gardner 1993). Although, *Arabidopsis* plants over-expressing wheat *wali5* did not exhibit tolerance to Al toxicity (Ezaki et al. 2000).

More recently, we also cloned a candidate gene *TaSTOP1* in bread wheat which belongs to a member of Cys2His2 zinc finger transcription factor family proteins (Garcia-Oliveira et al. 2013). *TaSTOP1* loci were localised on the long arm of homoeologous group 3 chromosomes [3AL (*TaSTOP1-A*), 3BL (*TaSTOP1-B*) and 3DL (*TaSTOP1-D*)]. Earlier, important loci associated for Al tolerance in wheat on long arm of homoeologous group 3, particularly 3BL, have been consistently reported in classical studies using chromosomal manipulation (Aniol and Gustafson 1984; Papernik et al. 2001), QTL mapping (Zhou et al. 2007; Cai et al. 2008; Dai et al. 2013), and genome-wide association analysis (Raman et al. 2010; Navakode et al. 2014). The role of a zinc finger transcription factor *ART1* identified through mutational analysis in rice has also been shown in natural variation of Al tolerance in rice, earlier which was suggested that it was not involved in Al tolerance (Famoso et al. 2011). Thus, in the light of recent findings, *TaSTOP1* and *wali5* seem to be strong candidates for Al tolerance in wheat, and it would be very interesting to map the location the these genes in bi-parental population or in association panel of wheat for further understanding on their role in the natural variation of Al tolerance in wheat.

## 5 Transgenic Approach

Transgenic approach offers unique opportunities for validating gene function in Al tolerance and could also be an alternative technology to increase crop production in acidic soils through development of Al-tolerant cultivars by genetic engineering. Since the first transgenic tobacco plant developed by using the bacterial citrate synthase gene (de la Fuente et al. 1997) for enhanced Al tolerance, transgenic approach is being pursued actively throughout the world to generate transgenic lines in different plant species including crop plants for improved tolerance to Al phytotoxicity. So far, a range of genes originating from bacteria, nematode and plants have been used to generate transgenic plants for enhanced level of Al tolerance (Garcia-Oliveira et al. 2015), but only few transgenic approaches in wheat were tested to date (Table 2).

It is general indication that plant genes induced by a particular stress often serve to protect against that stress. Considering this hypothesis, two Al stress responsive genes *Wali5* and *MnSOD1* from wheat were transferred in *Arabidopsis* and canola, respectively (Ezaki et al. 2000; Basu et al. 2001). Transgenic *Arabidopsis* over-expressing *TaWali5* expressed transcript at high levels, but appeared unlikely to be a good strategy for improving Al tolerance in plants. Contrarily, transgenic oilseed rape plants over-expressing *MnSOD1* gene had 1.5- to 2.5-fold higher SOD activity, and at least exhibited significantly better root growth under Al stress than the root growth of wild type plants. With the cloning of a wheat gene *ALMT1* which co-segregated with Al tolerance, the major emphasis of plant biotechnologists has been on engineering genes that encode organic anion transporter proteins such as *ALMT1* and *MATE1*. Wheat *ALMT1* was transferred in rice, but it was noticed that transgenic rice plants overexpressing the *TaALMT1* gene did not exhibit increases in Al resistance (Sasaki et al. 2004) and concluded that this was because of the natural high Al resistance observed in rice plants. However, the same gene exhibited most striking results in barley followed by wheat, and Arabidopsis which showed greater relative root growth by 20, 8, and 4-fold, respectively (Table 2).

Circumstantial evidences favour the role of regulatory genes (transcriptional factors) in plant tolerance to abiotic stresses because transcription factors are at the top of the cascade that triggers the expression of several genes involved in plant defences against Al toxicity. One of the best examples of the expression of a single transcriptional factor that triggers plant defences is the case of the rice *ART1* gene which activates several genes including *ALMT1* and *MATE1* (Yamaji et al. 2009). Recently, the contribution of *ART1* locus to the natural variation for Al tolerance in rice has also been identified by QTL analysis (Famoso et al. 2011). As mentioned

**Table 2** Details of studies for Al tolerance enhancement in plants by genetic engineering

| Name of the gene | Transgenic strategy | | | Relative root growth | Proposed mechanism | Reference(s) |
|---|---|---|---|---|---|---|
| | Source of gene | Recipient | Promoter | | | |
| I. Stress responsive | | | | | | |
| *Wali5* | Wheat | *Arabidopsis* | CMV35S | Not changed | Protection from oxidative stress | Ezaki et al. (2000) |
| *MnSOD1* | Wheat | Canola | CMV35S | 2.5-fold | Protection from oxidative stress | Basu et al. (2001) |
| II. Transporter(s) | | | | | | |
| *ALMT1* | Wheat | Barley | Ubiquitin | 20-fold | Malate efflux (↑) | Delhaize et al. (2004) |
| | Wheat | Rice | CMV35S | Not changed | Malate efflux (↑) | Sasaki et al. (2004) |
| | Wheat | Wheat | Ubiquitin | 8.0-fold | Malate efflux (↑) | Pereira et al. (2010) |
| | Wheat | *Arabidopsis* | CMV35S | 4.0-fold | Malate efflux (↑) | Ryan et al. (2011) |
| *AACT1/MATE1* | Barley | Wheat | Ubiquitin | 1.3-fold | Citrate efflux (↑) | Zhou et al. (2013) |

*Wali5*: Wheat Al induced (Bowman-Birk Protease inhibitor); *MnSOD1*: Manganese superoxide dismutase 1

earlier, we cloned the *TaSTOP1* from bread wheat in the recent past (Garcia-Oliveira et al. 2013), but still needs to confirm its function in wheat.

By far, in all the studies constitutive promoters either CMV35S or maize ubiquitin have been used which are expressed in most of plant parts. It is noteworthy that the site of Al phytotoxicity is the root apex; thus, in future; a wise approach would be to use the promoters that express transgenes only in this region, because it would avoid the waste of metabolic energy and undesirable phenotype due to ectopic expression. Furthermore, a desirable promoter for engineering Al tolerance would be highly expressed in the root apex and only in the presence of Al.

# 6 Screening Techniques for Al Tolerance

Screening of germplasm is a pre-requisite step in any plant improvement programme, and breeders require an easy and reliable phenotyping technique for a trait of interest. During the previous 50 years, considerable advances have been made in the screening techniques to efficiently discriminate the tolerant plants from sensitive to Al phytotoxicity in crop plants including wheat. These screening techniques can be broadly grouped into hydroponic and soil media-based techniques.

## *6.1 Hydroponic Techniques*

Under field conditions, it is difficult to isolate the factors affecting nutrition of plants and mechanisms of element uptake, due to spatial heterogeneity of soil chemical and physical properties that simultaneously impact plant development. Thus, hydroponic culture is an ideal and most useful screening methodology that allows to understand the Al phytotoxicity in plants including wheat, because it provides not only an easy access to the root systems for Al analyses, but also it enables the application of accurate concentrations of minerals coupled with tight control over nutrient availability and pH. Moreover, hydroponic culture also permits non-destructive measurements of Al tolerance based on root growth, and the interpretation of the specific Al effect on the plant at defined vegetative stage is highly accurate (Carver and Ownby 1995). For assessment of Al tolerance in wheat, most of researchers studied root growth parameters such as root tolerance index, relative root growth and net root growth due to the easiness to measure such parameters. Even faster scoring of wheat and other species can be achieved visually using different staining techniques such as hematoxylin, eriochrome cyanine R, morin, Schiff's reagent or Evans blue staining (Table 3). Among these histochemical assays, hematoxylin and eriochrome cyanine R staining are widely used to screen for Al tolerance in wheat because of their consistent reliability in the staining patterns. Since the first evidence of role of malate efflux in Al tolerance of

**Table 3** Techniques for assessment of Al toxicity effects and discrimination of Al-tolerant and -sensitive genotypes grown in hydroponic culture

| | Techniques | Methodology | Measurement | Equipment | Cost incurred | Reference(s) |
|---|---|---|---|---|---|---|
| Measurement of Al effects on root growth | | | | | | |
| 1 | Root tolerance index (RTI)/relative root growth (RRG) | $\frac{\text{Root growth under Al stress}}{\text{Root growth without Al stress}} \times 100$ | Quantitative | – | Very cheap | Ma et al. (2005, 2006); Zhou et al. (2007) |
| 2 | Net root growth (NRG) | Root length measured after Al stress – root length measured prior to Al stress | Quantitative | – | Very cheap | Cai et al. (2008) |
| 3 | Root re-growth assay (eriochrome cyanine R dye) | Part of the root which grows during the Al stress recovery process | Quantitative | – | Cheaper | Martins-Lopes et al. (2009) |
| 4 | Root hair development | Detects rhizosheath size and root hair length | Quantitative | Camera attached with microscope and image analyses software | Costly | Delhaize et al. (2012b) |
| Measurement of Al accumulation | | | | | | |
| 1 | Hematoxilin staining | Detects Al accumulation by forming colour complex between hematoxylin and root-bound Al | Semi-quantitative | Visual score | Cheaper | Polle et al. (1978); Ma et al. (2005); Zhou et al. (2007); Cai et al. (2008) |
| 2 | Elemental analysis | Detects tissue-specific Al concentration | Quantitative | ICP/AAS | Costly | Silva et al. (2010); Delhaize et al. (2012b) |
| Al localisation in root tissues | | | | | | |
| 1 | Morin assay (fluorescent dye) | Forms fluorescent complex with Al ion | Semi-quantitative | Fluorescent microscope | Costly | Tice et al. (1992); Garcia-Oliveira, unpublished |

| | | | | | | |
|---|---|---|---|---|---|---|
| Measurement of root injury/stress | | | | | | |
| 1 | Evans blue assay (semi-permeable dye) | Detects loss of plasma membrane's integrity and cell viability | Semi-quantitative | Visual/ spectrophotometer | Cheaper | Xu et al. (2012); Garcia-Oliveira, unpublished |
| 2 | Schiff's reagent | Detects lipidic peroxidation | Semi-quantitative | Visual/ spectrophotometer | Cheaper | Yamamoto et al. (2001); Garcia-Oliveira, unpublished |
| Measurement of root exudates | | | | | | |
| 1 | Organic anion efflux | Quantify OAs secreted by intact/ excised root and internal concentration in different tissues | Quantitative | Enzymatic assay/HPLC | Costly | Delhaize et al. (1993b); Ryan et al. (2009); Tovkach et al. (2013); Garcia-Oliveira et al. (2014) |
| 2 | Phenolic exudate | Quantify total and individual phenolic compounds secreted from roots and internal concentration in different tissues | Quantitative | Spectrophotometer/ HPLC | Costly | Tolrá et al. (2009) |

wheat provided by Delhaize et al. (1993b), stimulated studies quantify the amount of different organic acids efflux in Al-tolerant bread wheat genotypes. This resulted in the identification of role of citrate efflux in some Al-tolerant bread wheat genotypes (Ryan et al. 2009). At present, exudation of organic acids is the most promising mechanism of Al tolerance in wheat yet studied, which must be included as an essential part of screening methodology for Al tolerance in bread wheat.

### 6.2 Soil Bioassay

Screening using soil bioassay is the most suitable methodology for studying the long-term effect of Al phytotoxicity specifically on the aerial plant parts, because free Al is directly toxic to plant roots and, in most cases, is little absorbed or translocated to the aerial plant parts. Therefore, small pots either may be filled with pre-washed and air dried sands impregnated with nutrient solution or may be filled with acid soils collected from the target regions. However, the estimation of effect of Al phytotoxicity on economic yield of wheat is imperative, because the improved genotypes for Al tolerance will ultimately grow in those areas which have major problem of Al phytotoxicity. Furthermore, large population can be screened with relatively low cost and less efforts under field condition. Although, the soil matrix may manifest multiple biotic and abiotic stresses which can complicate or affect the output.

## 7 Breeding for Al Toxicity Tolerance

Last comprehensive physiological and genetic studies have contributed significantly to understanding of the role of genetic variability in wheat response to Al toxicity. Collaborative research, involving the Brazilian wheat breeding programme and CIMMYT (International Maize and Wheat Improvement Center), has not only facilitated the development of Al-tolerant lines/cultivars but also distributed worldwide which helped in the development of segregating populations and near-isogenic lines for Al tolerance in wheat.

### 7.1 Conventional Breeding

Historically, numerous Al-tolerant wheat cultivars, such as Fronteira, Surpresa, Minuano, Jesuita and Guarani were developed by Brazilian wheat breeders using Al-tolerant genotypes Alfredo Chaves and Polissu (Rajaram et al. 1988). Subsequently, several improved bread wheat cultivars tolerant to Al phytotoxicity such as BH1146 and Carazinho in Brazil and Atlas 66 in USA were developed using the same source of material (Alfredo Chaves 6-21) which are well known for their high level of tolerance to Al phytotoxicity and still remain today as the standard of performance for Al tolerance. Most of the modern bread wheat cultivars tolerant to Al phytotoxicity developed in different countries have Brazilian material in their pedigree.

### *7.2 Genomic Assisted Breeding*

Undoubtedly, molecular markers have potential to facilitate and accelerate the wheat breeding programme for Al tolerance through precise transfer of chromosomal regions tightly linked with a marker allele or a gene(s) governing Al tolerance. With the advances in marker technology, a number of random DNA markers based upon RFLP, AFLP, SSR and DArT have been developed and also tagged with Al tolerance loci using traditional mapping population in wheat (Table 1). Unfortunately, only limited candidate genes have been so far cloned in wheat. Among these candidate genes, *TaALMT1* and *TaMATE1* genes have been mapped that accounts for phenotypic variation for Al tolerance. Functional markers derived from these candidate genes have been also developed to screen the wheat germplasm (Sasaki et al. 2006; Raman et al. 2005, 2010; Garcia-Oliveira et al. 2014).

As mentioned earlier in this chapter, Raman et al. (2010) identified several loci using random DNA markers along with functional markers derived from candidate gene *TaALMT1* for Al tolerance in wheat. It is expected that candidate gene-based functional markers will be better suitable for genomic assisted breeding than random DNA markers associated with the trait of interest. Therefore, for pyramiding of Al tolerance loci/gene(s), functional markers derived from additional candidate genes need to be tested via association mapping approach to determine allelic diversity within the genes conditioning Al tolerance in wheat. Thus in the future, both the random as well as gene-specific markers validated in different genetic backgrounds will be better suitable for marker assisted selection to tap the novel alleles for Al tolerance in bread wheat.

## 8 Future Perspectives

Among cereals, rice serves as model plant species, and numerous candidate genes including regulatory and structural genes in rice for Al tolerance have been cloned and functionally validated. However, the identification of such candidate genes in bread wheat is far behind than in other cereal crops. Previous studies in bread wheat rely on only few genotypes like 'BH1146', 'Chinese Spring', 'Carazinho', 'Atlas 66', 'ET8' and 'FSW', which have Brazilian cultivars in their ancestry except 'Chinese Spring' 'and 'FSW' that originated from China.

Over the years, Al toxicity research has mainly focused on acidic soil conditions, and researchers have also gained a better understanding of its physiological, genetic and molecular basis (Kochian 1995; Delhaize and Ryan 1995). Compelling evidences indicate that Al phytotoxicity also restricts development of plant roots including wheat, even at high pH (Ma et al. 2003) which led the researchers to realise that Al phytotoxicity can occur both under acidic and alkaline conditions. Very recently, Brautigan et al. (2012) showed that as pH rose above 9.2, anionic species of Al became more prevalent and the phytotoxic effect increased due to this form, not the quantity of Al present because the amount of Al entering stems and leaves at high pH and low pH was similar. Considering the knowledge gained from the model plants and cereals including wheat, the following area of research on Al tolerance in wheat await future results:

1. By far most of studies indicate that Al tolerance in wheat is not a widespread trait and the most of donor parents are identified from Al toxicity prone areas. Recent physiology orientated studies clearly identified the new mechanism (citrate efflux) of Al tolerance in wheat (Ryan et al. 2009; Garcia-Oliveira et al. 2014). Thus, attention in future should focus on systematic efforts on the screening of accessions of bread wheat where Al toxicity is prevailed.
2. Most of studies to date reported low Al uptake in root of Al-tolerant bread wheat genotypes as a general mechanism of Al tolerance. However, information regarding genetic variation for Al transported from roots is lacking. This was observed as an important feature for Al tolerance in the wheat *Leymus racemosus* addition line (Mohammed et al. 2013) and in a wild species of Poaceae, *Andropogon virginicus* L. (Ezaki et al. 2013), that could be one of the useful strategies to confer Al tolerance by maintaining its toxic content below a critical level in roots.
3. Besides external detoxification of Al, internal detoxification could be another important mechanism conferring tolerance. Of special importance seem to be the sequestration of toxic metals in specific subcellular compartments of particular tissues whose nature depends on the plant species and organs involved. In the case of Al, accumulation in leaf spikes and secretion from the trichomes. especially after long-term exposure to Al stress may play a role.
4. Development of rapid and new screening techniques would be another area to be worked on, as most of studies in wheat to date rely on organic acids efflux particularly malate efflux by taking the species-dependent nature of response into consideration. Conversely, more recent experiments identified other dependable parameters such as citrate efflux which has much stronger chelation activity than malate. Although, OAs seem to be the most important contributor to detoxification of Al in bread wheat, there are clearly other mechanisms operating in wheat that do not rely on organic acids efflux such as root hair formation but to date little is known about such mechanisms.
5. Considering the high intra- and interspecific variability found in tribe Triticeae, such as rye has one of the most efficient groups of genes for Al tolerance that could offer alternate source of tolerance genes for wheat (*Triticum ssp.*). It had

been proposed that the ability of triticale to release citrate from the roots when exposed to Al stress is inherited from the rye parent, but it is the wheat parent that determines the rate of OAs exudation (Stass et al. 2008).

6. At the molecular level, recent evidences are supporting the importance of regulatory genes (transcription factors) in plant tolerance to Al phytotoxicity, as Al may trigger resistance mechanisms either by directly binding to transcription factors or by indirectly activating different signalling pathways. Fast induction by Al of changes in the cytosolic $Ca^{2+}$ levels (Rengel 1992), reactive oxygen species (Ramírez-Benítez et al. 2011), and phytohormone-mediated signal transduction in plants (Massot et al. 2002) need further studies to connect with the expression of Al resistance mechanisms.
7. The epigenetic control of the expression of transcription factors and other genes of the responsive network under Al stress conditions is still completely missing, especially the role of small RNAs has to be elucidated.
8. By far, numerous loci associated with Al tolerance have been identified using cytological and molecular approaches in bread wheat. An immediate challenge to molecular approaches is to identify genes directly involved through their allelic variation for Al tolerance underlying these QTL and their validation through segregation or mutation/transgenic approaches so that they can be used to complement and enhance traditional breeding programmes by gene pyramiding.
9. The role of protein phosphorylation in the transport activity of the wheat root malate efflux transporter TaALMT1 by extracellular Al has been clearly demonstrated by integrative pharmacology, electrophysiology and site-directed mutagenesis approaches. In future, studies on the structural and functional analysis of the ALMT-type transporters need to be focused to further clarify the molecular basis of Al activation in wheat.

## 9 Conclusions

The improvement in grain yield has always been the top priority in wheat breeding programmes because the global demand for wheat is increasing as a consequence of burgeoning population along with incremental growth in income and also an ever increasing demand for animal products. Being a staple food, roughly wheat production needs to be double to keep up with the large expected demand in 2050. As mentioned in the beginning of this chapter, wheat is grown in diverse environmental conditions including both acidic and alkaline soils where Al phytotoxicity appears to be one of the major constraints for higher productivity of this cereal. Thus, to accommodate the rising wheat demand, development of improved bread wheat cultivars having higher tolerance to Al phytotoxicity is likely to offer an additional opportunity for yield gain in such areas. Although substantial progress has also been made in genetically modifying wheat to enhance its Al tolerance through conventional breeding, biotechnologists need to develop new strategies to assist the conventional breeders for rapid characterisation of wheat gene pools along with efficiently utilising the genetic variation for Al tolerance in wheat. In conclusion, considerable progress has been made regarding understanding the physiological and genetic mechanisms of Al tolerance in wheat, but much still needs to be learned in relation to the molecular mechanisms which is far behind among the cereals particularly rice.

## References

Aniol A, Gustafson JP (1984) Chromosome location of genes controlling aluminum tolerance in wheat, rye, and triticale. Can J Genet Cytol 26:701–705

Archambault DJ, Zhang G, Taylor GJ (1996) Accumulation of Al in root mucilage of an Al-resistant and an Al-sensitive cultivar of wheat. Plant Physiol 112:1471–1478

Basu U, Basu A, Taylor GJ (1994) Differential exudation of polypeptides by roots of aluminum-resistant and aluminum-sensitive cultivars of *Triticum aestivum* L. in response to aluminium stress. Plant Physiol 106:151–158

Basu U, McDonald-Stephens JL, Archambault DJ, Good AG, Briggs KG, Aung T, Taylor GJ (1997) Genetic and physiological analysis of doubled-haploid, aluminum-resistant lines of wheat provide evidence for the involvement of a 23 kD, root exudates polypeptide in mediating resistance. Plant Soil 196:283–288

Basu U, Good AG, Aung T, Slaski JJ, Basu A, Briggs KG, Taylor GJ (1999) A 23-kDa, root exudates polypeptide co-segregates with aluminum resistance in *Triticum aestivum*. Physiol Plant 106:53–61

Basu U, Good AG, Taylor GJ (2001) Transgenic *Brassica napus* plants overexpressing aluminum-induced mitochondrial maganese superoxide dismutase cDNA are resistant to aluminium. Plant Cell Environ 24:1269–1278

Beckman I (1954) Sobre o cultivo e melhoramento do trigo (*Triticum vulgare*, Vill) no sul do Brasil. Agron Sul Rio Grandense 1:64–72 (in Portuguese)

Berzonsky WA, Kimber G (1986) Tolerance of *Triticum* species to Al. Plant Breed 97:275–278

Blamey FPC, Edmeades DC, Wheeler DM (1990) Role of root cation exchange capacity in different aluminum tolerance of Lotus species. J Plant Nutr 13:729–744

Blamey FPC, Robinson NJ, Asher CJ (1993) Interspecific differences in aluminium tolerance in relation to root cation-exchange capacity. Dev Plant Soil Sci 50:91–96

Brautigan DJ, Rengasamy P, Chittleborough DJ (2012) Aluminium speciation and phytotoxicity in alkaline soils. Plant Soil 360:187–196

Bray EA, Bailey-Serres J, Weretilnyk E (2000) Responses to abiotic stresses. In: Gruissem W, Buchannan B, Jones R (eds) Responses to abiotic stresses. American Society of Plant Physiologists, Rockville, MD, pp 1158–1249

Cai S, Bai GH, Zhang D (2008) Quantitative trait loci for aluminum resistance in Chinese wheat landrace FSW. Theor Appl Genet 117:49–56

Camargo CEO (1981) Wheat improvement. I. The heritability of tolerance to aluminum toxicity. Bragantia 40:33–45 (in Portuguese)

Camargo CEO (1984) Wheat improvement. VI. Heritability studies on aluminum tolerance using three concentrations of aluminum in nutrient solutions. Bragantia 43:279–291 (in Portuguese)

Campbell LG, Lafever HN (1981) Heritability of aluminum tolerance in wheat. Cereal Res Commun 9:281–287

Carver BF, Ownby JD (1995) Acid soil tolerance in wheat. Adv Agron 54:117–173

Chalmers KJ, Campbell AW, Kretschmer J, Karakousis A, Henschke PH, Pierens S, Harker N, Pallotta M, Cornish GB, Shariflou MR, Rampling LR, McLauchlan A, Daggard G, Sharp PJ,

Holton TA, Sutherland MW, Appels R, Langridge P (2001) Construction of three linkage maps in bread wheat, *Triticum aestivum*. Aust J Agric Res 52:1089–1119

Choudhry MA (1978) Genetic differences for aluminum tolerance in five wheat crosses. Agr Abstr Madison, Am Soc Agr, 151

Christiansen-Weniger CI, Groneman AF, van Veen JA (1992) Associative $N_2$ fixation and root exudation of organic acids from wheat cultivars of different aluminum tolerance. Plant Soil 139:167–174

Dai J, Bai G, Zhang D, Hong D (2013) Validation of quantitative trait loci for aluminum tolerance in Chinese wheat landrace FSW. Euphytica 192:171–179

de la Fuente JM, Ramırez-Rodrıguez V, Cabrera-Ponce JL, Herrera-Estrella L (1997) Aluminum tolerance in transgenic plants by alteration of citrate synthesis. Science 276:1566–1568

Delhaize E, Ryan PR (1995) Aluminum toxicity and tolerance in plants. Plant Physiol 107:315–321

Delhaize E, Ryan PR, Randall PJ (1993a) Aluminum tolerance in wheat (*Triticum aestivum* L.) (II. Aluminum-stimulated excretion of malic acid from root apices). Plant Physiol 103:695–702

Delhaize E, Craig S, Beaton CD, Bennet RJ, Jagadish VC, Randall PJ (1993b) Aluminum tolerance in wheat (*Triticum aestivum* L.) I. Uptake and distribution of aluminum in root apices. Plant Physiol 103:685–693

Delhaize E, Ryan PR, Hebb DM, Yamamoto Y, Sasaki T, Matsumoto H (2004) Engineering high level aluminium tolerance in barley with the ALMT1 gene. Proc Natl Acad Sci U S A 101:15249–15254

Delhaize E, James RA, Ryan PR (2012a) Aluminium tolerance of root hairs underlies genotypic differences in rhizosheath size of wheat (*Triticum aestivum*) grown on acid soil. New Phytol 195:609–619

Delhaize E, Ma JF, Ryan PR (2012b) Transcriptional regulation of aluminium tolerance genes. Trends Plant Sci 17:341–348

Dixon J, Braun HJ, Kosina P, Crouch J (2009) Wheat facts and futures – 2007. CIMMYT, Mexico City

Elmayan T, Tepfer M (1995) Evaluation in tobacco of the organ specificity and strength of the rolD promoter, domain A of the 35S promoter and the 35S2 promoter. Transgenic Res 4:388–396

Ezaki B, Gardner RC, Ezaki Y, Matsumoto H (2000) Expression of aluminum-induced genes in transgenic *Arabidopsis* plants can ameliorate Al stress and/or oxidative stress. Plant Physiol 122:657–665

Ezaki B, Jataran K, Higashi A, Takahashi K (2013) A combination of five mechanisms confers a high tolerance for aluminum to a wild species of Poaceae, *Andropogon virginicus* L. Environ Exp Bot 93:35–44

Famoso AN, Zhao K, Clark RT, Tung CW, Wright MH, Bustamante C, Kochian LV, McCouch SR (2011) Genetic architecture of aluminum tolerance in rice (*Oryza sativa*) determined through genome-wide association analysis and QTL mapping. PLoS Genet 7:e1002221

Foy CD, Fleming AL, Burns GR, Armiger WH (1967) Characterization of differential aluminum tolerance among varieties of wheat and barley. Soil Sci Soc Am Proc 31:513–521

Foy CD, Lee EH, Coradetti CA, Taylor GJ (1990) Organic acids related to differential aluminum tolerance in wheat (*Triticum aestivum*) cultivars. In: Beusichem ML (ed) Plant nutrition-physiology and application. Kluwer, Dordrecht, pp 381–389

Furuichi T, Sasaki T, Tsuchiya Y, Ryan PR, Delhaize E, Yamamoto Y (2010) Extracellular hydrophilic carboxy-terminal domain regulates the activity of TaALMT1, the aluminum-activated malate transport protein of wheat. Plant J 64:47–55

Furukawa J, Yamaji N, Wang H, Mitani N, Murata Y, Sato K, Katsuhara M, Takeda K, Ma JF (2007) An aluminum-activated citrate transporter in barley. Plant Cell Physiol 48:1081–1091

Garcia-Oliveira AL, Benito C, Prieto P, de Andrade MR, Rodriques-Pousada C, Guedes-Pinto H, Martins-Lopes P (2013) Molecular characterization of TaSTOP1 homoeologues and their response to aluminium and proton ($H^+$) toxicity in bread wheat (*Triticum aestivum* L.). BMC Plant Biol 13:134

Garcia-Oliveira AL, Martins-Lopes P, Tolrá R, Poschenrieder C, Tarquis M, Guedes-Pinto H, Benito C (2014) Molecular characterization of the citrate transporter gene *TaMATE1* and expression analysis of upstream genes involved in organic acid transport under Al stress in bread wheat (*Triticum aestivum* L.). Physiol Plant 152:441–452

Garcia-Oliveira AL, Chander S, Barceló J, Poschenrieder C (2015) Aluminium stress in crop plants. In: Yadav P, Kumar S, Jain V (eds) Recent advances in plant stress physiology. Astral Int Publ (in press)

Granados G, Pandey S, Ceballos H (1993) Response to selection for tolerance to acid soilsin tropical maize population. Crop Sci 33:936–940

Guerinot ML (2007) It's elementary: enhancing $Fe^{3+}$ reduction improves rice yield. Proc Natl Acad Sci U S A 104:7311–7312

Hamel F, Breton C, Houde M (1998) Isolation and characterization of wheat aluminum-regulated genes: possible involvement of aluminum as a pathogenesis response elicitor. Planta 205:531–538

Han C, Dai SF, Liu DC, Pu ZJ, Wei YM, Zheng YL, Wen DJ, Zhao L, Yan ZH (2013) TaALMT1 promoter sequence compositions, acid tolerance, and Al tolerance in wheat cultivars and landraces from Sichuan in China. Genet Mol Res 12:5602–5616

Henderson M, Ownby JD (1991) The role of root cap mucilage secretion in aluminum tolerance in wheat. Biochem Curr Topics Plant Physiol 10:134–141

Horst WJ, Wagner A, Marschner H (1982) Mucilage protects root meristems from aluminium injury. Z Pflanzenphysiol 105:435–444

Hu SW, Bai GH, Carver BF, Zhang DD (2008) Diverse origins of aluminum-resistance sources in wheat. Theor Appl Genet 118:29–41

Kerridge PC, Kronstad WE (1968) Evidence of genetic resistance to aluminium toxicity in wheat (*Triticum aestivum* vill., Host). Agron J 60:710–711

Kinraide TB (1988) Proton extrusion by wheat roots exhibiting severe aluminum toxicity symptoms. Plant Physiol 88:418–423

Kinraide TB, Ryan PR, Kochian LV (1992) Interactive effects of $Al^{3+}$, $H^+$ and other cations on root elongation considered in terms of cell-surface electrical potential. Plant Physiol 99:1461–1468

Kochian LV (1995) Cellular mechanisms of aluminium toxicity and resistance in plants. Annu Rev Plant Physiol Plant Mol Biol 46:237–260

Lafever HN, Campbell LG, Foy CD (1977) Differential response of wheat cultivars to Al. Agron J 69:563–568

Ligaba A, Kochian L, Pineros M (2009) Phosphorylation at S384 regulates the activity of the TaALMT1 malate transporter that underlies aluminum resistance in wheat. Plant J 60:411–423

Ligaba A, Dreyer I, Margaryan A, Schneider DJ, Kochian L, Piñeros M (2013) Functional, structural and phylogenetic analysis of domains underlying the Al sensitivity of the aluminum-activated malate/anion transporter, TaALMT1. Plant J 76:766–780

Ma G, Rengasamy P, Rathgen J (2003) Phytotoxicity of aluminium to wheat plants in high-pH solutions. Aust J Exp Agric 43:497–501

Ma HX, Bai GH, Carver BF, Zhou LL (2005) Molecular mapping of a quantitative trait locus for aluminum tolerance in wheat cultivar Atlas 66. Theor Appl Genet 112:51–57

Ma HX, Bai GH, Lu WZ (2006) Quantitative trait loci for aluminum resistance in wheat cultivar Chinese Spring. Plant Soil 283:239–249

Magalhães JV, Liu J, Guimarães CT, Lana UGP, Alves VMC, Wang YH, Schaffert RE, Hoekenga OA, Piñeros MA, Shaff JE, Klein PE, Carneiro NP, Coelho CM, Trick HN, Kochian LV (2007) A gene in the multidrug and toxic compound extrusion (MATE) family confers aluminum tolerance in sorghum. Nat Genet 39:1156–1161

Martins-Lopes P, Maças B, Guedes-Pinto H (2009) Portuguese bread wheat germplasm evaluation for aluminium tolerance. Cereal Res Commun 37:179–188

Massot N, Nicander B, Barceló J, Poschenrieder C, Tillberg EE (2002) A rapid increase in cytokinin levels and enhanced ethylene evolution precede $Al^{3+}$-induced inhibition of root growth in bean seedlings (*Phaseolus vulgaris* L.). Plant Growth Regul 37:105–112

Miller TE, Iqel N, Readers SM, Mahmood A, Cant KA, King IP (1997) A cytogenetic approach to the improvement of aluminium tolerance in wheat. New Phytol 137:93–98

Mohammed YSA, Eltayeb AE, Tsujimoto H (2013) Enhancement of aluminum tolerance in wheat by addition of chromosomes from the wild relative *Leymus racemosus*. Breeding Sci 63:407–416

Navakode S, Weidner A, Lohwasser U, Röder MS, Börner A (2009) Molecular mapping of quantitative trait loci (QTLs) controlling aluminium tolerance in bread wheat. Euphytica 166:283–290

Navakode S, Neumann K, Kobiljski B, Lohwasser U, Borner A (2014) Genome wide association mapping to identify aluminium tolerance loci in bread wheat. Euphytica 198:401–411

Osawa H, Matsumoto H (2001) Possible involvement of protein phosphorylation in aluminum-responsive malate efflux from wheat root apex. Plant Physiol 126:411–420

Ownby JD, Popham HR (1990) Citrate reverses the inhibition of wheat root growth caused by aluminum. J Plant Physiol 135:588–591

Papernik LA, Bethea AS, Singleton TE, Magalhães JV, Garvin DF, Kochian LV (2001) Physiological basis of reduced Al tolerance in ditelosomic lines of Chinese spring wheat. Planta 212:829–834

Parker DR (1995) Root growth analysis: an underutilised approach to understanding aluminium rhizotoxicity. Plant Soil 171:151–157

Pellet DM, Grunes DL, Kochian LV (1996) Multiple aluminum-resistance mechanisms in wheat. Roles for root apical phosphate and malate exudation. Plant Physiol 112:591–597

Pereira JF, Zhou G, Delhaize E, Richardson T, Ryan PR (2010) Engineering greater aluminium resistance in wheat by over-expressing TaALMT1. Ann Bot 106:205–214

Pinto-Carnide O, Guedes-Pinto H (1999) Aluminium tolerance variability in rye and wheat Portuguese germplasm. Genet Resour Crop Evol 46:81–85

Polle E, Konzak CF, Kittrick JA (1978) Visual detection of aluminium tolerance levels in wheat by hematoxylin staining of seedling roots. Crop Sci 18:823–827

Puthota V, Cruz-Ortega R, Jonson J, Ownby J (1991) An ultrastructural study of the inhibition of mucilage secretion in the wheat root cap by aluminum. In: Wright RJ, Baligar VC, Murrmann RP (eds) Plant-soil interactions at low pH. Kluwer, Dordrecht, pp 779–789

Rajaram S, Pfeiffer W, Singh R (1988) Developing bread wheat for acid soils through shuttle breeding. In: Kohli MM, Rajaram S (eds) Wheat breeding in acid soils. Review of Brazilian/CIMMYT Collaboration, 1974–1986, pp 51–58

Raman H, Zhang KR, Cakir M, Appels R, Garvin DF, Maron LG, Kochian LV, Moroni JS, Raman R, Imtiaz M, Drake-Brockman F, Waters I, Martin P, Sasaki T, Yamamoto Y, Matsumoto H, Hebb DM, Delhaize E, Ryan PR (2005) Molecular characterization and mapping of ALMT1, the aluminium-tolerance gene of bread wheat (*Triticum aestivum* L.). Genome 48:781–791

Raman H, Ryan PR, Raman R, Stodart BJ, Zhang K, Martin P, Wood R, Sasaki T, Yamamoto Y, Mackay M, Hebb DM, Delhaize E (2008) Analysis of TaALMT1 traces the transmission of aluminum resistance in cultivated common wheat (*Triticum aestivum* L.). Theor Appl Genet 116:343–354

Raman H, Stodart B, Ryan PR, Delhaize E, Emebiri L, Raman R, Coombes N, Milgate A (2010) Genome-wide association analyses of common wheat (*Triticum aestivum* L.) germplasm identifies multiple loci for aluminium resistance. Genome 53:957–966

Ramírez-Benítez JE, Muñoz-Sánchez JA, Becerril-Chi KM, Miranda-Ham ML, Castro-Concha LA, Hernández-Sotomayor SMT (2011) Aluminum induces changes in oxidative burst scavenging enzymes in Coffea arabica L. suspension cells with differential Al tolerance. J Inorg Biochem 105:1523–1528

Rengel Z (1992) Disturbance of $Cs^{2+}$ homeostasis as a primary trigger in the Al toxicity syndrome. Plant Cell Environ 15:931–938

Ribeiro-Carvalho C, Guedes-Pinto H, Heslop-Harrison JS, Schwarzacher T (2001) Introgression of rye chromatin on chromosome 2D in the Portuguese wheat landrace 'Barbela'. Genome 44:1122–1128

Riede CR, Anderson JA (1996) Linkage of RFLP markers to an aluminum tolerance gene in wheat. Crop Sci 36:905–909

Rincón M, Gonzales RA (1992) Aluminum partitioning in intact roots of aluminum-tolerant and aluminum-sensitive wheat (*Triticum aestivum* L.) cultivars. Plant Physiol 99:1021–1028

Ryan PR, Shaff JE, Kochian LV (1992) Correlation among ionic currents, ion fluxes, and root elongation in aluminum-sensitive and aluminum-tolerant wheat cultivars. Plant Physiol 99:1193–1200

Ryan PR, Delhaize E, Randall PJ (1995a) Characterization of Al stimulated efflux of malate from the apices of Al-tolerant wheat roots. Planta 196:103–110

Ryan PR, Delhaize E, Randall P (1995b) Malate efflux from root apices and tolerance to aluminium are highly correlated in wheat. Funct Plant Biol 22:531–536

Ryan PR, Delhaize E, Jones DL (2001) Function and mechanism of organic anion exudation from plants. Annu Rev Plant Physiol Plant Mol Biol 52:527–560

Ryan PR, Raman H, Gupta S, Horst WJ, Delhaize E (2009) A second mechanism for aluminum resistance in wheat relies on the constitutive efflux of citrate from roots. Plant Physiol 149:340–351

Ryan PR, Tyerman SD, Sasaki T, Furuichi T, Yamamoto Y, Zhang WH, Delhaize E (2011) The identification of aluminium-resistance genes provides opportunities for enhancing crop production on acid soils. J Exp Bot 62:9–20

Samuels TD, Kucukakyuz K, Rincon-Zachary M (1997) Al partitioning patterns and root growth as related to al sensitivity and al tolerance in wheat. Plant Physiol 113:527–534

Sasaki T, Yamamoto Y, Ezaki B, Katsuhara M, Ahn SJ, Ryan PR, Delhaize E, Matsumoto H (2004) A wheat gene encoding an aluminum-activated malate transporter. Plant J 37:645–653

Sasaki T, Ryan PR, Delhaize E, Hebb DM, Ogihara Y, Kawaura K, Noda K, Kojima T, Toyoda A, Matsumoto H, Yamamoto Y (2006) Sequence upstream of the wheat (Triticum aestivum L.) ALMT1 gene and its relationship to aluminum resistance. Plant Cell Physiol 47:1343–1354

Sears ER (1954) The aneuploids of common wheat. Missouri Agricultural Experimental Station Research Bulletin 572:1–58

Silva JP, Reboredo F, Guedes-Pinto H, Mello-Sampayo T (1991) 'Barbela', a bread wheat cultivar tolerant to aluminum. Brotéria-Genética 12(87):65–68

Silva S, Pinto-Carnide O, Martins-Lopes P, Matos M, Guedes-Pinto H, Santos C (2010) Differential aluminium changes on nutrient accumulation and root differentiation in an Al sensitive vs. tolerant wheat. Environ Exper Bot 68:91–98

Snowden KC, Gardner RC (1993) Five genes induced by aluminum in wheat (*Triticum aestivum* L.) roots. Plant Physiol 103:855–861

Stass A, Smit I, Eticha D, Oettler G, Horst WJ (2008) The significance of organic anion exudation for the aluminum resistance of primary triticale derived from wheat and rye parents differing in aluminum resistance. J Plant Nutr Soil Sci 171:634–642

Tang Y, Garvin DF, Kochian LV, Sorrells ME, Carver BF (2002) Physiological genetics of aluminum tolerance in the wheat cultivar Atlas 66. Crop Sci 42:1541–1546

Tice KR, Parker DR, DeMason DA (1992) Operationally defined apoplastic and symplastic aluminum fractions in root tips of aluminumin toxicated wheat. Plant Physiol 100:309–318

Tokizawa M, Kobayashi Y, Saito T, Kobayashi M, Iuchi S, Nomoto M, Tada Y, Yamamoto YY, Koyama H (2015) STOP1, CAMTA2 and other transcription factors are involved in aluminum-inducible AtALMT1 expression. Plant Physiol doi:10.1104/pp.114.256552

Tolrá R, Barcelo J, Poschenrieder C (2009) Constitutive and aluminium induced patterns of phenolic compounds in two maize varieties differing in aluminium tolerance. J Inorg Biochem 103:1486–1490

Tovkach A, Ryan PR, Richardson AE, Lewis DC, Rathjen TM, Ramesh S, Tyerman SD, Delhaize E (2013) Transposon mediated alteration of TaMATE1B expression in wheat confers constitutive citrate efflux from root apices. Plant Physiol 161:880–892

Wallace SU, Anderson IC (1984) Aluminum toxicity and DNA synthesis in wheat roots. Agron J 76:5–8

Whitten M (1997) Subsurface acidification: estimation lime requirements from lime dissolution rates in the field. In: Williamson DR (ed) Proceedings of the fourth triennial Western Australian soil science conference, African Reef Resort, Geraldton, Western Australia, pp 128–131

Xu FJ, Li G, Jin CW, Liu WJ, Zhang SS, Zhang YS, Lin XY (2012) Aluminum-induced changes in reactive oxygen species accumulation, lipid peroxidation and antioxidant capacity in wheat root tips. Biol Plantarum 56:89–96

Yamaguchi M, Sasaki T, Sivaguru M, Yamamoto Y, Osawa H, Ahn SJ, Matsumoto H (2005) Evidence for the plasma membrane localization of Al-activated malate transporter (ALMT1). Plant Cell Physiol 46:812–816

Yamaji N, Huang CF, Nagao S, Yano M, Sato Y, Nagamura Y, Ma JF (2009) A zinc finger transcription factor ART1 regulates multiple genes implicated in aluminum tolerance in rice. Plant Cell 21:3339–3349

Yamamoto Y, Kobayashi Y, Matsumoto H (2001) Lipid peroxidation is an early symptom triggered by aluminum, but not the primary cause of elongation inhibition in pea roots. Plant Physiol 125:199–208

Zhang G, Taylor GJ (1988) Effect of aluminum on growth and distribution of aluminum in tolerant and sensitive cultivars of *Triticum aestivum* L. Commun Soil Sci Plant Anal 19:1195–1205

Zhang G, Taylor GJ (1989) Kinetics of aluminum uptake by excised roots of aluminum-tolerant and aluminum-sensitive cultivars of *Triticum aestivum* L. Plant Physiol 91:1094–1099

Zhang G, Taylor GJ (1990) Kinetics of aluminum uptake in *Triticum aestivum* L. Identity of the linear phase of aluminum uptake by excised roots of aluminum-tolerant and aluminum-sensitive cultivars. Plant Physiol 94:577–584

Zhou LL, Bai GH, Ma HX, Carver BF (2007) Quantitative trait loci for aluminum resistance in wheat. Mol Breed 19:153–161

Zhou GF, Delhaize E, Zhou M, Ryan PR (2013) The barley MATE gene, HvAACT1, increases citrate efflux and Al tolerance when expressed in wheat and barley. Ann Bot 112:603–612

# Rice Arsenal Against Aluminum Toxicity

**Rafael Augusto Arenhart, Lauro Bucker-Neto, Rogerio Margis, Zhi-Yong Wang, and Marcia Margis-Pinheiro**

**Abstract** One of the major constraints on crop production is the ability of plants to grow in acidic soils, where aluminum (Al) is soluble in its toxic form ($Al^{3+}$). However, some plants can address this Al toxicity by utilizing different strategies such as exclusion (an external mechanism) and detoxification (an internal mechanism). Rice, an important food source, is one of the most Al-tolerant crops, but the mechanism of this tolerance is not well understood. In this review, we provide an overview of Al-tolerance mechanisms in rice and show that this species can employ several strategies that together provide tolerance to Al toxicity.

## 1 Introduction

Under acidic conditions, aluminum (Al), which is a major worldwide environmental concern, is solubilized in its trivalent ionic form ($Al^{3+}$) and interacts with plants. This phenomenon causes major problems in agriculture because Al is the most abundant metal in the soil and because up to 50 % of the world's arable land has acidic soil (Von Uexkull and Mutert 1995).

A plant's root apex plays a major role in Al recognition and response and also serves as the primary site for Al toxicity (Kollmeier et al. 2000). The first symptom of Al toxicity is the inhibition of root growth, which is caused by the inhibition of cell wall expansion and elongation and, after long exposures, cell wall division (Matsumoto 2000; Panda et al. 2009). The effects of these injuries include poor uptake of water and nutrients, the disruption of cytosolic calcium ($Ca^{2+}$) and proton activity ($H^+$), oxidative stress, and other problems; these consequences are discussed in detail in Kochian et al. (2005).

R.A. Arenhart • L. Bucker-Neto • M. Margis-Pinheiro (✉)
Departamento de Genética, Universidade Federal do Rio Grande do Sul, Avenida Bento Gonçalves 9500, sala 207, prédio 43312, 91501–970 Porto Alegre, Brasil
e-mail: marcia.margis@ufrgs.br

R. Margis
Departamento de Biofísica e Centro de Biotecnologia, Universidade Federal do Rio Grande do Sul, Porto Alegre, Brasil

Z.-Y. Wang
Department of Plant Biology, Carnegie Institution for Science, Stanford, CA 94305, USA

 
S.K. Panda, F. Baluška (eds.), *Aluminum Stress Adaptation in Plants*, Signaling and Communication in Plants 24, DOI 10.1007/978-3-319-19968-9_8

Nevertheless, through the course of evolution, plants have developed mechanisms that allow them to cope with Al in the soil. Two main types of mechanisms account for Al tolerance: external detoxification systems and internal detoxification systems. External detoxification, which is the better-documented mechanism, occurs via the efflux of organic acid anions (malate, citrate, and oxalate) from the roots; these anions form chelating complexes that prevent the entrance of Al into the cells. Internal detoxification mechanisms occur by the chelation of Al with organic acid anions and sequestration in vacuoles. For a detailed review of these Al-tolerance mechanisms, readers can consult (Kochian et al. 2002; Hoekenga and Magalhaes 2011).

Rice (*Oryza sativa*—Poaceae) is the most Al-tolerant crop under field conditions (Foy 1988) and the most Al-tolerant cereal grass under controlled conditions (Famoso et al. 2010). However, an explanation of the exact mechanism of this resistance is still lacking. Both Al internalization and Al exclusion seem to occur in rice (Xia et al. 2010; Yokosho et al. 2011). Indeed, rice is up to six times more tolerant of Al than other members of the Poaceae family, such as maize, wheat, and sorghum (Famoso et al. 2010), suggesting that Al tolerance among these grasses was acquired due to selection during the domestication process.

One can hypothesize that Al-tolerance traits could have been selected in these grasses according to their origin of domestication. For example, maize was domesticated in Mexico from *Balsas teosinte* approximately 8000 years ago (Doebley 2004; Doebley et al. 2006), whereas wheat was domesticated in the Fertile Crescent from *Triticum dicoccoides* approximately 10,000 years ago (Ozkan et al. 2002), and the domestication of cultivated rice occurred in southern China (Huang et al. 2012). A soil pH map[1] indicates that the regions where cultivated rice was domesticated have acidic soil and that the regions where maize and wheat were domesticated have neutral to basic soils (Fig. 1a).

Within *O. sativa*, two main subspecies from several genetically differentiated variety groups exist: japonica and indica (Garris et al. 2005). Based on genetic evidence, all rice varietal groups descended from the wild species *Oryza rufipogon* (Huang et al. 2012). *O. sativa* japonica rice was first domesticated from a specific population of *O. rufipogon* around the center of the Pearl River Basin in southern China, and *O. sativa* indica rice was subsequently developed from crosses between japonica rice and local wild rice as the initial cultivars spread into Southeast and southern Asia (Huang et al. 2012). *O. rufipogon* is, in general, very tolerant to Al (Nguyen et al. 2003; Cao et al. 2011), whereas *O. sativa* japonica and *O. sativa* indica are less tolerant. In all, Indica subspecies are less tolerant of Al than japonica subspecies (Ma et al. 2002; Famoso et al. 2011). Molecular evidence shows that the japonica group has less genetic diversity than the indica group (Garris et al. 2005).

[1] The soil pH map (from 2000 to 2010) was retrieved from www.globalsoilmap.com and does not represent the pH of the soil 10,000 years ago, during the period in which rice was domesticated. However, until the 1800s, most acidic soil remained untouched and under forest cover. There was only some encroachment into regions with acidic soils in densely populated regions of the world, such as East Asia (Von Uexkull and Mutert 1995). For the map source, please see Hengl (2009).

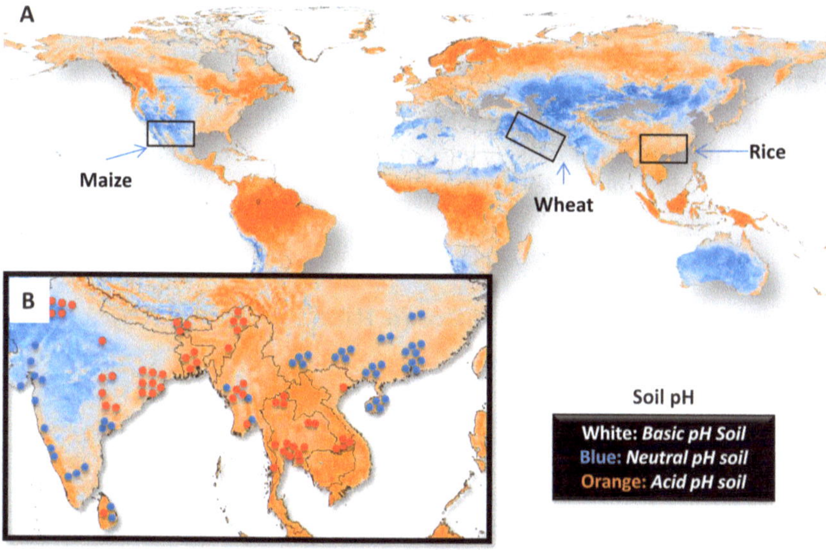


**Fig. 1** Global map of soil pH. (**a**) Map representing the mean soil pH, predicted using a correlation with world maps (Hengl 2009). The origins and location of maize, wheat, and rice cultivation are indicated by *black boxes*. Maize and wheat were first cultivated in soil with a neutral to basic pH, while rice was cultivated in acidic soils. (**b**) The *O. rufipogon* accessions that originated *O. Sativa* (Japonica) are shown as *blue spots*, and the *O. rufipogon* accessions that originated *O. Sativa* (*Indica*) are shown as *red spots*. Magnification of the map showing the soil pH in south Asia. The *spot dots* are based on data retrieved from Huang et al. (2012)

A good explanation for this phenomenon was proposed by Kovach et al. (2007). A climate change resulted in the return of glacial-like conditions across Northern Asia from 11,500 to 13,000 years ago (Higham 2002; Lu et al. 2002). The colder weather would have eliminated a large portion of the japonica-like wild ancestors in the Yangtze River Valley. Humans were forced to rely on a narrowing gene pool, forcing a more rapid movement toward the domestication of rice in this region. In contrast, the warmer tropics of South and Southeast Asia would have maintained larger, more diverse populations of indica-like *O. rufipogon*, which could have been foraged by humans for a longer time, resulting in a more gradual domestication process. Crosses between *O. rufipogon* and japonica from this region may have contributed to the less Al-tolerant indica group (Fig. 1b).

Another clue for this comes from Zhao et al. (2013). Nitrogen (N), which plays a crucial role in plant growth, is present in the soils as ammonium and nitrate (two major types of inorganic N sources), with ammonium being more available in acidic soils and nitrate being more available in neutral to basic soils. Indica rice cultivars are generally Al-sensitive and nitrate-preferring, while Japonica cultivars are Al-tolerant and relatively ammonium-preferring. This Al tolerance was negatively correlated with their nitrate preference, suggesting that Al tolerance in rice is antagonistic with nitrate preference and synergistic with ammonium preference under acidic conditions (Zhao et al. 2013). However, acidic soils are distributed in regions where both japonica and indica can grow, suggesting that other factors besides N and Al could have driven the differentiation of japonica and indica and should therefore be evaluated (Zhao et al. 2013).

Despite the lower degree of Al tolerance of the Indica group compared to the japonica group, subspecies of the indica group are still more Al-tolerant than other grasses. How indica subspecies became less tolerant to Al is still an open question.

## 2 Mechanisms of Al Tolerance in Rice

### *2.1 Organic Acid Release*

The ability of rice to tolerate toxic levels of Al began to be understood in the end of the 1980s (Foy 1988). At the beginning of the 2000s, rice researchers focused on searching for QTLs for Al tolerance (Ma et al. 2002; Nguyen et al. 2002, 2003). The first study regarding Al tolerance in rice demonstrated that the japonica variety was more Al-tolerant than the Indica variety and that japonica accumulated less Al in the root apex, indicating that Al exclusion rather than internal detoxification played major role in rice (Ma et al. 2002). However, recent evaluations of the Al concentration in the root apex showed no difference between japonica and indica varietal groups, demonstrating no relationship between Al exclusion and Al tolerance in rice (Famoso et al. 2010). Organic citrate is secreted in response to Al in both japonica and indica, and the cultivars do not differ significantly in the amount of secreted citrate; there is also no evidence of citrate secretion in response to increasing Al concentrations. Indeed, the amount of citrate that is secreted by rice was not significant, being one tenth of the citrate secreted by rye (Ma et al. 2002), which has a similar Al tolerance to rice (Famoso et al. 2010). This outcome implies that mechanisms to reduce or to mask the toxic forms of Al in the apoplast and symplast other than citrate release may determine the degree of Al tolerance in rice. Meanwhile, it was shown that the expression level of an Al-induced citrate transporter is positively correlated with the amount of citrate secretion in rice cultivars that differ in their Al tolerance (Yokosho et al. 2011).

This contradiction between rice citrate levels and Al tolerance may be explained by the following hypothesis: one portion of the citrate may function in Al exclusion, forming Al-citrate complexes in the rhizosphere, and another portion may function in Al sequestration into vacuoles. Even though the primary mechanism is Al exclusion, rice can tolerate certain intracellular amounts of Al. For example, an Al-tolerant rice cultivar (Pusa Basmati) accumulated more Al in the roots than an Al-sensitive cultivar (Vikas) (Meriga et al. 2004). Indeed, rice plants accumulate Al (Xia et al. 2010) and sequester it into vacuoles by a specific transporter (Huang et al. 2011). The Al hyperaccumulator *Hydrangea macrophylla* detoxifies Al by forming complexes with citrate in a 1:1 molar ratio in the leaves (Ma et al. 1997). In

*Fagopyrum esculentum*, another Al hyperaccumulator, Al is bound to oxalate ions in the roots and leaves and to citrate ions in the xylem (Ma et al. 1997; Zheng et al. 1998). The most studied Al hyperaccumulators usually use citrate–Al complexes in the xylem, raising the hypothesis that this is a major Al transport route (Grevenstuk and Romano 2013).

## 2.2 The Antioxidant Defense System, Cell Wall and Plasma Membrane Composition

Under adverse environmental conditions, plants produce increased reactive oxygen species (ROS), leading to the oxidation of biological macromolecules and, as consequence, to lipid peroxidation, membrane damage, and enzyme inactivation. To alleviate oxidative injury, nonenzymatic systems (such as reduced glutathione (GSH), ascorbic acid (AsA), carotenoids, and phenolics) and enzymatic systems [such as superoxide dismutase (SOD), ascorbate peroxidase (APx), catalase (CAT), peroxidase (POD), glutathione reductase (GR), and glutathione POD (GPX)] are activated (Ma et al. 2012). Al stress generates ROS in rice plants and consequently increases the levels of ROS scavenging proteins (Sharma and Dubey 2007; Pandey et al. 2013). Under low concentrations of Al (up to 40 μM), a rice Al-tolerant cultivar will show higher levels of ROS-scavenging enzymes compared to an Al-sensitive cultivar; this will be accompanied by the lignification of the roots in an Al-sensitive cultivar, most likely due to the failure to scavenge ROS products such as hydrogen peroxide ($H_2O_2$) (Ma et al. 2012).

Moreover, rice plants in which cytosolic APx1/2 is silenced show a higher tolerance of moderate Al concentrations (up to 150 μM) compared to wild type plants (Rosa et al. 2010). However, at high Al concentrations (up to 750 μM), APx1/2-silenced plants become Al sensitive because the low levels of cytosolic APx are not sufficient to compensate for more stressful conditions, resulting in a more Al-sensitive phenotype (Arenhart et al. 2013).

Lipid peroxidation, a symptom of Al toxicity, varies between Al-tolerant and Al-sensitive cultivars. Two scenarios have been proposed: (1) the lipid composition of plants may be variable, making some less susceptible to peroxidation, or (2) plants with highly effective Al exclusion mechanisms suffer less lipid peroxidation because less $Al^{3+}$ reaches the plasma membrane (Hoekenga and Magalhaes 2011). However, decreases in the activities of SOD and APX due to prolonged exposure to Al lead to DNA damage; this suggests that lipid peroxidation is a consequence rather than the cause of Al injury to plant roots (Meriga et al. 2004). The exogenous application of magnesium (Mg), calcium (Ca), salicylic acid (Sa) (Pandey et al. 2013), or nitric oxide (NO) reduced the toxicity of Al in rice seedlings by regulating the expression and activity of antioxidant enzyme activities and reducing ROS levels, but with the exception of NO, the relationship between rice Al tolerance and the endogenous concentrations of these molecules has not been described (Yang et al. 2013).

Nitric oxide, an important signaling molecule, alleviates Al-induced oxidative stress in *Phaseolus vulgaris* and *Cassia tora* (Wang and Yang 2005; Wang et al. 2010). However, endogenous NO responses to Al seem to be related to higher degrees of Al tolerance. For example, in Arabidopsis root apex transition zone, local release of large amounts of NO is blocked by Al treatment (Illés et al. 2006). Nevertheless, in rice, NO increases significantly after Al treatment (Yang et al. 2013). Moreover, rice seedlings that were pretreated with sodium nitroprusside (SNP, a NO donor) were more tolerant of Al treatment (Zhang et al. 2011). Despite the fact that genetic analyses have not implicated ROS scavenging genes or their regulators in natural variation in Al tolerance (Hoekenga and Magalhaes 2011), the ROS scavenging system appears to contribute to internal and external Al tolerance in rice.

Rice plants that were pretreated with SNP became more tolerant of Al and had lower pectin and hemicellulose levels, lower Al accumulation in the root tips and cell walls, a higher degree of methylation of pectin, and a lower cell wall Al-binding capacity than roots that were exposed to Al but not pretreated with SNP (Zhang et al. 2011). Consistent with this, the levels of cell wall polysaccharides (pectin, hemicellulose 1 and 2) in the root apex were reported to be significantly higher in an Al-sensitive cultivar than in an Al-tolerant cultivar in the absence of Al, and Al exposure increased the root apex hemicellulose content more significantly in an Al-sensitive cultivar (Yang et al. 2008). Furthermore, root cell wall pectin methyl esterase activity was constitutively higher in an Al-sensitive cultivar than in an Al-tolerant one, and this was accompanied by a higher proportion of demethylated pectins. The Al adsorption and desorption kinetics of the root tip cell wall also indicated that more Al was adsorbed, and the Al was more tightly bound in Al-sensitive plants. These data were consistent with Al content, pectin methylesterase activity, and pectin demethylesterification, suggesting that cell wall polysaccharides are important in Al exclusion, specifically from the rice root apex (Yang et al. 2008). Moreover, a rice cultivar that contains mutation affecting the root outer cell layers (epidermis, exodermis, and sclerenchyma) accumulated more Al than wild type rice (Huang et al. 2009b). In this mutant, the exodermal cells were changed into sclerenchyma-like cells, experiencing a decrease in cell size and a thickening of cell walls.

In addition to the cell wall, the composition of the plasma membrane seems to play a role in Al tolerance in rice. In rice and other Al-tolerant species, the membrane surface is less negatively charged than in Al-sensitive ones (Wagatsuma et al. 2005). This plasma membrane negative charge is one mechanism that may underlie variations in Al tolerance within species (Khan et al. 2009). An Al-sensitive rice cultivar showed increased plasma membrane permeability and greater Al uptake than an Al-tolerant cultivar. The lipid composition of the plasma membrane differed between these cultivars, with the Al-tolerant cultivar presenting a lower ratio of phospholipids to major neutral lipid $\Delta^5$-sterol than the sensitive cultivar, suggesting that the plasma membrane of the Al-tolerant cultivar is less negatively charged and has reduced permeability compared to that of the Al-sensitive cultivar (Khan et al. 2009).

In another study, four rice cultivars that differ in their Al tolerance were compared, and a decrease in the lipid and fatty acid content was observed in the sensitive cultivars. In the roots of the susceptible cultivars, the levels of phospholipids such as phosphatidylcholine decreased, whereas the amount of lipid remained unchanged in the tolerant cultivars. The study suggests that the stability of lipid composition and the capacity to maintain lipid biosynthetic activities may help rice under Al stress (Huynh et al. 2012).

## 2.3 *The Role of the Root Border Cells*

Root border cells (RBCs) are special living cells that are attached to the root apex and play key roles in plant development. RBCs are released from the root tip and secrete a mucilage to protect plants from environmental factors. Recently, Driouich et al. (2013) published detailed review of RBCs and their involvement with plant responses to stress. Mucilage also protects *P. vulgaris* from Al toxicity because the physical removal of RBCs from the root tips resulted in a higher Al accumulation in the root tips and a more severe inhibition of root elongation (Miyasaka and Hawes 2001). In rice, the physical removal of the RBCs from root tips resulted in a more severe inhibition of root elongation and a higher Al accumulation in the root tips in an Al-sensitive cultivar than in an Al-tolerant cultivar (Cai et al. 2011). Furthermore, the Al content of the root tips was lower in roots surrounded by RBCs than that in roots deprived of RBCs, and cell viability and Al-induced mucilage exudation were always higher in the RBCs from the Al-tolerant cultivar than from the Al-sensitive cultivar (Cai et al. 2011).

The dissociation of root cap cells from root tissue is essential for RBC separation. For this, pectate lyases act to depolymerize pectic polysaccharides by cleaving internal linkages. However, these enzymes have little activity on methylated pectin, which predominates in plant cell walls (O'Neill et al. 1990). Pectin methylesterase de-esterifies pectin by removing methoxyl groups to produce methanol and PGA, a substrate that is susceptible to degradation by pectate lyase. In fact, the release of RBCs is dependent on pectin methylesterase activity (Stephenson and Hawes 1994; Wen et al. 1999).

RBCs attached to roots seem to help rice avoid Al toxicity, and one possibility is that Al-tolerant rice cultivars have more attached RBCs than the sensitive cultivars because the Al-sensitive cultivars have a higher degree of pectin methylesterase activity under Al stress (Yang et al. 2008), which may cause premature RBC release. RBCs are most likely one of the first Al barriers in the roots and should be considered as one component of high Al tolerance in rice.

## 2.4 The Al-Tolerance Genes in Rice

Searches for rice Al-tolerance genes have been the focus of many studies over the last few years, and many candidate genes have been found using various distinct approaches: differential display reverse transcription-PCR (DDRT-PCR) (Zhang et al. 2007), cDNA amplified fragment length polymorphism (cDNA-AFLP) (Mao et al. 2004), proteomic analysis (Yang et al. 2007, 2013), semi-quantitative and real-time polymerase chain reaction (Zhang et al. 2010), microarrays (Yamaji et al. 2009; Tsutsui et al. 2012), genome-wide associations and QTLs (Famoso et al. 2011), and RNA-Seq (Arenhart et al. 2014). Despite the volume of data that has been generated, the comparison of all these data to identify Al-tolerance genes is not an easy task, because different cultivars, time, and Al concentrations were used. Moreover, many of the Al-responsive genes represent reactions to Al-toxicity and are not actually involved in the Al-tolerance mechanisms. Nevertheless, some important rice Al-tolerance genes have been characterized over the past years. For a review of the genes that are involved in Al-tolerance mechanisms in important crop species, please see Delhaize et al. (2012).

At least two transcription factors seem to play a role in Al tolerance in rice: ART1 (Aluminum Resistance Transcription Factor 1) and ASR5 (Aba, Stress and Ripening). ART1 is constitutively expressed in the roots and is not induced by Al. However, an art1-knockout mutant is highly Al-sensitive (Yamaji et al. 2009), indicating a central role for ART1 in the Al-tolerance mechanisms in rice. In the same way, ASR5 is also expressed in the roots, and transcript levels are increased in response to Al (Arenhart et al. 2013). In addition, ASR5-silenced plants are Al-sensitive. Furthermore, both ART1 and ASR5 regulate genes that are important in the response to Al (Yamaji et al. 2009; Arenhart et al. 2014).

Seven ART1-regulated genes have been characterized so far; Nrat1 (Nramp aluminum transporter 1) is a specific Al transporter that is involved in the uptake of Al to cells for sequestration to vacuoles. The silencing of Nrat1 resulted in decreased Al uptake, increased Al binding to the cell wall, and enhanced Al sensitivity (Xia et al. 2010). ALS1 (Aluminum Sensitive 1), a half-size ABC transporter, is also involved in the sequestration of Al to the vacuoles (Huang et al. 2011). MGT1 (Magnesium Transporter 1), a Mg transporter, confers Al tolerance in rice by increasing the concentration of Mg in the cell and decreasing Al-binding sites (Chen et al. 2012). CDT3 (cadmium tolerance 3) anchors to the plasma membrane and functions as a chelator, binding Al and preventing its entrance into the cells (Xia et al. 2013). Silencing CDT3 results in decreased Al accumulation in the root plasma membrane and cell wall but increased Al concentration in the cell sap. FRDL4 (Ferric Reductase Defective3-like 4), a citrate transporter, secretes citrate from roots, chelating Al in the rhizosphere (Yokosho et al. 2011). Finally, STAR1 and STAR2 (sensitive to aluminum rhizotoxicity 1 and 2), an ATP-binding domain and a transmembrane domain, respectively, transport UDP-glucose, which is implicated in cell wall modifications that mask Al-binding sites in the cell wall (Huang et al. 2009a). ASR5 regulates at least 36 genes in response to Al in rice but only STAR1 is well characterized so far (Arenhart et al. 2014).

# 3 Concluding Remarks

Under acidic conditions in the soil, $Al^{3+}$ (the rhizotoxic form) is formed and incorporated by plants, subsequently inhibiting root elongation and expansion and resulting in several other injuries to the plants. Due to the reactivity of Al, the nucleus, cell wall, plasma membrane, and cytoskeleton can be targets of Al injury (Kochian et al. 2004). A considerable number of different Al-tolerance mechanisms have been proposed, and it is likely that multiple Al-tolerance mechanisms are employed by different plant species (Kochian et al. 2004).

Al transport systems in the plasma membrane/tonoplast of rice appear to be key contributors to the Al tolerance of rice compared to other crops because even the most rice Al-sensitive aus lines, which possess a functionally deficient NRAT1 transporter (Li et al. 2014), are still more Al-tolerant than other cereal species, including maize, sorghum, and wheat (Famoso et al. 2010). Rice, one of the most Al-tolerant crops, seems to employ several mechanisms in response to Al that, when combined, result in a greater tolerance of Al (Fig. 2). A summary of these mechanisms is listed below; note that these mechanisms do not necessarily occur in this order:

## 3.1 External Detoxification Mechanisms

- *Mucilage secretion*: Rice root border cells attached to the root apices secrete a thicker mucilage that binds to Al and prevents Al entrance into the root cells (Cai et al. 2011).
- *Regulation of plasma membrane lipid composition*: A lower ratio of phospholipids to major neutral lipid $\Delta^5$-sterol leads to a decreased negative charge and a reduced permeability of the plasma membrane, preventing Al entrance into the symplast (Khan et al. 2009). Furthermore, the stability of the plasma membrane lipid composition (e.g., phosphatidylcholine) may act in plasma membrane-mediated prevention of Al binding (Huynh et al. 2012).
- *Regulation of cell wall composition*: a lower content of cell wall polysaccharides such as pectin and Hemicellulose 1 and 2 as well as a higher degree of methylesterification result in fewer carboxylic groups that serves as Al-binding sites, also preventing Al entrance into the root cells (Yang et al. 2008).
- *Regulation of nitric oxide*: After Al exposure, NO levels increase, and NO acts as an antioxidant molecule, helping in the ROS defense system (Yang et al. 2013) and blocking the increase in pectin content (Zhang et al. 2011).

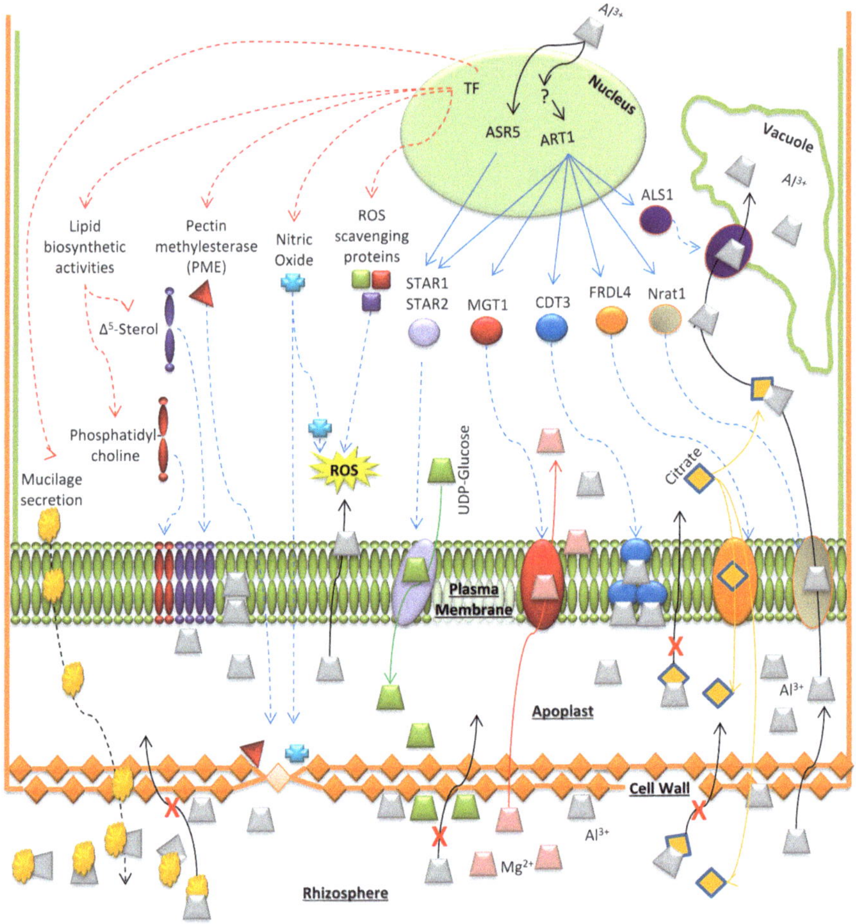


**Fig. 2** Summary of rice mechanisms to cope with Al stress. In response to Al, rice utilizes several mechanisms to prevent the entrance of Al to cells. These mechanisms include the production of a mucilage in the attached root border cells to bind Al; the modification of the plasma membrane and cell wall components; increases in the levels of nitric oxide which acts as an antioxidant molecule in ROS defense system and as signaling molecule to block increases in pectin content; the release of citrate into the apoplast and rhizosphere; the release of UDP-Glucose in cell wall to mask Al-binding sites; the regulation of magnesium content in cell to decrease Al-binding sites, and the chelation of Al in plasma membrane by the CDT3 protein. Because some Al may enter the symplast and cause an increase in ROS content, rice can alleviate the toxicity of Al by regulating the expression of ROS scavenging genes. Finally, a portion of the Al in the symplast is transported and accumulated in vacuoles. *Black arrows*: aluminum efflux; *orange arrow*: citrate efflux; *red arrow*: magnesium efflux; *green arrow*: UDP-Glucose efflux; *red dashed arrows*: genes that are possibly regulated by ASR5, ART1, or another transcription factor (TF); *blue arrows*: ART1- and ASR5-regulated genes; *blue dashed arrows*: protein localization in the cell; *black dashed arrow*: mucilage induced by Al in the root border cells

- *Release of citrate:* The release of citrate into the apoplast and rhizosphere leads to Al-citrate complex formation, reducing Al entrance into the root cells (Yokosho et al. 2011).
- *Use of UDP-Glucose in the cell wall:* UDP-Glucose binds to the cell wall, masking potential Al-binding sites (Huang et al. 2009a).
- *Regulation of the magnesium content in the cell sap:* Increasing the Mg content also decreases Al-binding sites on a cell (Chen et al. 2012).
- *Al chelation in the plasma membrane*: A specialized plasma membrane protein is able to chelate Al, preventing its entrance into the cell (Xia et al. 2013).

## 3.2 Internal Detoxification

- *Induction of ROS scavenging systems*: As consequence of Al toxicity, ROS injury is alleviated by increasing ROS scavenging proteins (Ma et al. 2012).
- *Sequestration of Al into vacuoles*: Specific proteins transport Al to less toxic compartments such as vacuoles (Xia et al. 2010, 2013).

Despite some Al sequestration into vacuoles, rice is essentially an Al-excluder because most of the mechanisms employed by rice attempt to prevent the entrance of Al into the cell. In conclusion, there are many mechanisms that act synergistically to protect rice from Al exposure. Furthermore, with new technologies for broad analysis of data, more Al-regulated genes should be discovered that might help to elucidate this puzzling and complex mechanism.

**Acknowledgments** This work was supported by the Coordenação de Aperfeiçoamento de Pessoal de Nível Superior (CAPES: www.capes.gov.br), Fundação de apoio a Pesquisa do Rio Grande do Sul (FAPERGS), and the Brazilian National Council of Technological and Scientific Development (CNPq). This research was partially supported by a grant from the NIH (R01GM066258) to Z.-Y. Wang.

## References

Arenhart RA, De Lima JC, Pedron M, Carvalho FEL, Silveira JAG, Rosa SB, Caverzan A, Andrade CMB, Schünemann M, Margis R et al (2013) Involvement of ASR genes in aluminium tolerance mechanisms in rice. Plant Cell Environ 36:52–67

Arenhart RA, Bai Y, Oliveira LF, Neto LB, Schunemann M, Maraschin F, Mariath J, Silverio A, Martins G, Margis R et al (2014) New insights into aluminum tolerance in rice: the ASR5 protein binds the STAR1 promoter and other aluminum-responsive genes. Mol Plant 7:709–721

Cai M, Zhang S, Xing C, Wang F, Ning W, Lei Z (2011) Developmental characteristics and aluminum resistance of root border cells in rice seedlings. Plant Sci 180:702–708

Cao Y, Lou Y, Han Y, Shi J, Wang Y, Wang W, Ming F (2011) Al toxicity leads to enhanced cell division and changed photosynthesis in *Oryza rufipogon* L. Mol Biol Rep 38:4839–4846

Chen ZC, Yamaji N, Motoyama R, Nagamura Y, Ma JF (2012) Up-regulation of a magnesium transporter gene OsMGT1 is required for conferring aluminum tolerance in rice. Plant Physiol 159:1624–1633
Delhaize E, Ma JF, Ryan PR (2012) Transcriptional regulation of aluminium tolerance genes. Trends Plant Sci 17:341–348
Doebley J (2004) The genetics of maize evolution. Ann Rev Genet 38:37–59
Doebley JF, Gaut BS, Smith BD (2006) The molecular genetics of crop domestication. Cell 127:1309–1321
Driouich A, Follet-Gueye M-L, Vicré-Gibouin M, Hawes M (2013) Root border cells and secretions as critical elements in plant host defense. Curr Opin Plant Biol 16:489–495
Famoso AN, Clark RT, Shaff JE, Craft E, McCouch SR, Kochian LV (2010) Development of a novel aluminum tolerance phenotyping platform used for comparisons of cereal aluminum tolerance and investigations into rice aluminum tolerance mechanisms. Plant Physiol 153:1678–1691
Famoso AN, Zhao K, Clark RT, Tung C-W, Wright MH, Bustamante C, Kochian LV, McCouch SR (2011) Genetic architecture of aluminum tolerance in rice (*Oryza sativa*) determined through genome-wide association analysis and QTL mapping. PLoS Genet 7:e1002221
Foy C (1988) Plant adaptation to acid, aluminum-toxic soils. Soil Sci Plant Anal 19:959–987
Garris AJ, Tai TH, Coburn J, Kresovich S, McCouch S (2005) Genetic structure and diversity in *Oryza sativa* L. Genetics 169:1631–1638
Grevenstuk T, Romano A (2013) Aluminium speciation and internal detoxification mechanisms in plants: where do we stand? Metallomics 5:1584–1594
Hengl T (2009) A practical guide to geostatistical mapping. Creative Commons Attribution-Noncommercial-No Derivative Works 3.0 License, pp 291
Higham CFW (2002) Languages and farming dispersals: Austroasiatic languages and rice cultivation. In: Bellwood P, Renfrew C (eds) Examining the farming/language dispersal hypothesis. McDonald Institute for Archaeological Research: Cambridge, pp 223–232
Hoekenga OA, Magalhaes JV (2011) Mechanisms of aluminum tolerance. Springer, Berlin
Huang C, Yamaji N, Mitani N, Yano M, Nagamura Y, Ma JF (2009a) A bacterial-type ABC transporter is involved in aluminum tolerance in rice. Plant Cell 21:655–667
Huang C-F, Yamaji N, Nishimura M, Tajima S, Ma JF (2009b) A rice mutant sensitive to Al toxicity is defective in the specification of root outer cell layers. Plant Cell Physiol 50:976–985
Huang C, Yamaji N, Chen Z, Ma JF (2011) A tonoplast-localized half-size ABC transporter is required for internal detoxification of aluminum in rice. Plant J 69:857–867
Huang X, Kurata N, Wei X, Wang Z-X, Wang A, Zhao Q, Zhao Y, Liu K, Lu H, Li W et al (2012) A map of rice genome variation reveals the origin of cultivated rice. Nature 490:497–501
Huynh V-B, Repellin A, Zuily-Fodil Y, Pham-Thi A-T (2012) Aluminum stress response in rice: effects on membrane lipid composition and expression of lipid biosynthesis genes. Physiol Plant 146:272–284
Illés P, Schlicht M, Pavlovkin J, Lichtscheidl I, Baluska F, Ovecka M (2006) Aluminium toxicity in plants: internalization of aluminium into cells of the transition zone in Arabidopsis root apices related to changes in plasma membrane potential, endosomal behaviour, and nitric oxide production. J Exp Bot 57:4201–4213
Khan MSH, Tawaraya K, Sekimoto H, Koyama H, Kobayashi Y, Murayama T, Chuba M, Kambayashi M, Shiono Y, Uemura M et al (2009) Relative abundance of Delta(5)-sterols in plasma membrane lipids of root-tip cells correlates with aluminum tolerance of rice. Physiol Plant 135:73–83
Kochian LV, Pence NS, Letham DLD, Pineros MA, Magalhaes JV, Hoekenga OA, Garvin DF (2002) Mechanisms of metal resistance in plants: aluminum and heavy metals. Plant Soil 247:109–119
Kochian LV, Hoekenga OA, Pineros MA (2004) How do crop plants tolerate acid soils? Mechanisms of aluminum tolerance and phosphorous efficiency. Annu Rev Plant Biol 55:459–493

Kochian LV, Piñeros MA, Hoekenga OA (2005) The physiology, genetics and molecular biology of plant aluminum resistance and toxicity. Plant Soil 274:175–195

Kollmeier M, Felle HH, Horst WJ (2000) Genotypical differences in aluminum resistance of maize are expressed in the distal part of the transition zone. Is reduced basipetal auxin flow involved in inhibition of root elongation by aluminum? Plant Physiol 122:945–956

Kovach MJ, Sweeney MT, McCouch SR (2007) New insights into the history of rice domestication. Trends Genet 23:578–587

Li J-Y, Liu J, Dong D, Jia X, McCouch SR, Kochian LV (2014) Natural variation underlies alterations in Nramp aluminum transporter (NRAT1) expression and function that play a key role in rice aluminum tolerance. Proc Natl Acad Sci U S A 111:6503–6508

Lu H, Liu Z, Wu N, Berne S, Saito Y, Liu B, Wang L (2002) Rice domestication and climatic change: phytolith evidence from East China. Boreas 31:378–385

Ma J, Hiradate S, Nomoto K, Iwashita T, Matsumoto H (1997) Internal detoxification mechanism of Al in Hydrangea. Plant Physiol 113:1033–1039

Ma JF, Shen R, Zhao Z, Wissuwa M, Takeuchi Y, EbitaniT YM (2002) Response of rice to Al stress and identification of quantitative trait Loci for Al tolerance. Plant Cell Physiol 43:652–659

Ma B, Gao L, Zhang H, Cui J, Shen Z (2012) Aluminum-induced oxidative stress and changes in antioxidant defenses in the roots of rice varieties differing in Al tolerance. Plant Cell Rep 31:687–696

Mao C, Yi K, Yang L, Zheng B, Wu Y, Liu F, Wu P (2004) Identification of aluminium-regulated genes by cDNA-AFLP in rice (*Oryza sativa* L.): aluminium-regulated genes for the metabolism of cell wall components. J Exp Bot 55:137–143

Matsumoto H (2000) Cell biology of aluminium toxicity and tolerance in higher plants. Int Rev Cytol 200:1–46

Meriga B, Reddy BK, Rao KR, Reddy LA, Kishor PBK (2004) Aluminium-induced production of oxygen radicals, lipid peroxidation and DNA damage in seedlings of rice (*Oryza sativa*). J Plant Physiol 161:63–68

Miyasaka SC, Hawes MC (2001) Possible role of root border cells in detection and avoidance of aluminum toxicity. Plant Physiol 125:1978–1987

Nguyen VT, Nguyen BD, Sarkarung S, Martinez C, Paterson AH, Nguyen HT (2002) Mapping of genes controlling aluminum tolerance in rice: comparison of different genetic backgrounds. Mol Genet Genom 267:772–780

Nguyen BD, Brar DS, Bui BC, Nguyen TV, Pham LN, Nguyen HT (2003) Identification and mapping of the QTL for aluminum tolerance introgressed from the new source, Oryza Rufipogon Griff., into indica rice (Oryza sativa L.). Theor Appl Genet 106:583–593

O'Neill M, Albersheim P, Darvill A (1990) The pectic polysaccharides of primary cell walls. Meth Plant Biochem 2:415–441

Ozkan H, Brandolini A, Schafer-Pregl R, Salamini F (2002) AFLP analysis of a collection of tetraploid wheats indicates the origin of emmer and hard wheat domestication in southeast Turkey. Mol Biol Evol 19:1797–1801

Panda SK, Baluska F, Matsumoto H (2009) Aluminum stress signaling in plants. Plant Signal Behav 4:592–597

Pandey P, Srivastava RK, Dubey RS (2013) Salicylic acid alleviates aluminum toxicity in rice seedlings better than magnesium and calcium by reducing aluminum uptake, suppressing oxidative damage and increasing antioxidative defense. Ecotoxicology 22:656–670

Rosa SB, Caverzan A, Teixeira FK, Lazzarotto F, Silveira JAG, Ferreira-Silva SL, Abreu-Neto J, Margis R, Margis-Pinheiro M (2010) Cytosolic APx knockdown indicates an ambiguous redox responses in rice. Phytochemistry 71:548–558

Sharma P, Dubey RS (2007) Involvement of oxidative stress and role of antioxidative defense system in growing rice seedlings exposed to toxic concentrations of aluminum. Plant Cell Rep 26:2027–2038

Stephenson MB, Hawes MC (1994) Correlation of pectin methylesterase activity in root caps of pea with root border cell separation. Plant Physiol 106:739–745

Tsutsui T, Yamaji N, Huang CF, Motoyama R, Nagamura Y, Ma JF (2012) Comparative genome-wide transcriptional analysis of Al-responsive genes reveals novel Al tolerance mechanisms in rice. PLoS One 7:e48197

Von Uexkull HR, Mutert E (1995) Global extent, development and economic impact of acid soils. Plant Soil 171:1–15

Wagatsuma T, Khan SH, Rao IM, Wenzl P, Tawaraya K, Yamamoto T, Kawamura T, Hosogoe K, Ishikawa S (2005) Methylene blue stainability of root-tip protoplasts as an indicator of aluminum tolerance in a wide range of plant species, cultivars and lines. Soil Sci Plant Nutr 51:991–998

Wang Y-S, Yang Z-M (2005) Nitric oxide reduces aluminum toxicity by preventing oxidative stress in the roots of *Cassia tora* L. Plant Cell Physiol 46:1915–1923

Wang H-H, Huang J-J, Bi Y-R (2010) Nitrate reductase-dependent nitric oxide production is involved in aluminum tolerance in red kidney bean roots. Plant Sci 179:281–288

Wen F, Zhu Y, Hawes MC (1999) Effect of pectin methylesterase gene expression on pea root development. Plant Cell 11:1129–1140

Xia J, Yamaji N, Kasai T, Ma JF (2010) Plasma membrane-localized transporter for aluminum in rice. Proc Natl Acad Sci U S A 107:18381–18385

Xia J, Yamaji N, Ma JF (2013) A plasma membrane-localized small peptide is involved in rice aluminum tolerance. Plant J 76:345–355

Yamaji N, Huang CF, Nagao S, Yano M, Sato Y, Nagamura Y, Ma JF (2009) A zinc finger transcription factor ART1 regulates multiple genes implicated in aluminum tolerance in rice. Plant Cell 21:3339–3349

Yang Q, Wang Y, Zhang J, Shi W, Qian C, Peng X (2007) Identification of aluminum-responsive proteins in rice roots by a proteomic approach: cysteine synthase as a key player in Al response. Proteomics 7:737–749

Yang JL, Li YY, Zhang YJ, Zhang SS, Wu YR, Wu P, Zheng SJ (2008) Cell wall polysaccharides are specifically involved in the exclusion of aluminum from the rice root apex. Plant Physiol 146:602–611

Yang L, Tian D, Todd CD, Luo Y, Hu X (2013) Comparative proteome analyses reveal that nitric oxide is an important signal molecule in the response of rice to aluminum toxicity. J Proteom Res 12:1316–1330

Yokosho K, Yamaji N, Ma JF (2011) An Al-inducible MATE gene is involved in external detoxification of Al in rice. Plant J 68:1061–1069

Zhang J, He Z, Tian H, Zhu G, Peng X (2007) Identification of aluminium-responsive genes in rice cultivars with different aluminium sensitivities. J Exp Bot 58:2269–2278

Zhang J, Yin Y, Wang Y, Peng X (2010) Identification of rice Al-responsive genes by semi-quantitative polymerase chain reaction using sulfite reductase as a novel endogenous control. J Integr Plant Biol 52:505–514

Zhang Z, Wang H, Wang X, Bi Y (2011) Nitric oxide enhances aluminum tolerance by affecting cell wall polysaccharides in rice roots. Plant Cell Rep 30:1701–1711

Zhao XQ, Guo SW, Shinmachi F, Sunairi M, Noguchi A, Hasegawa I, Shen RF (2013) Aluminium tolerance in rice is antagonistic with nitrate preference and synergistic with ammonium preference. Ann Bot 111:69–77

Zheng SJ, Ma JF, Matsumoto H (1998) High aluminum resistance in buckwheat. Plant Physiol 117:745–751

# The Molecular Physiology and Regulation of Aluminum Resistance in Higher Plants

**Hiroyuki Koyama, Yuriko Kobayashi, Sanjib K. Panda, and Gregory J. Taylor**

**Abstract** Plants have evolved a variety of aluminum (Al)-resistance mechanisms that are regulated by complex biological systems. Two distinct categories of Al resistance were proposed in the late 1980s, namely "exclusion" of Al from the symplasm and "internal tolerance." Exclusion mechanisms reduce the amount of rhizotoxic Al ($Al^{3+}$) in the symplasm of cells and internal tolerance mechanisms reduce Al toxicity, and the resulting damage occurs once Al has entered the cytosol. Since these concepts were introduced, many studies have identified physiological and genetic mechanisms of Al resistance that provide support for "exclusion" and "internal tolerance" at the molecular level. Excretion of organic anions (OA) from root cells, which detoxify Al by chelation, appears to be the most common mechanism of Al exclusion in plants. In addition, modification of the chemical properties of the plasma membrane and cell wall contribute to a reduction of Al rhizotoxicity in the root tips. Sequestration of Al in the vacuole, translocation of Al to the shoot, and enhanced capacity to cope with Al-inducible reactive oxygen species are important mechanisms of internal Al tolerance. Various genes that control these traits, such as genes encoding OA transporters, have been identified in plants. Studies of the transcriptional regulation of these genes by STOP1/ART1-type zinc finger transcription factors show that multiple Al-resistance genes are likely co-regulated by the same signal transduction pathway in different plant species. In addition, regulation of Al-resistance mechanisms is coordinated with resistance to other stress factors associated with the acid soil syndrome.

H. Koyama (✉) • Y. Kobayashi
Faculty of Applied Biological Sciences, Gifu University, 501-1193 Gifu, Japan
e-mail: koyama@gifu-u.ac.jp; k_yuriko@gifu-u.ac.jp

S.K. Panda
Faculty of Life Science and Bioinformatics, Assam University, Silchar, India
e-mail: drskpanda@gmail.com

G.J. Taylor
Faculty of Sciences, Department of Biological Sciences, University of Alberta, Edmonton T6G2E1, Canada
e-mail: gregory.taylor@ualberta.ca



S.K. Panda, F. Baluška (eds.), *Aluminum Stress Adaptation in Plants*, Signaling and Communication in Plants 24, DOI 10.1007/978-3-319-19968-9_9

# 1 Molecular Physiology of $Al^{3+}$ Rhizotoxicity

Many crops grow poorly on acid soils due to acid soil syndrome. This syndrome consists of multiple stress factors, including toxicities of aluminum ($Al^{3+}$), manganese ($Mn^{2+}$), and protons ($H^+$) and deficiencies of calcium ($Ca^{2+}$) and phosphates ($PO_4^{2-}$; Taylor 1991; Kochian et al. 2004). While these stress factors induce complex stress conditions, $Al^{3+}$ rhizotoxicity is considered to be the most harmful in terms of yield loss, particularly under conditions of drought stress. If a plant is incapable of developing an adequate root system as a result of $Al^{3+}$ rhizotoxicity, it becomes susceptible to drought because it cannot access water in the subsoil (Foy 1992; Lynch and Wojciechowski 2015).

Molecular mechanisms of Al rhizotoxicity have been reported in several studies making use of the model plant, Arabidopsis (*Arabidopsis thaliana*). The harmful effects of $Al^{3+}$ appear at growing root tips because $Al^{3+}$ disturbs essential processes involved in cell division, elongation (Matsumoto 2000), and genotoxicity (Nezames et al. 2012b). A study with corn demonstrated that cells in the distal transition zone (DTZ), where cells are preparing to undergo elongation, is one of the most sensitive parts of the root to $Al^{3+}$ toxicity (Sivaguru and Horst 1998). One of the mechanisms that regulates this process has been clarified by a molecular physiological study of Arabidopsis. By combining reverse genetics of Al-inducible genes and characterization of ethylene signaling mutants, Yang et al. (2014) concluded that inhibition of root growth is driven by a localized increase in IAA biosynthesis in the DTZ, which in turn is regulated by ethylene signaling. However, a loss of function mutation in the cell cycle checkpoint regulator *TANMEI* (*ATAXIA TELANGIECTASIA-MUTATED AND RAD3-RELATED*) conferred Al resistance (Nezames et al. 2012b), suggesting that $Al^{3+}$ is also genotoxic and targets cell division in the root tip.

The bulk of Al in neutral and basic soils exists in nontoxic forms such as Al oxide and aluminosilicates. In acid soils, Al is mobilized as rhizotoxic $Al^{3+}$, the most toxic of the monomeric ions (Kinraide 2003). Aluminum toxicity in acid soils can be amended by liming, which neutralizes soil acidity (Kinraide 1998). Aluminum in soil solution is controlled at sufficiently safe levels for sensitive root cells if the soil pH is >5.5, but overtreatment with liming may induce harmful effects through alkalinity. Application of gypsum ($CaSO_4$) to partially neutralized soil (e.g., pH 5.0–5.5) eliminates the harmful effects of Al in the soil. $CaSO_4$ is a neutral salt (i.e., no neutralization of acidity) that can increase Ca concentrations in the soil solution more effectively than $CaCO_3$ (Sumner et al. 1986). A molecular physiological study of Arabidopsis provided experimental evidence that explains the complexity of Al rhizotoxicity in acid soils, including the ameliorative effects of $CaSO_4$ (Kobayashi et al. 2013b; Fig. 1).

Electrostatic studies have modeled the rhizotoxicity of Al for growing roots of wheat and other plant species. These studies predicted that rhizotoxicity is determined by $Al^{3+}$ activity at the plasma membrane (PM) surface ($\{Al^{3+}\}_{PM}$), as opposed to $Al^{3+}$ activity in the bulk-phase solution ($\{Al^{3+}\}_{bulk}$; Kinraide 1994,

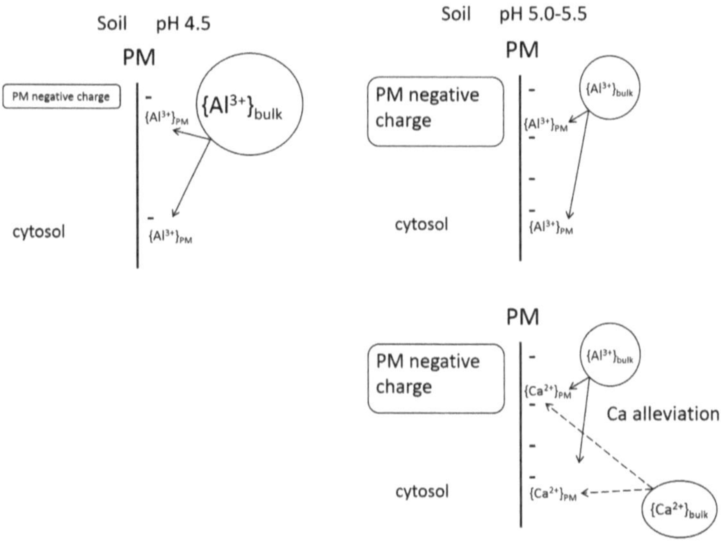


**Fig. 1** Mechanism of Al rhizotoxicity and Ca alleviation in weakly acidic soils. Aluminum rhizotoxicity is determined by $\{Al^{3+}\}$ at the plasma membrane (PM) surface ($\{Al^{3+}\}_{PM}$). At pH > 5.0, $\{Al^{3+}\}$ in the bulk-phase soil solution decreases compared with that at pH 4.5. By contrast, the negative charge of weakly acidic ligands (phospholipids) increases at pH > 5.0 compared with lower pH. Through the balance of these two factors, $\{Al^{3+}\}_{PM}$ is maintained at pH 5.0–5.5 and shows rhizotoxicity. The toxicity can be efficiently negated by increasing $\{Ca^{2+}\}_{PM}$ by application of a soluble Ca fertilizer such as gypsum (see Kobayashi et al. 2013b)

1998). The activity of $\{Al^{3+}\}_{PM}$, the $Al^{3+}$ attracted to the PM surface, is determined by both the $\{Al^{3+}\}_{bulk}$ and the electronegativity at the PM surface. In the pH range of 5.0–5.5, $\{Al^{3+}\}_{bulk}$ is decreased compared with that at lower pH (e.g., pH 4.5). In contrast, the electronegativity of the PM increases owing to dissociation of $H^+$ from weakly acidic phospholipids. Within this pH range, Al-sensitive mutants of Arabidopsis show Al-induced inhibition of growth that completely fits predictions of the electrostatic model. Application of a sufficient amount of $Ca^{2+}$ removes $\{Al^{3+}\}_{PM}$ by masking negative ligands on the PM surface by increasing $\{Ca^{2+}\}_{PM}$ (Ryan et al. 1993). This may account for the alleviative effects of gypsum on Al rhizotoxicity in partially neutralized soils. In fact, an Arabidopsis mutant (the double mutant of *phosphatidate phosphohydrolase 1* and *2*) that has a greater negative charge on the PM surface, because of its inability to convert phospholipids to electrically neutral lipids (Eastmond et al. 2010), is more susceptible to Al rhizotoxicity than the wild type in partially neutralized acid soil. Growth of *pah1pah2* is recovered by application of gypsum (Kobayashi et al. 2013b).

# 2 Molecular Mechanisms of Al Resistance

Two distinct categories of Al resistance were proposed in the late 1980s, namely "exclusion" of Al from the symplasm and "internal tolerance" (Taylor 1987, 1988, 1991). These categories broadly reflected monographs by Levitt (1980) that categorized resistance mechanisms as either "avoidance" mechanisms or "tolerance" mechanisms. In the context of Al, exclusion mechanisms were defined as those that reduce the amount of rhizotoxic Al ($Al^{3+}$) in the symplasm of cells. Internal tolerance mechanisms were defined as those that reduce Al toxicity and the resulting damage that occurs once Al has entered the cytosol. A more modern view might be that exclusion mechanisms are those that reduce the amount of rhizotoxic Al ($Al^{3+}$) in the symplasm and at sensitive sites within the apoplasm (e.g., the plasma membrane surface), while tolerance mechanisms reduce Al toxicity and the resulting damage that occurs once Al has access to these sensitive sites.

Since these concepts were introduced, many studies have identified physiological and genetic mechanisms of Al resistance that provide support for "exclusion" and "tolerance" at the molecular level, including a series of studies with Arabidopsis and rice that have identified molecular mechanisms of Al resistance that protect sensitive root tips from Al toxicity. In this section, we review our current understanding of molecular mechanisms of organic anion (OA) excretion and internal Al tolerance.

## *2.1 Molecular Mechanisms of Organic Anion Excretion from Roots*

Excretion of OA from roots plays a critical role in protection of root tips from Al rhizotoxicity in many plant species (Kochian et al. 2004). Organic anions such as citrate, malate, and oxalate can detoxify Al in the rhizosphere by chelation, since chelated forms are less toxic than $Al^{3+}$. Differences in both the form and amount of OAs that are excreted determine varietal differences in resistance to Al among many plant species (Ma et al. 1998; Zheng et al. 1998; Wenzl et al. 2001; Kobayashi et al. 2005). The amount of OAs excreted from roots is usually not sufficient to detoxify $\{Al^{3+}\}_{bulk}$, but it may protect sensitive root tips by reducing $\{Al^{3+}\}_{PM}$. A modeling study suggested that excretion of OAs is likely optimized to avoid unnecessary carbon loss, while still reducing Al uptake across the PM (Kinraide et al. 2005). This process is controlled by transcriptional regulation of transport genes that mediate OA excretion and activation of OA transporters by Al.

Genes that encode OA transporters capable of mediating Al resistance have been isolated from a variety of plant species during the last decade. A gene for a malate transporter involved in Al-induced malate excretion was first isolated from wheat (Sasaki et al. 2004). This gene, *TaALMT1* (*Triticum aestivum ALUMINUM ACTIVATED MALATE TRANSPORTER1*), encodes a malate transporter localized at the plasma membrane, which is activated by exogenous addition of Al. Similar patterns of Al-induced excretion have been observed for citrate excretion. Genes that encode the citrate-transporting MULTIDRUG AND TOXIC COMPOUNDS EXTRUSION (MATE) transporters were first isolated from barley (Furukawa et al. 2007) and sorghum (Magalhaes et al. 2007). Functional orthologous genes have been isolated from Arabidopsis (Liu et al. 2009), various crop species (Liu et al. 2013; Yokosho et al. 2011), and woody plant species (Sawaki et al. 2013). The expression pattern of specific OA transport genes determines a plant's overall pattern of OA excretion. For example, the expression level of the *ALMT1* from Arabidopsis (*AtALMT1*) is higher than that of *AtMATE*, which explains why it excretes a larger amount of malate than citrate (Sawaki et al. 2009).

Aluminum-inducible excretion of OAs is regulated at both the transcriptional and posttranslational levels. Although *TaALMT1* was reported to be a constitutively expressed gene (Sasaki et al. 2004), expression of orthologues of both *ALMT1* and *MATE* in a variety of plant species is Al-inducible. In Arabidopsis, pharmacological studies using inhibitors of protein phosphatases and kinases suggest that Al-inducible expression of *AtALMT1* involves protein phosphorylation/dephosphorylation (Kobayashi et al. 2007). A protein kinase inhibitor (staurosporine) and a protein phosphatase inhibitor (calyculin A) both inhibit Al-induced expression of *AtALMT1* and a citrate-transporting *MATE* in eucalyptus (Sawaki et al. 2013). Specific protein kinases and protein phosphatases involved in this response are yet to be identified, but these results strongly suggest that protein phosphorylation is involved in the activation of OA transporter proteins by Al.

Studies of *Ta*ALMT1 demonstrate that exogenous Al can trigger malate-transport activity (Sasaki et al. 2004). When *TaALMT1* is expressed in *Xenopous* oocytes, addition of Al to the incubation solution activates malate transport. Substitution of negatively charged amino acids of ALMT1 that are conserved among a variety of species inactivates transport activity, suggesting that binding of $Al^{3+}$ to specific negatively charged amino acids in the protein is essential for malate excretion (Sasaki et al. 2014). In addition, studies with Arabidopsis suggest that protein phosphorylation plays a critical role in Al activation of malate transport by ALMT1 (Kobayashi et al. 2007). Normally, Al-induced malate excretion abruptly ceases if Al is removed from the incubation solution. However, addition of a protein phosphatase inhibitor results in continued malate excretion when plants are moved to an Al-free solution. This suggests that protein dephosphorylation is required for the inactivation of *At*ALMT1. In addition, mutation of several amino acids in putative target sites for protein phosphorylation altered the Al activation of *Ta*ALMT1 (Ligaba et al. 2009). These reports suggest that Al activation of ALMT1 for malate excretion is regulated in a complex manner that involves chemical (i.e., Al binding) and biological (i.e., protein phosphorylation) processes.

In Arabidopsis, Al-induced expression of *AtALMT1* shows time- and dose-dependent responses across a wide dynamic range (e.g., 100 times that of control). This suggests that expression of *AtALMT1* is regulated by multiple transcription factors (Kobayashi et al. 2013a; Tokizawa et al. 2015). In fact, a series of transcription factors have now been identified (Tokizawa et al. 2015) (Fig. 2). A zinc finger

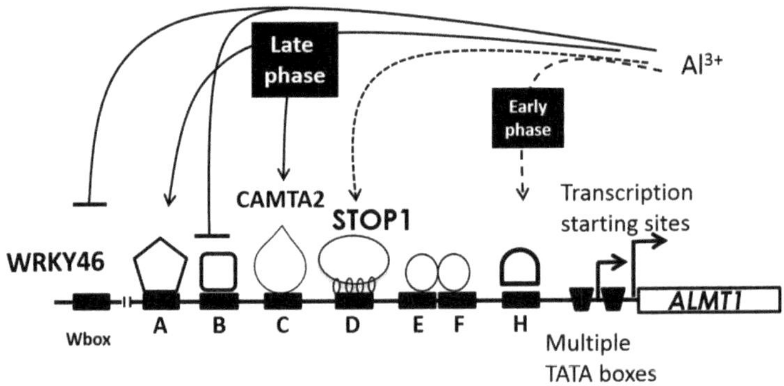


**Fig. 2** Promoter structure and transcription factors controlling *AtALMT1* for Al-inducible gene expression. STOP1 transcription factors are critical for *AtALMT1* expression and are involved in the early response to Al. CAMTA2 (a transcription activator) and WRKY46 (a transcription repressor) regulate the late phase of the Al response. Expression of these genes is suppressed (*WRKY46*) and induced (*CAMTA2*) by Al. Multiple transcription starting sites are regulated by corresponding TATA boxes

protein SENSITIVE TO PROTON RHIZOTOXICITY1 (STOP1) is essential for induction of *AtALMT1* expression (Iuchi et al. 2007, 2008). A WRKY-DOMAIN CONTAINING TRANSCRIPTION FACTOR 46 (WRKY46) functions as a transcriptional repressor (Ding et al. 2013), and a CALMODULIN-BINDING TRANSCRIPTION ACTIVATOR2 (CAMTA2) functions as a transcriptional activator (Tokizawa et al. 2015). Expression of *WRKY46* is repressed by Al, while *CAMTA2* expression is induced. The *AtALMT1* promoter possesses *cis*-elements that interact with unidentified repressors and activators that determine its specific expression in the root tip. In addition, the *AtALMT1* promoter carries multiple TATA boxes, which is a common characteristic of stress-responsive genes that show a wide-dynamic range of expression (Yamamoto et al. 2011; Fig. 2). Such activation is highly sensitive to $\{Al^{3+}\}_{PM}$. For example, *AtALMT1* expression is induced at a lower $\{Al^{3+}\}_{PM}$ than that required to induce toxicity in the *Atalmt1* mutant, which is hypersensitive to Al. This finding suggests that the Al-sensing system for activation of *AtALMT1* is sufficiently sensitive to provide protection from Al toxicity at physiological relevant concentrations (Kobayashi et al. 2013b).

Excretion of OAs is likely linked to the biosynthesis and metabolism of OAs. It has been shown that overexpression of citrate synthase confers Al resistance by enhancing citrate excretion (de la Fuente et al. 1997; Anoop et al. 2003). Overexpression also improves phosphate acquisition from sparingly soluble Al-phosphate present in soil (Koyama et al. 1999, 2000). Immobilization of phosphate by Al is a common problem in acidic soil, and thus this approach should be useful for molecular breeding of acidic soil-tolerant varieties that show both Al resistance and efficient utilization of P from Al-phosphate. Other metabolic engineering procedures, such as overexpression of phospho*enol*pyruvate carboxylase, have been also reported in various plant species (Trejo-Téllez et al. 2010). However, the interaction of OA metabolism and regulation of OA transporters remains unclear.

## 2.2 Internal Al Tolerance: Sequestration, Efflux of Al, and Adaptation to Al-Induced ROS Damage

Notwithstanding its polyvalent nature, Al quickly enters the cytoplasm after Al is attracted to the PM surface. A study using the $^{26}Al$ isotope to measure short-term Al influx in *Chara corallina* determined that Al can enter the cytoplasm within a few minutes (Taylor et al. 2000). Subsequently, Al present in the cytoplasm was sequestrated into the vacuole. Recently, a vacuolar Al transporter and a plasma membrane Al transporter were identified as key components of Al resistance in rice. Aluminum-resistant cultivars showed a higher expression level of *NRAMP ALUMINUM TRANSPORTER1* (*NRAT1*) than sensitive cultivars (Li et al. 2014). Enhanced uptake of Al is ultimately linked to higher sequestration in the vacuole and may reduce the amount of Al at the PM surface and/or cell wall (Li et al. 2014). These results suggest that sequestration of Al into the vacuole is an important strategy for Al resistance. A homologue of ALUMINUM SENSITIVE 1 (ALS1) in rice (*Os*ALS1; Huang et al. 2012) has been characterized as a vacuolar Al transporter. Aluminum uptake by NRAT1 and sequestration into the vacuole by *Os*ALS1 might be coordinately regulated. The vacuolar sequestration of Al appears to be dependent on the activity of the vacuolar $H^+$-ATPase. Dysfunction of the vacuolar $H^+$-ATPase in yeast results in an Al-hypersensitive phenotype (Hamilton et al. 2001a, b).

Aluminum efflux (export) from the cytoplasm is one mechanism of internal Al tolerance. An Al-sensitive mutant of *Arabidopsis*, *aluminum sensitive 3* (*als3*), accumulates Al in the root tips and decreases amounts of Al in the shoots (Larsen et al. 2005). These results suggest that ALS3 is involved in translocation of Al from roots to shoots. *ALS3* encodes a half-type ABC transporter (bacterial type), and the homologue *SENSITIVE TO ALUMINUM RHIZOTOXICITY2* (*STAR2*) has been isolated by positional cloning of a rice Al-sensitive mutant (Huang et al. 2009). In rice, the STAR1–STAR2 complex transports UDP-glucose rather than Al. This suggests that ALS3-like proteins transport UDP-glucose and are associated with Al efflux from the cell by a currently unknown mechanism (Fig. 3).

Treatment with Al triggers production of free radicals, and a subsequent antioxidant response is induced in a wide range of plant species. Production of reactive oxygen species (ROS) by Al is seemingly induced in a complex manner. In tobacco BY-2 cells, Al induces ROS production, which in turn initiates membrane lipid peroxidation. The harmful effects of Al are sharply enhanced in the presence of $Fe^{2+}$ (Yamamoto et al. 1997, 2002). This suggests that production of free radicals might be enhanced biochemically as a result of Fenton-like reaction mechanisms. This is supported by studies of solution chemistry in which a pro-oxidant activity of Al is observed where Al activates the Fenton reaction in conjunction with Fe (Mujika et al. 2011; Ruipérez et al. 2012). In addition, Al treatment induces a series of physiological events that increase ROS production. For example, Al treatment transiently increases the $Ca^{2+}$ concentration in the cytosol (Bhuja et al. 2004), and it can activate NADPH oxidase of the PM, resulting in production

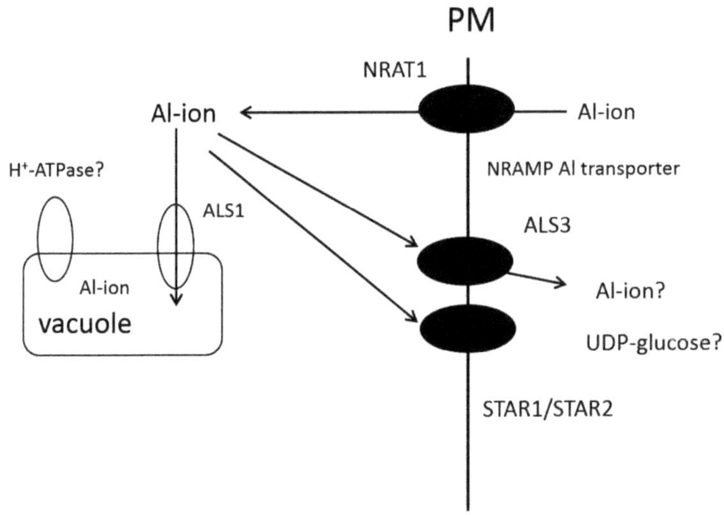


**Fig. 3** Sequestration and export of Al from the cytosol by multiple transporters. An NRAMP-type Al transporter (NRAT1) transports Al and possibly maintains $\{Al^{3+}\}_{PM}$ at a low level. In the cytosol, ALS1 and other vacuolar transporters sequester Al into the vacuole. Research on Arabidopsis suggests that ALS3 exports Al from the cytoplasm, but the rice homologue STAR2 (by interaction with STAR1) is reported to transport UDP-glucose, but not Al

of superoxide radicals ($O_2^{\bullet -}$) (Sagi and Fluhr 2001). Histochemical analysis of transgenic Arabidopsis expressing a gene for a pH-sensitive green fluorescent protein showed that Al acidifies the cytosol (Moseyko and Feldman 2001) and disturbs redox homeostasis through inactivation of the reduction capacity of thiol reductants at pH < 7.0. Taken together, these findings indicate that Al induces ROS production in the cytoplasm after its attraction to the PM.

Plant cells express a variety of genes that encode ROS-scavenging enzymes in response to treatment with Al (Richards et al. 1998). Research on transgenic plants reveals that overexpression of these ROS-responsive genes confers Al resistance. For example, Ezaki et al. (2000) demonstrated that ectopic expression of *NtPox* (tobacco peroxidase) and *parB* (tobacco glutathione *S*-transferase) confers Al resistance in transgenic Arabidopsis. Basu et al. (2001) reported that overexpression of *MnSOD* (superoxide dismutase) confers Al resistance in *Brassica napus*. It has been suggested that enhanced transcription of genes for ROS-scavenging enzymes is involved in Al resistance (Basu et al. 2001). By analogy to work yeast, other components of the ROS-scavenging system such as phospholipid hydroperoxide glutathione peroxidases (PHGPX) and UREidosuccinate transport (URE2) might also play a role in mediating resistance (Basu et al. 2004) (Fig. 4).

Transcriptome analyses indicate that many metabolic pathways are altered in response to treatment with Al, including the nitrogen and sulfur assimilation pathways. These responses suggest that metabolic reprogramming might play an

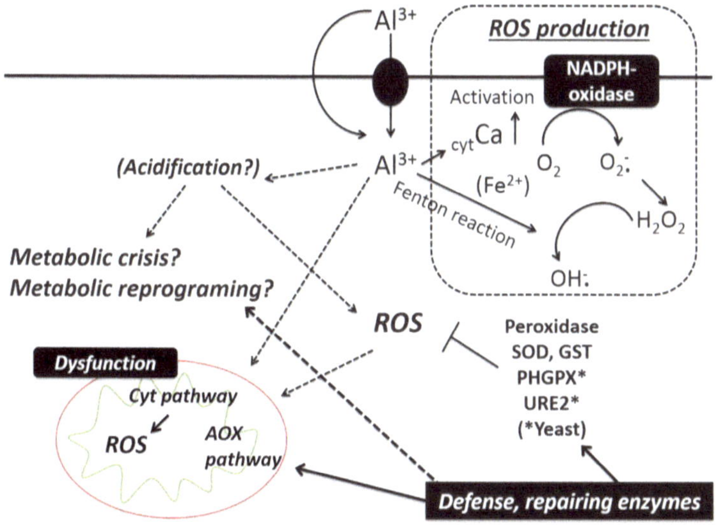


**Fig. 4** Aluminum-induced reactive oxygen species (ROS) stress. Aluminum increases cytosolic Ca concentrations, which triggers activation of NADP-oxidase. This process produces $H_2O_2$, which then amplifies ROS production by a Fenton-like reaction. Aluminum is known to acidify the cytosol, and acidification is known to be a factor that induces metabolic crisis. Aluminum also shifts mitochondrial respiration to an alternative pathway. Overexpression of genes for peroxidase, alternative oxidase (AOX), and other enzymes confers Al resistance in some plant species and yeast

important role in adaptation to Al toxicity. Metabolic reprogramming helps to detoxify ROS and maintain the cellular energy status under stress conditions (Baena-González et al. 2007). Overexpression of a plastid-localized malate dehydrogenase, which performs roles in malate metabolism and redox homeostasis, confers Al resistance in alfalfa (Tesfaye et al. 2001). Recently, an Al-sensitive mutant of Arabidopsis that synthesizes lower amounts of polyamines (including spermine) than the wild type has been isolated (Nezames et al. 2012a). Polyamine synthesis plays roles in adaptation to ROS stress in a wide range of organisms (Alcázar et al. 2010).

Aluminum-induced ROS production results in mitochondrial dysfunction, which is a critical event in the inhibition of cell growth in tobacco cell lines (Yamamoto et al. 1997). More recent studies have shown that ROS signaling activates the alternative oxidase (AOX) pathway, which has a protective role in the stress responses in plants (Panda et al. 2008). The contribution of *AOX* upregulation to Al resistance is further supported by the finding that tobacco cells overexpressing *AOX* show improved resistance to Al (Panda et al. 2013). Upregulation of genes encoding components of ROS scavenging pathways plays a role in other mechanisms active in Al resistance, in particular recovery from oxidative damage. Altogether, we infer that alteration of metabolic pathways involving metabolic reprogramming is involved in ROS-mediated transcriptomic adaptation.

## 3 The STOP1/ART1 System Controls Expression of Al-Resistance Genes

The zinc finger transcription factor SENSITIVE TO PROTON RHIZOTOXICITY1 (STOP1) was identified by positional cloning of an $H^+$-sensitive mutant of Arabidopsis (Iuchi et al. 2007). Root growth of the *stop1* mutant was inhibited by low pH ($H^+$-rhizotoxicity). Interestingly, the mutant also showed hypersensitivity to Al. Furthermore, *AtALMT1* expression was completely suppressed. Further investigation showed that multiple Al-resistance and $H^+$-resistance genes are co-regulated by the same regulatory system (Fig. 5). For example, the major Al-resistance genes *ALS3* and *AtMATE* were co-repressed in the *stop1* mutant (Sawaki et al. 2009). A rice orthologue of *STOP1*, *ALUMINUM-RESISTANCE TRANSCRIPTION FACTOR1* (*ART1*), was also identified by positional cloning of an Al-sensitive mutant (Yamaji et al. 2009). The *art1* mutant shows repression of homologues of *ALS3* (rice; *STAR2*) and *AtMATE* (rice; *OsFRDL4*). In addition, a magnesium transporter (Chen et al. 2012) and a plasma-membrane-localized cysteine-rich peptide (Xia et al. 2013) are critical for Al resistance in rice, and their expression is regulated by the ART1 transcription factor.

It has been suggested that the STOP1/ART1 pathway (Fig. 5) is shared by a wide range of plant species. This suggestion has been further analyzed in various plant

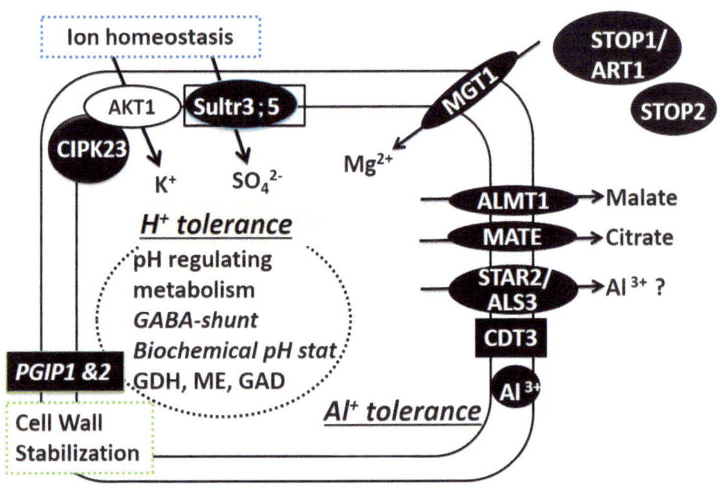


**Fig. 5** STOP1/ART1 regulatory genes involved in resistance to Al and proton toxicity. Functional analyses of STOP1 in Arabidopsis and ART1 in rice show that multiple genes controlling Al and proton resistance are coordinately regulated by STOP1/ART1 zinc finger transcription factors. MGT1, magnesium transporter 1; ALMT, aluminum-activated malate transporter; MATE, citrate-transporting MATE; STAR2/ALS3, aluminum-sensitive 3 protein; GDH, glutamate dehydrogenase; ME, malic enzyme; GAD, gamma butyric acid decarboxylase; PGIP, polygalacturonase inhibitor protein; AKT1, Arabidopsis $K^+$-transporter; CIPK23, CBL-interacting protein kinases (regulator of AKT1); CDT3 (Cys rich small protein in rice)

species, including woody plants [e.g., *Eucalyptus*, Sawaki et al. (2014)], monocots [e.g., wheat, Garcia-Oliveira et al. (2013), and bryophytes (Ohyama et al. (2013)]. Knockdown of the *STOP1* orthologue in the moss *Physcomitrella patens* suppresses Al resistance (Ohyama et al. 2013). These findings indicate that the STOP1/ART1 system may be ubiquitous among land plant species.

Reverse genetics of STOP1 orthologues provides a powerful approach to uncover molecular mechanisms of Al resistance in plant species. Knockdown of the STOP1 orthologue in tobacco results in an Al-sensitive phenotype. Aluminum-responsive citrate excretion plays a critical role in Al resistance in this plant species and is transcriptionally regulated by a citrate-transporting MATE that is regulated by the STOP1 orthologue. A similar approach may be useful to clarify unidentified Al-resistance mechanisms at the molecular level. For example, knockdown in buckwheat and tea plants may enable identification of the gene encoding the oxalate transporter that contributes strongly to Al resistance of these plants (Zheng et al. 1998).

The STOP1/ART1 system involves other transcription factors in the Al signaling process (Fig. 5). Arabidopsis carries a unique homologue of STOP1, namely STOP2, whose expression is regulated by STOP1 (Kobayashi et al. 2013c). Functional analyses show that STOP2 cannot activate transcription of *AtALMT1*, but can activate *AtMATE* and ALS3. ABSCISIC ACID, STRESS AND RIPENING 5 (ASR5) has been identified in rice and coordinately regulates expression of *STAR1* with ART1 (Arenhart et al. 2014). Interestingly, a genome-wide BLAST search indicates that Arabidopsis does not possess an *ASR5* orthologue, but target *cis*-elements that have been coordinately identified in the promoter of *AtALMT1* with the *cis*-element of STOP1 (Tokizawa et al. 2015). In wheat, the same combination is conserved in the promoter of *TaALMT1*, which suggests that a series of orthologous transcription factors coordinately regulate Al-resistance genes in various plant species.

A recent study utilizing a systems biology approach clarified that in the *stop1* mutant, many genes encoding proteins that contribute to $H^+$ resistance are suppressed (Sawaki et al. 2009). For example, the regulatory proteins of the major $K^+$-transporter *ARABIDOPSIS K-TRANSPORTER1* (*AKT1*) and sulfate transporter *SULFATE TRANSPORTER 3.1* (*Surtr 3.1*), which play an important role in ion homeostasis, are repressed in the *stop1* mutant. As another example, genes encoding enzymes that belong to pH-regulatory pathways such as the biochemical pH-stat and GABA (γ-amino butyric acid shunt)-shunt (Bouche and Fromm 2004) are repressed in the *stop1* mutant. In addition, two genes that are involved in stabilization of the cell wall pectic-polysaccharide network, *POLYGALACTURONASE INHIBITING PROTEIN 1* and *2* (*PGIP1* and *PGIP2*; Spadoni et al. 2006), are also repressed in the *stop1* mutant. These genes are controlled by STOP2, and the growing root tips of *pgip1* and *pgip2* show enhanced cellular damage in low-pH solution that is associated with the stability of the pectin network. Co-regulation of Al- and $H^+$-resistance genes by the same transduction pathway is reasonable because $H^+$ rhizotoxicity is apparent in naturally acidic soils in which Al becomes soluble.

In Arabidopsis, a variety of Al-resistance genes are also inducible by $H^+$. We suggest that $Al^{3+}$ and $H^+$ generate an identical signal, although the sensory mechanisms (e.g., the receptor protein) of each stressor remain unknown. However, Al-resistance genes play pleiotropic roles in other stress responses. For example, malate excreted via the functioning of *At*ALMT1 recruits beneficial rhizobacteria to the root surface. The recruitment of rhizobacteria may activate systemic-induced resistance. Infection of aerial tissues by pathogenic bacteria or the FLG22 peptide (a conserved peptide pattern in bacterial flagella) induced expression of *AtALMT1* in roots (Kobayashi et al. 2013a). These results reflect the dual function of AtALMT1 in Al resistance and plant immune responses. Similarly complex regulation has been observed for ALMT1 in soybean. Transcription of that gene is activated by Al- and P deficiency (Liang et al. 2013). This finding is consistent with the proposed role of OA excretion in increased P acquisition.

Membrane-binding proteins, such as receptors or other proteins that can induce signals (e.g., changes in pH and $Ca^{2+}$ concentration in the cytoplasm, $H_2O_2$ production), may be involved in the initial step of Al signaling. Further research is needed to identify the molecular mechanisms underlying these processes. In addition, longer exposure to Al can initiate ABA signaling because depressed root growth can lead to water deficiency in aerial tissues, thereby inducing ABA signaling (Zhang et al. 2006). As well, longer exposure to Al can initiate IAA accumulation as a result of ethylene-mediated signaling at the root tips by Al (Yang et al. 2014). ROS-mediated signals and ROS-induced disruption of respiration and metabolic pathways may also trigger complex signals. For example, specific metabolites such as Fe–citrate can activate transcription of several genes involved in primary metabolism (Finkemeier et al. 2013).

As described above, many genes and proteins that are directly involved in mediating resistance to Al and genes and proteins involved in regulation of resistance have been reported in last decade. These findings have enriched our knowledge of the molecular basis of Al resistance in a wide variety of plants. Collectively, these studies suggest that various resistance mechanisms are coordinately regulated in a complex manner. Further research is required to elucidate the complex signaling systems that mediate Al resistance and the mechanisms to repair Al-induced damage in plants.

## References

Alcázar R, Altabella T, Marco F, Bortolotti C, Reymond M, Koncz C, Carrasco P, Tiburcio AF (2010) Polyamines: molecules with regulatory functions in plant abiotic stress tolerance. Planta 231:1237–1249

Anoop VM, Basu U, McCammon MT, McAlister-Henn L, Taylor GJ (2003) Modulation of citrate metabolism alters aluminum tolerance in yeast and transgenic canola overexpressing a mitochondrial citrate synthase. Plant Physiol 132:2205–2217

Arenhart RA, Bai Y, Valter de Oliveira LF, Bucker Neto L, Schunemann M, Maraschin FD, Mariath J, Silverio A, Sachetto-Martins G, Margis R, Wang ZY, Margis-Pinheiro M (2014)

New insights into aluminum tolerance in rice: the ASR5 protein binds the STAR1 promoter and other aluminum-responsive genes. Mol Plant 7:709–721

Baena-González E, Rolland F, Thevelein JM, Sheen J (2007) A central integrator of transcription networks in plant stress and energy signalling. Nature 448:938–942

Basu U, Good AG, Taylor GJ (2001) Transgenic *Brassica napus* plants overexpressing aluminium-induced mitochondrial manganese superoxide dismutase cDNA are resistant to aluminium. Plant Cell Environ 24:1278

Basu U, Southron JL, Stephens JL, Taylor GJ (2004) Reverse genetic analysis of the glutathione metabolic pathway suggests a novel role of *PHGPX* and *URE2* genes in aluminum resistance in *Saccharomyces cerevisiae*. Mol Genet Genom 271:627–637

Bhuja P, McLachlan K, Stephens J, Taylor GJ (2004) Accumulation of 1,3-β-D-glucans, in response to aluminum and cytosolic calcium in *Triticum aestivum*. Plant Cell Physiol 45:543–549

Bouche N, Fromm H (2004) GABA in plants: just a metabolite? Trends Plant Sci 9:110–115

Chen ZC, Yamaji N, Motoyama R, Nagamura Y, Ma J (2012) Up-regulation of a magnesium transporter gene *OsMGT1* is required for conferring aluminum tolerance in rice. Plant Physiol 159:1624–1633

de la Fuente JM, Ramirez-Rodriguez V, Cabrera-Ponce JL, Herrera-Estrella L (1997) Aluminum tolerance in transgenic plants by alteration of citrate synthesis. Science 276:1566–1568

Ding ZJ, Yan JY, Xu XY, Li GX, Zheng SJ (2013) WRKY46 functions as a transcriptional repressor of *ALMT1*, regulating aluminum-induced malate secretion in Arabidopsis. Plant J 76:825–835

Eastmond PJ, Quettier AL, Kroon JT, Craddock C, Adams N, Slabas AR (2010) Phosphatidic acid phosphohydrolase 1 and 2 regulate phospholipid synthesis at the endoplasmic reticulum in Arabidopsis. Plant Cell 22:2796–2811

Ezaki B, Gardner RC, Ezaki Y, Matsumoto H (2000) Expression of aluminum-induced genes in transgenic Arabidopsis plants can ameliorate aluminum stress and/or oxidative stress. Plant Physiol 122:657–665

Finkemeier I, König A, Heard W, Nunes-Nesi A, Pham PA, Leister D, Fernie AR, Sweetlove LJ (2013) Transcriptomic analysis of the role of carboxylic acids in metabolite signaling in Arabidopsis leaves. Plant Physiol 162:239–253

Foy CD (1992) Soil chemical factors limiting plant root growth. In: Hatfield JL, Stewart BA (eds) Limitations to plant root growth, vol 19, Advances in soil science. Springer, New York, NY, pp 97–149

Furukawa J, Yamaji N, Wang H, Mitani N, Murata Y, Sato K, Katsuhara M, Takeda K, Ma JF (2007) An aluminum-activated citrate transporter in barley. Plant Cell Physiol 48:1081–1091

Garcia-Oliveira AL, Benito C, Prieto P, de Andrade MR, Rodrigues-Pousada C, Guedes-Pinto H, Martins-Lopes P (2013) Molecular characterization of *TaSTOP1* homoeologues and their response to aluminium and proton ($H^+$) toxicity in bread wheat (*Triticum aestivum* L.). BMC Plant Biol 13:134

Hamilton CA, Good AG, Taylor GJ (2001a) Vacuolar $H^+$-ATPase, but not mitochondrial $F_1F_0$-ATPase, is required for aluminum resistance in *Saccharomyces cerevisiae*. FEMS Microbiol Lett 205:231–236

Hamilton CA, Good AG, Taylor GJ (2001b) Induction of vacuolar ATPase and mitochondrial ATP synthase by aluminum in an aluminum-resistant cultivar of wheat. Plant Physiol 125:2068–2077

Huang CF, Yamaji N, Mitani N, Yano M, Nagamura Y, Ma JF (2009) A bacterial-type ABC transporter is involved in aluminum tolerance in rice. Plant Cell 21:655–667

Huang C, Yamaji N, Chen Z, Ma JF (2012) A tonoplast-localized half-size ABC transporter is required for internal detoxification of aluminum in rice. Plant J 69:857–867

Iuchi S, Koyama H, Iuchi A, Kobayashi Y, Kitabayashi S, Kobayashi Y, Ikka T, Hirayama T, Shinozaki K, Kobayashi M (2007) Zinc finger protein STOP1 is critical for proton tolerance in

Arabidopsis and coregulates a key gene in aluminum tolerance. Proc Natl Acad Sci U S A 104:9900–9905
Iuchi S, Kobayashi Y, Koyama H, Kobayashi M (2008) STOP1, a Cys2/His2 type zinc-finger protein, plays critical role in acid soil tolerance in Arabidopsis. Plant Signal Behav 3:128–130
Kinraide TB (1994) Use of a Gouy-Chapman-Stern model for membrane-surface electrical potential to interpret some features of mineral rhizotoxicity. Plant Physiol 106:1583–1592
Kinraide TB (1998) Three mechanisms for the calcium alleviation of mineral toxicities. Plant Physiol 118:513–520
Kinraide TB (2003) Toxicity factors in acidic forest soils: attempts to evaluate separately the toxic effects of excessive $Al^{3+}$ and $H^+$ and insufficient $Ca^{2+}$ and $Mg^{2+}$ upon root elongation. Eur J Soil Sci 54:323–333
Kinraide TB, Parker DR, Zobel RW (2005) Organic acid secretion as a mechanism of aluminium resistance: a model incorporating the root cortex, epidermis, and the external unstirred layer. J Exp Bot 56:1853–1865
Kobayashi Y, Furuta Y, Ohno T, Hara T, Koyama H (2005) Quantitative trait loci controlling aluminium tolerance in two accessions of *Arabidopsis thaliana* (Landsberg *erecta* and Cape Verde Islands). Plant Cell Environ 28:1516–1524
Kobayashi Y, Hoekenga OA, Itoh H, Nakashima M, Saito S, Shaff JE, Maron LG, Pineros MA, Kochian LV, Koyama H (2007) Characterization of *AtALMT1* expression in aluminum-inducible malate release and its role for rhizotoxic stress tolerance in Arabidopsis. Plant Physiol 145:843–852
Kobayashi Y, Kobayashi Y, Sugimoto M, Lakshmanan V, Iuchi S, Kobayashi M, Bais HP, Koyama H (2013a) Characterization of the complex regulation of *AtALMT1* expression in response to phytohormones and other inducers. Plant Physiol 162:732–740
Kobayashi Y, Kobayashi Y, Watanabe T, Shaff JE, Ohta H, Kochian L, Wagatsuma T, Kinraide TB, Koyama H (2013b) Molecular and physiological analysis of $Al^{3+}$ and $H^+$ rhizotoxicities at moderately acidic conditions. Plant Physiol 163:180–192
Kobayashi Y, Ohyama Y, Kobayashi Y, Ito H, Iuchi S, Fujita M, Zhao C, Tanveer T, Ganesan M, Kobayashi M, Koyama H (2013c) STOP2 activates transcription of several genes for Al- and low pH-tolerance that are regulated by STOP1 in Arabidopsis. Mol Plant 7:311–322
Kochian LV, Hoekenga OA, Piñeros MA (2004) How do crop plants tolerate acid soils? Mechanisms of aluminum tolerance and phosphorous efficiency. Annu Rev Plant Biol 55:459–493
Koyama H, Takita E, Kawamura A, Hara T, Shibata D (1999) Over expression of mitochondrial citrate synthase gene improves the growth of carrot cells in Al-phosphate medium. Plant Cell Physiol 40:482–488
Koyama H, Kawamura A, Kihara T, Hara T, Takita E, Shibata D (2000) Overexpression of mitochondrial citrate synthase in *Arabidopsis thaliana* improved growth on a phosphorus-limited soil. Plant Cell Physiol 41:1030–1037
Larsen PB, Geisler MJ, Jones CA, Williams KM, Cancel JD (2005) *ALS3* encodes a phloem-localized ABC transporter-like protein that is required for aluminum tolerance in Arabidopsis. Plant J 41:353–363
Levitt J (1980) Responses of plants to environmental stresses. Volume I. Chilling, freezing, and high temperature stresses. Volume II. Water, radiation, salt, and other stresses. Academic Press, New York, NY
Li JY, Liu J, Dong D, Jia X, McCouch SR, Kochian LV (2014) Natural variation underlies alterations in Nramp aluminum transporter (*NRAT1*) expression and function that play a key role in rice aluminum tolerance. Proc Natl Acad Sci U S A 111:6503–6508
Liang C, Pineros MA, Tian J, Yao Z, Sun L, Liu J, Shaff J, Coluccio A, Kochian LV, Liao H (2013) Low pH, aluminum, and phosphorus coordinately regulate malate exudation through *GmALMT1* to improve soybean adaptation to acid soils. Plant Physiol 161:1347–1361
Ligaba A, Kochian L, Piñeros M (2009) Phosphorylation at S384 regulates the activity of the *TaALMT1* malate transporter that underlies aluminum resistance in wheat. Plant J 60:411–423

Liu J, Magalhaes JV, Shaff J, Kochian LV (2009) Aluminum-activated citrate and malate transporters from the MATE and ALMT families function independently to confer Arabidopsis aluminum tolerance. Plant J 57:389–399

Liu MY, Chen WW, Xu JM, Fan W, Yang JL, Zheng SJ (2013) The role of *VuMATE1* expression in aluminium-inducible citrate secretion in rice bean (*Vigna umbellata*) roots. J Exp Bot 64:1795–1804

Lynch JP, Wojciechowski T (2015) Opportunities and challenges in the subsoil: pathways to deeper rooted crops. J Exp Bot. doi:10.1093/jxb/eru508

Ma JF, Hiradate S, Matsumoto H (1998) High aluminum resistance in buckwheat. II. Oxalic acid detoxifies aluminum internally. Plant Physiol 117:753–759

Magalhaes JV, Liu J, Guimaraes CT, Lana UG, Alves VM, Wang Y, Schaffert RE, Hoekenga OA, Pineros MA, Shaff JE (2007) A gene in the multidrug and toxic compound extrusion (MATE) family confers aluminum tolerance in sorghum. Nat Genet 39:1156–1161

Matsumoto H (2000) Cell biology of aluminum toxicity and tolerance in higher plants. Int Rev Cytol 200:1–46

Moseyko N, Feldman LJ (2001) Expression of pH-sensitive green fluorescent protein in *Arabidopsis thaliana*. Plant Cell Environ 24:557–563

Mujika J, Ruiperez F, Infante I, Ugalde J, Exley C, Lopez X (2011) Pro-oxidant activity of aluminum: stabilization of the aluminum superoxide radical ion. J Phys Chem A 115:6717–6723

Nezames CD, Ochoa V, Larsen PB (2012a) Mutational loss of Arabidopsis *SLOW WALKER2* results in reduced endogenous spermine concomitant with increased aluminum sensitivity. Funct Plant Biol 40:67–78

Nezames CD, Sjogren CA, Barajas JF, Larsen PB (2012b) The Arabidopsis cell cycle checkpoint regulators TANMEI/ALT2 and ATR mediate the active process of aluminum-dependent root growth inhibition. Plant Cell 24:608–621

Ohyama Y, Ito H, Kobayashi Y, Ikka T, Morita A, Kobayashi M, Imaizumi R, Aoki T, Komatsu K, Sakata Y, Satoshi I, Koyama H (2013) Characterization of *AtSTOP1* orthologous genes in tobacco and other plant species. Plant Physiol 162:1937–1946

Panda SK, Yamamoto Y, Kondo H, Matsumoto H (2008) Mitochondrial alterations related to programmed cell death in tobacco cells under aluminium stress. C R Biol 331:597–610

Panda SK, Sahoo L, Katsuhara M, Matsumoto H (2013) Overexpression of alternative oxidase gene confers aluminum tolerance by altering the respiratory capacity and the response to oxidative stress in tobacco cells. Mol Biotechnol 54:551–563

Richards KD, Schott EJ, Sharma YK, Davis KR, Gardner RC (1998) Aluminum induces oxidative stress genes in *Arabidopsis thaliana*. Plant Physiol 116:409–418

Ruipérez F, Mujika J, Ugalde J, Exley C, Lopez X (2012) Pro-oxidant activity of aluminum: promoting the Fenton reaction by reducing Fe (III) to Fe (II). J Inorg Biochem 117:118–123

Ryan P, Kinraide T, Kochian L (1993) $Al^{3+}$-$Ca^{2+}$ interactions in aluminum rhizotoxicity. Planta 192:98–103

Sagi M, Fluhr R (2001) Superoxide production by plant homologues of the $gp91^{phox}$ NADPH oxidase. Modulation of activity by calcium and by tobacco mosaic virus infection. Plant Physiol 126:1281–1290

Sasaki T, Yamamoto Y, Ezaki B, Katsuhara M, Ahn SJ, Ryan PR, Delhaize E, Matsumoto H (2004) A wheat gene encoding an aluminum-activated malate transporter. Plant J 37:645–653

Sasaki T, Tsuchiya Y, Ariyoshi M, Ryan PR, Furuichi T, Yamamoto Y (2014) A domain-based approach for analyzing the function of aluminum-activated malate transporters from wheat (*Triticum aestivum*) and *Arabidopsis thaliana* in *Xenopus* oocytes. Plant Cell Physiol 55:2126–2138

Sawaki Y, Iuchi S, Kobayashi Y, Kobayashi Y, Ikka T, Sakurai N, Fujita M, Shinozaki K, Shibata D, Kobayashi M, Koyama H (2009) STOP1 regulates multiple genes that protect Arabidopsis from proton and aluminum toxicities. Plant Physiol 150:281–294

Sawaki Y, Kihara-Doi T, Kobayashi Y, Nishikubo N, Kawazu T, Kobayashi Y, Koyama H, Sato S (2013) Characterization of Al-responsive citrate excretion and citrate-transporting MATEs in *Eucalyptus camaldulensis*. Planta 237:979–989

Sawaki Y, Kobayashi Y, Kihara-Doi T, Nishikubo N, Kawazu T, Kobayashi M, Kobayashi Y, Iuchi S, Koyama H, Sato S (2014) Identification of a STOP1-like protein in *Eucalyptus* that regulates transcription of Al tolerance genes. Plant Sci 223:8–15

Sivaguru M, Horst WJ (1998) The distal part of the transition zone is the most aluminum-sensitive apical root zone of maize. Plant Physiol 116:155–163

Spadoni S, Zabotina O, Di Matteo A, Mikkelsen JD, Cervone F, De Lorenzo G, Mattei B, Bellincampi D (2006) Polygalacturonase-inhibiting protein interacts with pectin through a binding site formed by four clustered residues of arginine and lysine. Plant Physiol 141:557–564

Sumner M, Shahandeh H, Bouton J, Hammel J (1986) Amelioration of an acid soil profile through deep liming and surface application of gypsum. Soil Sci Soc Am J 50:1254–1258

Taylor GJ (1987) Exclusion of metals from the symplasm: a possible mechanism of metal tolerance in higher plants. J Plant Nutr 10:1213–1222

Taylor GJ (1988) The physiology of aluminum phytotoxicity. In: Sigel H (ed) Aluminum and its role in biology, vol 24, Metal ions in biological systems. Marcel Dekker, New York, NY, pp 123–163

Taylor GJ (1991) Current views of the aluminum stress response: the physiological basis of tolerance. Curr Top Plant Biochem Physiol 10:57–93

Taylor GJ, McDonald-Stephens JL, Hunter DB, Bertsch PM, Elmore D, Rengel Z, Reid RJ (2000) Direct measurement of aluminum uptake and distribution in single cells of *Chara corallina*. Plant Physiol 123:987–996

Tesfaye M, Temple SJ, Allan DL, Vance CP, Samac DA (2001) Overexpression of malate dehydrogenase in transgenic alfalfa enhances organic acid synthesis and confers tolerance to aluminum. Plant Physiol 127:1836–1844

Tokizawa M, Kobayashi Y, Saito T, Kobayashi M, Iuchi S, Nomoto M, Tada Y, Yamamoto YY, Koyama H (2015) SENSITIVE TO PROTON RHIZOTOXICITY1, CALMODULIN BINDING TRANSCRIPTION ACTIVATOR2, and other transcription factors are involved in *ALUMINUM-ACTIVATED MALATE TRANSPORTER1* expression. Plant Physiol 167:991–1003

Trejo-Téllez L, Stenzel R, Gómez-Merino F, Schmitt J (2010) Transgenic tobacco plants overexpressing pyruvate phosphate dikinase increase exudation of organic acids and decrease accumulation of aluminum in the roots. Plant Soil 326:187–198

Wenzl P, Patino GM, Chaves AL, Mayer JE, Rao IM (2001) The high level of aluminum resistance in signal grass is not associated with known mechanisms of external aluminum detoxification in root apices. Plant Physiol 125:1473–1484

Xia J, Yamaji N, Ma JF (2013) A plasma membrane-localized small peptide is involved in rice aluminum tolerance. Plant J 76:345–355

Yamaji N, Huang CF, Nagao S, Yano M, Sato Y, Nagamura Y, Ma JF (2009) A zinc finger transcription factor ART1 regulates multiple genes implicated in aluminum tolerance in Rice. Plant Cell 21:3339–3349

Yamamoto Y, Hachiya A, Matsumoto H (1997) Oxidative damage to membranes by a combination of aluminum and iron in suspension-cultured tobacco cells. Plant Cell Physiol 38:1333–1339

Yamamoto Y, Kobayashi Y, Devi SR, Rikiishi S, Matsumoto H (2002) Aluminum toxicity is associated with mitochondrial dysfunction and the production of reactive oxygen species in plant cells. Plant Physiol 128:63–72

Yamamoto Y, Yoshioka Y, Hyakumachi M, Maruyama K, Yamaguchi-Shinozaki K, Tokizawa M, Koyama H (2011) Prediction of transcriptional regulatory elements for plant hormone responses based on microarray data. BMC Plant Biol 11:39

Yang ZB, Geng X, He C, Zhang F, Wang R, Horst WJ, Ding Z (2014) TAA1-regulated local auxin biosynthesis in the root-apex transition zone mediates the aluminum-induced inhibition of root growth in Arabidopsis. Plant Cell 26:2889–2904

Yokosho K, Yamaji N, Ma JF (2011) An Al-inducible MATE gene is involved in external detoxification of Al in rice. Plant J 68:1061–1069

Zhang J, Jia W, Yang J, Ismail AM (2006) Role of ABA in integrating plant responses to drought and salt stresses. Field Crops Res 97:111–119

Zheng SJ, Ma JF, Matsumoto H (1998) High aluminum resistance in buckwheat: I. Al-induced specific secretion of oxalic acid from root tips. Plant Physiol 117:745–751

# Physiological and Molecular Regulation of Aluminum Resistance in Woody Plant Species

**Marjorie Reyes-Díaz, Claudio Inostroza-Blancheteau, and Zed Rengel**

**Abstract** Aluminum (Al) is the main limiting factor for plant growth in acid soils. Woody plant species are well adapted to acid soils with high $Al^{3+}$ concentration. The external resistance mechanisms comprise Al immobilization in the rhizosphere (Al excluders) and internal ones include complexation of Al in cells (Al accumulators). This chapter provides a critical analysis of the physiological and molecular regulation of Al-resistance mechanisms in woody plant species.

## 1 Introduction

Acid soils are prevalent in many regions of the world and represent one-third of the available terrestrial land worldwide; moreover, the acidity problem is being aggravated due to the extensive use of ammonium fertilizers (von Uexkull and Mutert 1995). The wide distribution of acid soils is specific to boreal and the tropic regions that correspond with location of woody plant species. Acid soils typically have pH in water <5.0–5.5 and are associated with several stress factors, such as toxicity of Al, $Mn^{2+}$, and $H^{+}$ and deficiency of P and $Ca^{2+}$. However, ionic aluminum (Al) is

M. Reyes-Díaz (✉)
Departamento de Ciencias Químicas y Recursos Naturales, Facultad de Ingeniería y Ciencias, Universidad de La Frontera, P.O. Box 54-D, Temuco, Chile

Center of Plant, Soil Interaction and Natural Resources Biotechnology, Scientific and Technological Bioresource Nucleus (BIOREN), Universidad de La Frontera, Temuco, Chile
e-mail: marjorie.reyes@ufrontera.cl

C. Inostroza-Blancheteau
Núcleo de Investigación en Producción Alimentaria (NIPA), Facultad de Recursos Naturales, Universidad Católica de Temuco, P.O. Box 56-D, Temuco, Chile

Escuela de Agronomía, Facultad de Recursos Naturales, Universidad Católica de Temuco, Temuco, Chile

Z. Rengel
Soil Science and Plant Nutrition, School of Earth and Environment, The University of Western Australia, Crawley, WA 6009, Australia

 
S.K. Panda, F. Baluška (eds.), *Aluminum Stress Adaptation in Plants*, Signaling and Communication in Plants 24, DOI 10.1007/978-3-319-19968-9_10

considered as the main limiting factor for growth and productivity of plants (Kochian et al. 2004; Mora et al. 2004; Osawa et al. 2011). Aluminum is the third most abundant element in the Earth's crust, after oxygen and silicon. Acidity solubilizes Al from nontoxic silicate or oxide forms into phytotoxic Al (mainly trivalent ion $Al^{3+}$) (Tahara et al. 2008a). Aluminum can inhibit root growth in many plant species at micromolar concentrations, suggesting Al interferes with dividing and expanding root cells to inhibit root elongation (Mora et al. 2005; Tamás et al. 2006). This inhibition limits uptake of water and nutrients and increases plant susceptibility to other stresses in the root zone, i.e., low pH, $Ca^{2+}$ imbalance, etc. (Koyama et al. 2001; Grisel et al. 2010; Inostroza-Blancheteau et al. 2012).

Most studies regarding Al stress are focused on roots, where Al effects are manifested first (Yamamoto et al. 2001; Barceló and Poschenrieder 2002). Inhibition of shoot growth, in turn, appears to be a secondary response to Al, being mediated through interactions between root Al and nutrient translocation to shoots (Lidon et al. 1999). In fact, restricted shoot growth becomes evident only after root growth was limited by exposure to toxic concentrations of Al ions (Rengel 1996).

Despite decades of extensive efforts to decipher the mechanism(s) of Al phytotoxicity, the primary cause of Al remains largely speculative (Horst 1995; Kochian 1995; Rengel 1996; Matsumoto 2000; Barceló and Poschenrieder 2002). Aluminum has a strong binding affinity for oxygen donor compounds such as inorganic phosphate, nucleotides, RNA, DNA, proteins, carboxylic acids, and phospholipids (Ma 2000). Aluminum cannot catalyze redox reactions, but instead triggers lipid peroxidation and production of reactive oxygen species (ROS) in roots (Yamamoto et al. 2002; Tamás et al. 2006). Most of Al associated with roots is located in the cell wall, having displaced calcium (Rengel 1992, 1996; Rengel and Zhang 2003). Calcium is also present on the surface of the plasma membrane, interacting with proteins and phospholipids and affecting membrane fluidity and thereby its functionality (Hanson 1984; Kaus 1987); Al can displace Ca from the plasma membrane binding sites as well as provoke a disruption of Ca homeostasis in the cytosol (Rengel and Zhang 2003). However, more evidence is needed to prove any of these mechanisms operating in woody plant species.

Several Al-resistance mechanisms (external and internal) have been proposed (Watanabe and Osaki 2002; Ryan and Delhaize 2010; Inostroza-Blancheteau et al. 2012). External resistance mechanisms comprise Al immobilization in the cell wall, selective permeability of the plasma membrane, root-induced increase in the rhizosphere pH, and exudation of chelating ligands (Wagatsuma and Yamasaka 1985; Delhaize et al. 1993; Basu et al. 1994; Emmanuel and Peter 1995; Pellet et al. 1995). Internal resistance mechanisms include complexation of Al by proteins and other ligands in the cytosol, compartmentalization in the vacuoles, evolution of Al-tolerant enzymes, elevation of enzyme activity, and induction of biosynthesis of specific proteins (Keltjens and Ulden 1987; Kasai et al. 1992; Taylor et al. 1997).

Plant species that have internal Al tolerance mechanisms can sometimes accumulate Al at high concentrations (Al-accumulators), ranging from 0.2 to 40 g Al $kg^{-1}$ DW (Table 1). In this context, Al-accumulator species are those accumulating over 1 g Al $kg^{-1}$ DW in their leaves (Jansen et al. 2003). However, until now,

**Table 1** Al accumulation in leaves several woody plant species

| Plant species | Al accumulation ($g\ kg^{-1}$ DW) | Classification | References |
|---|---|---|---|
| *Vaccinium corymbosum* | 0.4 | Non-accumulator | Reyes-Diaz et al. (2009, 2010) |
| *Ugni molinae* | 0.3 | Non-accumulator | Reyes-Diaz et al. unpublished |
| *Diospyros sumatrana* | 1.0 | Accumulator | Masunaga et al. (1998) |
| *Ganua mottleyana* | 1.0 | Accumulator | Masunaga et al. (1998) |
| *Palaquium abovatum* | 1.0 | Accumulator | Masunaga et al. (1998) |
| *Canthium confertum* | 1.8 | Accumulator | Jansen et al. (2003) |
| *Danais fragrans* | 9.0 | Accumulator | Jansen et al. (2003) |
| *Gouldia terminalis* | 1.9 | Accumulator | Jansen et al. (2003) |
| *Miconia albicans* | 6.0 | Accumulator | Haridasan (2008) |
| *Miconia pohliana* | 5.0 | Accumulator | Haridasan (2008) |
| *Qualea parviflora* | 4.0 | Accumulator | Haridasan (2008) |
| *Populus hibridus* | 0.2 | Accumulator | Wannaz et al. (2012) |
| *Eucalyptus rostrata* | 0.7 | Accumulator | Wannaz et al. (2012) |
| *Pinus spp* | 0.1 | Accumulator | Wannaz et al. (2012) |
| *Camellia sinensis* | 30 | Hyperaccumulator | Matsumoto et al. (1976) |
| *Miconia albicans* | 11 | Hyperaccumulator | De Medeiros and Haridasan (1985) |
| *Faramea marginata* | 16 | Hyperaccumulator | Britez et al. (1997) |
| *Melastoma malabathricum* | 10 | Hyperaccumulator | Watanabe et al. (1998) |
| *Faramea marginata* | 18 | Hyperaccumulator | Britez et al. (2002) |
| *Coptosapelta olaciformis* | 10 | Hyperaccumulator | Jansen et al. (2003) |
| *Craterispermum laurinum* | 30 | Hyperaccumulator | Jansen et al. (2003) |
| *Vochysia rufa* | 28 | Hyperaccumulator | Haridasan (2008) |
| *Vochysia thyrsoidea* | 36 | Hyperaccumulator | Haridasan (2008) |
| *Vochysia tucanorum* | 40 | Hyperaccumulator | Haridasan (2008) |
| *Salvertia convallariodora* | 28 | Hyperaccumulator | Haridasan (2008) |
| *Qualea multiflora* | 20 | Hyperaccumulator | Haridasan (2008) |
| *Conostegia xalapensis* | 19 | Hyperaccumulator | González-Santana et al. (2012) |

there is no clarity on the concept of Al-hyperaccumulator species, considering the published evidence and the wide range of species that accumulate Al (Table 1). We suggest that leaf Al concentration of more than 10 g Al $kg^{-1}$ DW is a suitable criterion for defining Al-hyperaccumulator plant species (Table 1).

Even though plants can be classified as Al excluders or Al accumulators, depending on whether Al is chelated by organic anions in the rhizosphere, or is taken up by roots and transported to shoots (Jansen et al. 2003), it is important to bear in mind that plant responses to Al toxicity are highly dependent of the species or cultivar under study (Kochian 1995; Rout et al. 2001; Barceló and Poschenrieder 2002). Also, wild species are frequently more resistant to Al stress and can accumulate higher concentrations of Al in their leaves than cultivated species (Kochian 1995; Piñeros et al. 2005).

## 2 Physiological Aspects of Aluminum Toxicity and Resistance in Woody Plants

At the physiological level, several studies showed that Al can affect negatively photosynthesis, photoprotective compounds, water content, mineral nutrition, etc. Aluminum negatively affected net photosynthesis in several plants species (Moustakas et al. 1995; Pereira et al. 2003; Chen 2006). In *Quercus glauca*, Akaya and Takenaka (2001) found leaf concentration of Al increasing and that of P decreasing, accompanied by a decrease in water absorption by roots, suggesting that Al may affect uptake and transport of water and nutrients.

The most abundant organic acid anion exuded from roots of woody species in response to Al toxicity is citrate, followed by oxalate and to a lesser extent malate (Table 2). Mitochondrial metabolism is a key in the regulation of organic acid biosynthesis in plants under Al stress (Nunes-Nesi et al. 2014). For example, the citrate synthase activity in *Secale cereale* roots can be increased by 30 % after 6 h of exposure to Al stress just before an increase in citrate efflux (Li et al. 2000). Similarly, in the Al-resistant tree species *Paraserianthes falcataria*, the activity of citrate synthase in roots was increased together with the quantity of transcript when the tree was exposed to Al toxicity (Osawa and Kojima 2006).

In woody species, root exudation of phenolic compounds into the rhizosphere was proposed as an Al-exclusion mechanism (Kidd et al. 2001; Barceló and Poschenrieder 2002). However, complexation of Al by phenolic compounds may be less important that the formation of complexes with organic acid anions because phenolics are less efficient than, e.g., citrate in complexing Al (Ofei-Manu et al. 2001).

A new Al-binding ligand was recently described in *Eucalyptus camaldulensis* roots; it is hydrolyzable tannin (oenothein B) with many adjacent phenolic hydroxyl groups (Tahara et al. 2014). Oenothein B was not detected in other woody plants

**Table 2** Aluminum-activated release of organic acid anions from roots of different woody plant species

| Plant species | Organic acid anion | Tissue from which exudation was measured | References |
|---|---|---|---|
| Horticultural species | | | |
| *Citrus junos* | Citrate | Whole roots | Deng et al. (2009) |
| *Camellia sinensis* | Oxalate | Root tips | Morita et al. (2011) |
| *Citrus grandis* | Citrate, malate | Root tips | Yang et al. (2011) |
| *Citrus sinensis* | Citrate, malate | Root tips | Yang et al. (2011) |
| Forest species | | | |
| *Picea abies* | Succinate, oxalate | Whole roots | Heim et al. (2001, 2003) |
| *Pinus sylvestris* | Oxalate | Whole roots | Ahonen-Jonnarth et al. (2003) |
| *Pinus densiflora* | Citrate | Whole roots | Tahara et al. (2005) |
| *Cryptomeria japonica* | Citrate, oxalate | Root tips | Hirano et al. (2012) |
| *Eucalyptus camaldulensis* | Citrate | Root tips | Ikka et al. (2013) |
| *Eucalyptus globulus* | Citrate, malate | Root tips | Silva et al. (2004) |
| *Eucalyptus grandis* | Citrate | Root tips | Silva et al. (2004) |
| *Eucalyptus euophylla* | Citrate, malate, oxalate | Root tips | Silva et al. (2004) |
| *Populus tremula* | Citrate oxalate | Root tips | Qin et al. (2007) |
| *Populus tremuloides* | Citrate, malate, oxalate, succinate | Root tips | Naik et al. (2009) |
| *Populus trichocarpa* | Citrate, malate, oxalate, succinate | Root tips | Naik et al. (2009) |
| *Pinus thunbergii* | Citrate, oxalate | Root tips | Hirano et al. (2012) |
| Ornamental species | | | |
| *Acacia auriculiformis* | Citrate, oxalate | Whole roots | Nguyen et al. (2003) |
| *Melaleuca cajuputi* | Citrate, oxalate | Whole roots | Nguyen et al. (2003) |
| *Melaleuca leucadendra* | Citrate | Whole roots | Nguyen et al. (2003) |
| *Melastoma malabathricum* | Citrate | Whole roots | Watanabe and Osaki (2002) |
| *Lespedeza bicolor* | Citrate, malate | Whole roots | Dong et al. (2008) |
| *Paraserianthes falcataria* | Citrate | Root tips | Osawa and Kojima (2006) |

(continued)

**Table 2** (continued)

| Plant species | Organic acid anion | Tissue from which exudation was measured | References |
|---|---|---|---|
| *Melaleuca bracteata* | Citrate | Root tips | Tahara et al. (2008b) |
| *Cinnamomum camphora* | Citrate | Root tips | Osawa et al. (2011) |

such as *Melaleuca bracteata* and *Populus nigra* and neither in the model herbaceous species Arabidopsis.

For tree species, callose induction after exposure to Al has been reported in the seedlings of *Castanea sativa* (Hirano et al. 2006), *Picea abies* (Jorns et al. 1991; Wissemeier et al. 1998; Hirano et al. 2004), *Populus spp.* (Qin et al. 2007; Smith et al. 2011), and some tropical species (Tahara et al. 2005). Callose is a defense-related cell wall polysaccharide (1,3-[β]-glucan) in root apices; it has been used as a physiological indicator in differentiating Al-sensitive and Al-resistant crop genotypes because its formation is induced within a few hours after Al exposure (Wissemeier et al. 1987; Horst et al. 1997; Eticha et al. 2005). In the coniferous *Pinus thunbergii*, callose was induced after 1-day exposure to Al and was distributed mainly in the root apex (Jones et al. 2006). However, no induction of callose was recorded in *Camellia japonica* after the exposure to Al in contrast to other tree species (Tahara et al. 2005; Stass et al. 2008). This could be related to Al distribution within the *Camellia japonica* roots, i.e., little Al in the inner root cells such as cortical cells and stele, which differed from that in other tree species. These results raised a possible additional mechanism of Al resistance (based on Al distribution in root cells) in tree species adapted to acid soils. However, more evidence is eagerly awaited.

## 3 Aluminum Effects on Photosynthesis in Woody Plants

Comparatively less is known about the effects of Al in leaves than roots. There is relatively little Al translocation from roots to the above-ground parts. Nevertheless, Al toxicity induces stunted, dark-green leaves; purpling of stems, leaves, and veins; and collapse of growing points or petioles, which in some cases is due to $Ca^{2+}$ deficiency or to a reduced $Ca^{2+}$-transport problem (Rout et al. 2001). Aluminum toxicity may lead also to malformations in chloroplast, even though detectable amounts of Al may not be observed, indicating indirect effects on chloroplast functioning (Moustakas et al. 1995). Aluminum toxicity decreases both total chlorophyll content and photosynthetic rate in some species (Akaya and Takenaka 2001; Rout et al. 2001; Chen et al. 2005a, b; Chen 2006). Aluminum decreases $CO_2$ assimilation in many plant species including *Sorghum bicolor* (Peixoto et al. 2002), *Zea maize* (Lidon et al. 1999), and the woody plant species *Citrus* (Pereira et al. 2000; Chen et al. 2005a, b; Jiang et al. 2008, 2009).

In *Citrus*, Al decreased net photosynthesis by inhibiting $CO_2$ assimilation (Chen et al. 2005a). A decrease in the rate of $CO_2$ assimilation in this species could be associated with structural damage of thylakoids and a decrease in some photochemical parameters such as the rate of variable fluorescence (Fv) to initial fluorescence (F0) (Pereira et al. 2000). Similar behavior was observed in net photosynthesis in *Quercus glauca*, but no changes were observed in the chlorophyll fluorescence and chlorophyll content with and without Al exposure (Akaya and Takenaka 2001; Chen et al. 2005a).

Yang et al. (2012) reported that *Citrus grandis* exposed to Al showed higher or similar intercellular $CO_2$ concentration as the control plants, indicating that an Al-induced decrease in $CO_2$ assimilation was primarily caused by non-stomatal factors, as previously reported for another species of *Citrus* (Chen et al. 2005a, b; Jiang et al. 2008, 2009). In *Citrus grandis*, despite decreased $CO_2$ assimilation, Al increased or did not affect non-structural carbohydrates in leaves. This may be due to a decreased demand for reduced C in growing sink tissues and the less dilution due to growth inhibition (Yang et al. 2012).

Moustakas et al. (1995) point out that Al stress inhibits photosynthesis as a result of a partial inhibition of photosynthetic electron transport in photosystem II (PSII) and closure of PSII reaction centers. In addition, chloroplast elemental loss and intra-thylakoid acidification have also been observed (Lidon et al. 1999). Reyes-Diaz et al. (2009) showed that a short-term exposure to Al differentially affected photochemical efficiency of photosystem II in *Vaccinium corymbosum* cultivars. Photochemical parameters decreased substantially in the Al treatments in Bluegold cultivar (up to 98 % inhibition) and Legacy cultivar (up to 80 % inhibition) without total recovery. In contrast, Brigitta cultivar showed a better PSII performance and root growth than the other cultivars, suggesting that Brigitta is the best cultivar for use in acid soils with Al toxicity, followed by Legacy, whereas Bluegold was the most sensitive. These results were confirmed by Reyes-Díaz et al. (2010), with the photochemical parameters being affected more in Bluegold than Legacy cultivar. In *Citrus* rootstocks treated with Al in the nutrient solution, Pereira et al. (2003) found that leaf area and dry mass of leaves decreased, suggesting a lower production of photoassimilates and less plant growth. In accordance, Pereira et al. (2000) indicated that reduced photosynthesis rate by Al caused a decrease in leaf area.

As mentioned above, Al toxicity negatively influences total chlorophyll in woody plants species. However, Wannaz et al. (2012) found an increase in total chlorophyll content concomitant with an increase in leaf Al concentration. Total chlorophyll was similar between deciduous tree species and was greater in conifers. In this study, chlorophyll degradation parameters were also measured (Phe-a/Chl-a), showing the highest values in *Pinus* spp. A positive correlation in *Pinus* needles was found between Al concentration and total chlorophyll ($r = 0.37$) and a negative correlation between total chlorophyll and Phe-a/Chl-a ($r = -0.48$). On the contrary, Yang et al. (2012) indicated that Al affected chlorophyll content less than $CO_2$ assimilation, suggesting that a decrease in chlorophyll content by Al is probably not the primary factor limiting $CO_2$ assimilation. This suggestion was previously reported for *Fagus sylvatica* (Ridolfi and Garrec 2000), *Citrus reshni*

(Chen et al. 2005b), and *C. grandis* (Jiang et al. 2008, 2009). According to Pereira et al. (2000) and Chen et al. (2005a), only a fraction of the absorbed light energy was used in electron transport in Al-treated *Citrus* leaves. That resulted in excess excitation energy with respect to the control leaves, which was not totally dissipated as heat by the antenna pigment complexes of the PSII as indicated by the lower NPQ (non-photochemical quenching). NPQ is highly correlated with the concentration of antheraxanthin (A) + zeaxanthin (Z) (Demmig-Adams and Adams 1996). However, some studies have shown no correlation between NPQ and A + Z concentration (Förster et al. 2001; Cousins et al. 2002).

Although the effects of many environmental stresses (water, temperature, nutrients, and salt) on xanthophyll cycle-dependent thermal energy dissipation have been examined in some detail (Niyogi et al. 1997; Logan et al. 1998; Ruban and Horton 1999; Adams et al. 2004; Zúñiga et al. 2006), little is known about the response of xanthophyll cycle-dependent thermal energy dissipation to Al toxicity. Yang et al. (2012) observed lower Fv/Fm in *C. grandis* leaves treated with Al compared with control ones, indicating that photoinhibitory damage to PSII complexes occurred (Maxwell and Johnson 2000). A decrease in Fv/Fm was due to an increase in Fo and a decrease in Fm. The higher Fo may be caused by both the damage of the oxygen-evolving complex and the inactivation of some reaction center of the PSII as indicated by Yamane et al. (1997). It is also supported by the finding that Fv was decreased in Al-treated leaves along an increase in Fo, which is the characteristic of photoinhibitory damage to the PSII acceptor side (Setlik et al. 1990). Hence, multiple lines of evidence suggest that Al could affect the photosynthetic apparatus; however, exact mechanisms are yet to be completely elucidated.

## 4 Molecular and Transcriptional Regulation of Aluminum Resistance in Woody Plant Species

At molecular level, several genes of crop and model plant species that were linked to better adaptation to elevated Al concentration in acid soil have been isolated and characterized. For example, the first malate transporter gene was identified in wheat (*Triticum aestivum* Al-activated malate transporter, *TaALMT1*) in response to Al toxicity (Sasaki et al. 2004). Indeed, much effort and progress has been made in understanding the physiological and molecular mechanisms underlying Al toxicity in herbaceous plants (e.g., Inostroza-Blancheteau et al. 2012; Soto-Cerda et al. 2015).

In herbaceous species, a release of organic acid anions into the rhizosphere in response to Al toxicity is mediated by *ALMT* genes that encode an anion channel specifically induced by Al, allowing malate efflux from roots (Schroeder et al. 2013). These genes have been characterized in different species such as: *Triticum aestivum*, *Secale cereale*, *Oryza sativa*, *Zea mays*, *Brassica napus*,

*Arabidopsis thaliana*, and others. In addition, genes associated with release of citrate from roots are multidrug and toxic compound extrusion transporter in *Sorghum bicolor* (*SbMATE*) and Al-activated citrate transporter 1 in *Hordeum vulgare* (*HvAACT1*), both belonging to the MATE family (Delhaize et al. 2012). However, little is known about the transcriptional regulation of Al-resistant genes in woody plants. Recently, four genes highly homologous to citrate-transporting multidrug and toxic compounds extrusion gene were isolated in *Eucalyptus camaldulensis*. One of the homologues was named *EcMATE1*; this gene was expressed more strongly in roots than shoots in response to Al toxicity and low pH (Sawaki et al. 2013). In addition, an analysis of transcriptome in *Populus tremula* revealed a total of 175 significantly upregulated and 69 downregulated genes. Two genes showed strong induction in roots and were closely related to Arabidopsis Al tolerance genes *ALS3* (for *Al-sensitive 3*) and *MATE*, suggesting an important role in Al tolerance in *Populus tremula* (Grisel et al. 2010).

In *Vaccinium corymbosum*, two cDNA libraries were established using Al-resistant and Al-sensitive genotypes with an aim of understanding the mechanisms of Al resistance in highbush blueberry (Inostroza-Blancheteau et al. 2011). In this study, a cDNA-amplified fragment length polymorphism (cDNA-AFLP) method was used to identify differential gene expression in Brigitta (Al-resistant) and Bluegold (Al-sensitive) cultivars. Seventy transcript-derived fragments (TDFs) were identified as being Al responsive, 31 of which showed significant homology to genes with known or putative functions. Several genes associated with stress responses such as *glutathione S-transferase*, *S-adenosylmethionine decarboxylase*, *aldehyde dehydrogenase*, *vacuolar* $H^+$*-pyrophosphatase*, and others were detected, but not the specific genes such as *ALMTs* in response to Al toxicity. However, this situation could be due to the limitations of the technique used.

We reported the first cloning and characterization of a calmodulin gene, *VcCaM1* (*Vaccinium corymbosum Calmodulin 1*), in this woody shrub induced by Al toxicity (Inostroza-Blancheteau et al. 2013). According to qRT-PCR and enzymatic analysis, it appeared that *VcCaM1* was not directly involved in Al resistance, but might be involved in improving plant performance under Al toxicity through regulation of $Ca^{2+}$ homeostasis and antioxidant systems in leaves.

Recently, in a leguminous tree *Acacia mangium*, 44 full-length sequences were identified and cloned using differential display and semi-quantitative RT-PCR and other molecular techniques. These include MATE- and ATP-binding cassette transporters and a plasma membrane ATPase gene that has been associated with responses to $H^+$ and Al toxicity in other species (Mizuno et al. 2014).

In herbaceous species (e.g., wheat, barley, rice, maize, sorghum, rye, and Arabidopsis), only *ALMT* and *MATE* genes have so far been described as major genes in response to Al toxicity. However, some studies suggest that regulatory genes coding for the transcription factors have an important role in Al resistance (Sawaki et al. 2009; Yamaji et al. 2009). Studies performed in Arabidopsis showed that a zinc-finger protein STOP1 (Sensitive TO Proton rhizotoxicity 1) is a critical transcription factor for proton ($H^+$) and Al tolerance in acid conditions (Iuchi et al. 2007). In tobacco plants, *NtSTOP1* regulated Al tolerance concomitant with the upregulation of *MATE* gene and release of citrate in response to Al stress (Ohyama et al. 2013). More recently, a STOP1-like protein that regulated transcription of *MATE* and *ALS3* gene was identified in *Eucalyptus* roots (Sawaki et al. 2014). Even though the activity of STOP-like protein has not been evaluated yet, bioinformatics tools and database searches showed that other woody species (e.g., conifers) contain a putative orthologue of this protein. This finding suggests the STOP1-like protein and the genes it regulates are ancestral and might be shared by a wide range of plant species. These advances together with progress in tree biotechnology (e.g., cloning, gene transformation, and overexpression) could be useful for molecular breeding of *Eucalyptus* and other tree species, with the capacity to produce transgenic plants shortening the genetic improvement cycle in these species.

## 5 Future Direction

In recent years, a wide range of studies described the identification and characterization of candidate genes and transcription factors that were upregulated by Al toxicity. With the recent advances in genetic engineering, there is an opportunity to enhance Al resistance of sensitive genotypes through the overexpression of appropriate endogenous genes or introduction of foreign genes. However, little information exists on the complete sequence of genes associated with Al resistance in woody plant species. Hence, it is indispensable to conduct more research at the molecular level to characterize the genome of woody species. Finally, it is necessary for breeders and researches to develop new strategies and protocols of genetic transformation to increase Al resistance using biotechnological tools.

## References

Adams WWIII, Zarter CR, Ebbert V, Demmig-Adams B (2004) Photoprotective strategies of overwintering evergreens. Bioscience 54:41–49

Ahonen-Jonnarth U, Göransson A, Finlay RD (2003) Growth and nutrient uptake of ectomycorrhizal *Pinus sylvestris* seedlings in a natural substrate treated with elevated Al concentrations. Tree Physiol 23:157–167

Akaya M, Takenaka C (2001) Effects of aluminum stress on photosynthesis of *Quercus glauca* Thumb. Plant Soil 237:137–146

Barceló J, Poschenrieder C (2002) Fast root growth responses, root exudates, and internal detoxification as clues to the mechanisms of aluminium toxicity and resistance: a review. Environ Exp Bot 48:75–92

Basu U, Godbold D, Tayor GJ (1994) Aluminum resistance in *Triticum aestivum* associated with enhanced exudation of malate. J Plant Physiol 144:747–753

Britez RM, Reismann CB, Silva SM, Athayde SF, Lima RX, de Quadros RMB (1997) Chemical characterization of two forests on the coastal plain of the Ilha do Mel, Paraná, Brazil. In:

Ando T, Fujita K, Mae T, Matsumoto H, Mori S, Sekiya J (eds) Plant nutrition – for sustainable food production and environment. Kluwer, Dordrecht, pp 461–462

Britez RM, Watanabe T, Jansen S, Reissmann CB, Osaki M (2002) The relationship between aluminium and silicon accumulation in leaves of *Faramea marginata* (Rubiaceae). New Phytol 156:445–456

Chen LS (2006) Physiological responses and tolerance of plant shoot to aluminum toxicity. J Plant Physiol Mol Biol 32:143–155

Chen LS, Qi Y-P, Smith BR, Liu X-H (2005a) Aluminium-induced decrease in $CO_2$ in citrus seedlings is unaccompanied by decreased activities of key enzymes involved in $CO_2$ assimilation. Tree Physiol 25:317–324

Chen LS, Qi Y-P, Liu X-H (2005b) Effects of aluminum on light energy utilization and photoprotective systems in citrus leaves. Ann Bot 96:35–41

Cousins AB, Adam NR, Wall GW, Kimball BA, Pinter PJ Jr, Ottman MJ, Leavitt SW, Webber AN (2002) Photosystem II energy use, non-photochemical quenching and the xanthophyll cycle in Sorghum bicolor grown under drought and free-air $CO_2$ enrichment (FACE) conditions. Plant Cell Environ 25:1551–1559

De Medeiros RA, Haridasan M (1985) Seasonal variations in the foliar concentrations of nutrients in some aluminium accumulating and non-accumulating species of the cerrado region of central Brazil. Plant Soil 88:433–436

Delhaize E, Ryan PR, Randall PJ (1993) Aluminum tolerance in wheat (*Triticum aestivum* L.). II. Aluminum stimulated excretion of malic acid from root apices. Plant Physiol 103:695–702

Delhaize E, Ma JP, Ryan PR (2012) Transcriptional regulation of aluminium tolerance genes. Trends Plant Sci 17:341–347

Demmig-Adams B, Adams WW III (1996) The role of xantophyll cycle carotenoids in the protection of photosynthesis. Trends Plants Sci 1:21–26

Deng W, Luo K, Li Z, Yang Y, Hu N, Wu Y (2009) Overexpression of *Citrus junos* mitochondrial citrate synthase gene in *Nicotiana benthamiana* confers aluminium tolerance. Planta 230:355–365

Dong XY, Shen RF, Chen RF, Zhu ZL, Ma JF (2008) Secretion of malate and citrate from roots is related to high Al-resistance in *Lespedeza bicolor*. Plant Soil 306:139–147

Emmanuel D, Peter PR (1995) Aluminum toxicity and tolerance in plants. Plant Physiol 107:315–321

Eticha D, The C, Welcker C, Narro L, Staß A, Horst WJ (2005) Aluminium-induced callose formation in root apices: inheritance and selection trait for adaptation of tropical maize to acid soils. Field Crops Res 93:252–263

Förster B, Osmond CB, Boynton JE (2001) Very high light resistant mutants of *Chlamydomonas reinhardtii*: responses of photosystem II, nonphotochemical quenching and xanthophyll pigments to light and $CO_2$. Photosynth Res 67:5–15

González-Santana IH, Márquez-Guzmán J, Cram-Heydrich S, Cruz-Ortega R (2012) *Conostegia xalapensis* (Melastomataceae): an aluminum accumulator plant. Physiol Plant 144:134–145

Grisel N, Zoller S, Künzli-Gontarczyk M, Lampart T, Münsterkötter M, Brunner I, Bovet L, Métraux JP, Sperisen C (2010) Transcriptome responses to aluminum stress in roots of aspen (*Populus tremula*). Plant Biol 10:185

Hanson JB (1984) The functions of calcium in plant nutrition. In: Tinker PB, Lauchli A (eds) Advances in plant nutrition, vol 1. Praeger Scientific, New York, NY, pp 149–208

Haridasan M (2008) Nutritional adaptations of native plants of the cerrado biome in acid soils. Rev Brasil Fisiol Veg 20:183–195

Heim A, Brunner I, Frey B, Frossard E, Luster J (2001) Root exudation, organic acids, and element distribution in roots of Norway spruce seedlings treated with aluminium in hydroponics. J Plant Nutr Soil Sci 164:519–526

Heim A, Brunner I, Frossard E, Luster J (2003) Aluminum effects on *Picea abies* at low solution concentrations. Soil Sci Soc Am J 67:895–898

Hirano Y, Graf-Pannatier E, Zimmermann S, Brunner I (2004) Induction of callose in roots of Norway spruce seedlings after short-term exposure to aluminum. Tree Physiol 24:1279–1283
Hirano Y, Walthert L, Brunner I (2006) Callose in root apices of European chestnut seedlings: a physiological indicator of aluminum stress. Tree Physiol 26:431–440
Hirano Y, Frey B, Brunner I (2012) Contrasting reactions of roots of two coniferous tree species to aluminum stress. Environ Exp Bot 77:2–18
Horst WJ (1995) The role of the apoplast in aluminium toxicity and resistance of higher plants: a review. Zeitschrift für Pflanzenernährung und Bodenkunde 158:419–428
Horst WJ, Püschel AK, Schmohl N (1997) Induction of callose formation is a sensitive marker for genotypic aluminium sensitivity in maize. Plant Soil 192:23–30
Ikka T, Ogawa T, Li D, Hiradate S, Morita A (2013) Effect of aluminum on metabolism of organic acids and chemical forms of aluminum in root tips of *Eucalyptus camaldulensis* Dehnh. Phytochemistry 94:142–147
Inostroza-Blancheteau C, Aquea F, Reyes-Díaz M, Alberdi M, Arce-Johnson P (2011) Identification of aluminum-regulated genes by cDNA-AFLP analysis of roots in two contrasting genotypes of highbush blueberry (*Vaccinium corymbosum* L.). Mol Biotechnol 49:32–41
Inostroza-Blancheteau C, Rengel Z, Alberdi M, Mora ML, Aquea F, Arce-Johnson P, Reyes-Díaz M (2012) Molecular and physiological strategies to increase aluminum resistance in plants. Mol Biol Rep 39:2069–2079
Inostroza-Blancheteau C, Aquea F, Loyola R, Slovin J, Josway S, Rengel Z, Reyes-Díaz M, Alberdi M, Arce-Johnson P (2013) Molecular characterisation of a calmodulin gene, *VcCaM1*, that is differentially expressed under aluminium stress in highbush blueberry. Plant Biol 15:1013–1018
Iuchi S, Koyama H, Iuchi A, Kobayashi Y, Kitabayashi S, Kobayashi Y, Ikka T, Hirayama T, Shinozaki K, Kobayashi M (2007) Zinc finger protein STOP1 is critical for proton tolerance in Arabidopsis and coregulates a key gene in aluminum tolerance. Proc Natl Acad Sci U S A 14:9900–9905
Jansen S, Watanabe T, Dessein S, Smets E, Robbrecht E (2003) A comparative study of metal levels in leaves of some Al-accumulating Rubiaceae. Ann Bot 91:657–663
Jiang HX, Chen LS, Zheng JG, Han S, Tang N, Smith B (2008) Aluminum-induced effects on photosystem II photochemistry in *Citrus* leaves assessed by the chlorophyll a fluorescence transient. Tree Physiol 28:1863–1871
Jiang H-X, Tang N, Zheng J-G, Li Y, Chen L-S (2009) Phosphorus alleviates aluminum-induced inhibition of growth and photosynthesis in *Citrus grandis* seedlings. Physiol Plant 137:298–311
Jones DL, Blancaflor EB, Kochian LV, Gilroy S (2006) Spatial coordination of aluminium uptake, production of reactive oxygen species, callose production and wall rigidification in maize roots. Plant Cell Environ 29:1309–1318
Jorns AC, Hecht-Buchholz C, Wissemeier AH (1991) Aluminum-induced callose formation in root tips of Norway spruce (*Picea abies* (L.) Karst.). Z Pflanzenernähr Bodenk 154:349–353
Kasai M, Sasaki M, Yamamoto Y, Matsumoto H (1992) Al increases $K^+$ efflux and activities of ATP-dependent and PPi-dependent $H^+$ pumps of tonoplast-enriched vesicles from barley roots. Plant Cell Physiol 33:1035–1039
Kaus H (1987) Some aspects of calcium dependent regulation in plant metabolism. Annu Rev Plant Physiol 38:47–72
Keltjens WG, Ulden PSR (1987) Effect of Al on nitrogen ($NH_4^+$ and $NO_3^-$) uptake, nitrate reductase activity and proton release in two sorghum cultivars different in Al tolerance. Plant Soil 104:227–234
Kidd PS, Llugany M, Poschenrieder C, Gunsé B, Barceló J (2001) The role of root exudates in aluminium resistance and silicon-induced amelioration of aluminium toxicity in three varieties of maize (*Zea mays* L.). J Exp Bot 52:1339–1352
Kochian LV (1995) Cellular mechanisms of aluminum toxicity and resistance in plants. Annu Rev Plant Physiol Plant Mol Biol 46:237–260

Kochian LV, Hoekenga OA, Pineros MA (2004) How do crop plants tolerate acid soils? Mechanisms of aluminum tolerance and phosphorous efficiency. Annu Rev Plant Biol 55:459–493
Koyama H, Toda T, Hara T (2001) Brief exposure to low-pH stress causes irreversible damage to the growing root in *Arabidopsis thaliana*: pectin-Ca interaction may play an important role in proton rhizotoxicity. J Exp Bot 52:361–368
Li XF, Ma JF, Matsumoto H (2000) Pattern of aluminum-induced secretion of organic acids differs between rye and wheat. Plant Physiol 123:1537–1544
Lidon FC, Barreiro MG, Ramalho JDC, Lauriano JA (1999) Effects of aluminium toxicity on nutrient accumulation in maize shoots: implications on photosynthesis. J Plant Nutr 22:397–416
Logan BA, Demmig-Adams B, Adams W III (1998) Antioxidant and xanthophylls cycle-dependent energy dissipation in *Cucurbita pepo* L. and *Vinca major* L. upon a sudden increase in growth PPFD in the field. J Exp Bot 49:1881–1888
Ma JF (2000) Role of organic acids in detoxification of aluminum in higher plants. Plant Cell Physiol 41:383–390
Masunaga T, Kubota T, Hotta M, Wakatsuki T (1998) Mineral composition of leaves and bark in aluminum accumulators in a tropical rain forest in Indonesia. Soil Sci Plant Nutr 44:347–358
Matsumoto H (2000) Cell biology of aluminium toxicity and tolerance in higher plants. Int Rev Cytol 200:1–46
Matsumoto H, Hirasawa E, Morimura S, Takahashi E (1976) Localization of aluminium in tea leaves. Plant Cell Physiol 17:627–631
Maxwell K, Johnson GN (2000) Chlorophyll fluorescence – a practical guide. J Exp Bot 51:659–668
Mizuno S, Ayabe S, Uchiyama H (2014) Expression of genes encoding transporters and enzyme proteins in response to low-pH and high-aluminum treatments in *Acacia mangium*, a stress-tolerant legume. Plant Biotechnol 31:61–66
Mora ML, Alfaro M, Williams PH, Stehr W, Demanet R (2004) Effect of fertilizer input on soil acidification in relation to growth and chemical composition of a pasture and animal production. J Soil Sci Plant Nutr (Chile) 4:29–40
Mora ML, Demanet R, Vistoso E, Gallardo F (2005) Influence of sulfate concentration in mineral solution on ryegrass grown at different pH and aluminium levels. J Plant Nutr 28:1–16
Morita A, Yanagisawa O, Maeda S, Takatsu S, Ikka T (2011) Tea plant (*Camellia sinensis* L.) roots secrete oxalic acid and caffeine into medium containing aluminium. Soil Sci Plant Nutr 57:796–802
Moustakas M, Ouzounidou G, Eleftherios PE, Lannoye R (1995) Aluminum effect on photosynthesis and elemental uptake in an aluminum-tolerant and non-tolerant wheat cultivar. J Plant Nutr 18:669–683
Naik D, Smith E, Cumming JR (2009) Rhizosphere carbon deposition, oxidative stress and nutritional changes in two poplar species exposed to aluminum. Tree Physiol 29:423–436
Nguyen NT, Nakabayashi K, Thompson J, Fujita K (2003) Role of exudation of organic acids and phosphate in aluminum tolerance of four tropical woody species. Tree Physiol 23:1041–1050
Niyogi KK, Björkman O, Grossman AR (1997) The roles of specific xanthophylls in photoprotection. Proc Natl Acad Sci U S A 94:14162–14167
Nunes-Nesi A, Brito DS, Inostroza-Blancheteau C, Fernie AR, Araújo WL (2014) The complex role of mitochondrial metabolism in plant aluminum resistance. Trends Plant Sci 19:399–407
Ofei-Manu P, Wagatsuma T, Ishikawa S, Tawaraya K (2001) The plasma membrane strength of the root tip cells and root phenolic compounds are correlated with Al tolerance in several common woody plants. Soil Sci Plant Nutr 47:359–375
Ohyama Y, Ito H, Kobayashi Y, Ikka T, Morita A, Kobayashi M, Imaizumi R, Aoki T, Komatsu K, Sakata Y, Iuchi S, Koyama H (2013) Characterization of *AtSTOP1* orthologous genes in tobacco and other plant species. Plant Physiol 162:1937–1946

Osawa H, Kojima K (2006) Citrate-release-mediated aluminum resistance is coupled to the inducible expression of mitochondrial citrate synthase gene in *Paraserianthes falcataria*. Tree Physiol 26:565–574
Osawa H, Endo I, Hara Y, Matsushima Y, Tange T (2011) Transient proliferation of proanthocyanidin-accumulating cells on the epidermal apex contributes to highly aluminum-resistant root elongation in camphor tree. Plant Physiol 155:433–446
Peixoto PHP, Da Matta FM, Cambraia J (2002) Responses of the photosynthetic apparatus to aluminum stress in two sorghum cultivars. J Plant Nutr 25:821–832
Pellet DM, Grunes DL, Kochian LV (1995) Organic acid exudation as an aluminum tolerance mechanism in maize (*Zea mays* L.). Planta 196:788–795
Pereira WE, de Siqueira DL, Martinez CA, Puiatti M (2000) Gas exchange and chlorophyll fluorescence in four citrus rootstocks under aluminum stress. J Plant Physiol 157:513–520
Pereira WE, Lopes de Siqueira D, Puiatti M, Martínez CA, Salomão LCC, Cecon PR (2003) Growth of citrus rootstocks under aluminium stress in hydroponics. Sci Agric 60:31–41
Piñeros MA, Shaff JE, Manslank S, Carvalho A, Vera M, Kochian LV (2005) Aluminum resistance in maize cannot be solely explained by root organic exudation. A comparative physiological study. Plant Physiol 137:231–241
Qin R, Hirano Y, Brunner I (2007) Exudation of organic acid anions from poplar roots after exposure to Al, Cu and Zn. Tree Physiol 27:313–320
Rengel Z (1992) Role of calcium in aluminium toxicity. New Phytol 121:499–513
Rengel Z (1996) Uptake of aluminium by plant cells. New Phytol 134:389–406
Rengel Z, Zhang W-H (2003) Role of dynamics of intracellular calcium in aluminium-toxicity syndrome. New Phytol 159:295–314
Reyes-Diaz M, Alberdi M, Mora ML (2009) Short-term aluminum stress differentially affects the photochemical efficiency of photosystem II in highbush blueberry genotypes. J Am Soc Hort Sci 134:14–21
Reyes-Díaz M, Inostroza-Blancheteau C, Millaleo R, Cruces E, Wulff-Zottele C, Alberdi M, Mora ML (2010) Long-term aluminum exposure effects on physiological and biochemical features of highbush blueberry cultivars. J Am Soc Hort Sci 135:212–222
Ridolfi M, Garrec JP (2000) Consequences of an excess Al and a deficiency in Ca and Mg for stomatal functioning and net carbon assimilation of beech leaves. Ann For Sci 57:209–218
Rout GR, Samantaray S, Das P (2001) Aluminium toxicity in plants: a review. Agronomie 21:3–21
Ruban AV, Horton P (1999) The xanthophyll cycle modulates the kinetics of nonphotochemical energy dissipation in isolated light-harvesting complexes, intact chloroplasts, and leaves of spinach. Plant Physiol 119:531–542
Ryan PR, Delhaize E (2010) The convergent evolution of aluminum resistance in plants exploits a convenient currency. Funct Plant Biol 37:275–284
Sasaki T, Yamamoto Y, Ezaki B, Katsuhara M, Ahn SJ, Ryan PR, Delhaize E, Matsumoto H (2004) A wheat gene encoding an aluminum-activated malate transporter. Plant J 37:645–653
Sawaki Y, Iuchi S, Kobayashi Y, Ikka T, Sakurai N, Fujita M, Shinozaki K, Shibata D, Kobayashi M, Koyama H (2009) *STOP1* regulates multiple genes that protect Arabidopsis from proton and aluminum toxicities. Plant Physiol 150:281–294
Sawaki Y, Kihara-Doi T, Kobayashi Y, Nishikubo N, Kawazu T, Kobayashi Y, Koyama H, Sato S (2013) Characterization of Al-responsive citrate excretion and citrate-transporting MATEs in *Eucalyptus camaldulensis*. Planta 237:979–989
Sawaki Y, Kobayashi Y, Kihara-Doic T, Nishikubo N, Kawazuc T, Kobayashi M, Kobayashi Y, Iuchi S, Koyama H, Sato S (2014) Identification of a STOP1-like protein in *Eucalyptus* that regulates transcription of Al tolerance genes. Plant Sci 223:8–15
Schroeder JI, Delhaize E, Frommer WB, Guerinot ML, Harrison MJ, Herrera-Estrella L, Horie T, Kochian LV, Munns R, Nishizawa NK, Tsay YF, Sanders D (2013) Using membrane transporters to improve crops for sustainable food production. Nature 497:60–66
Setlik I, Allakhverdiev SI, Nedbal L, Setlikova E, Klimov VV (1990) Three types of Photosystem II photoinactivation. I. Damaging process on the acceptor side. Photosynth Res 23:39–48

Silva IR, Novais RF, Jham GN, Barros NF, Gebrim FO, Nunes FN (2004) Responses of eucalypt species to aluminum: the possible involvement of low molecular weight organic acids in the Al tolerance mechanism. Tree Physiol 24:1267–1277
Smith E, Naik D, Cumming JR (2011) Genotypic variation in aluminum resistance, cellular aluminum fractions, callose and pectin formation and organic acid accumulation in roots of *Populus* hybrids. Environ Exp Bot 72:182–193
Soto-Cerda BJ, Inostroza-Blancheteau C, Mathias M, Penaloza E, Zuñiga J, Muñoz G, Rengel Z, Salvo-Garrido H (2015) Marker-assisted breeding for *TaALMT1*, a major gene conferring aluminium tolerance to wheat. Biol Plant 59(1):83–91
Stass A, Smit I, Eticha D, Oettler G, Horst WJ (2008) The significance of organic anion exudation for the aluminum resistance of primary triticale derived from wheat and rye parents differing in aluminum resistance. J Plant Nutr Soil Sci 171:634–682
Tahara K, Norisada M, Tange T, Yagi H, Kojima K (2005) Ectomycorrhizal association enhances Al tolerance by inducing citrate secretion in *Pinus densiflora*. Soil Sci Plant Nutr 51:397–403
Tahara K, Yamanoshita T, Norisada M, Hasegawa I, Kashima H, Sasaki S, Kojima K (2008a) Aluminum distribution and reactive oxygen species accumulation in root tips of two Melaleuca trees differing in aluminum resistance. Plant Soil 307:167–178
Tahara K, Norisada M, Yamanoshita T, Kojima K (2008b) Role of binding ligands in aluminum resistance of *Eucalyptus camaldulensis* and *Melaleuca cajuputi*. Plant Soil 302:175–187
Tahara K, Hashida K, Otsuka Y, Ohara S, Kojima K, Shinohara K (2014) Identification of a hydrolyzable tannin, oenothein B, as an aluminum-detoxifying ligand in a highly aluminum-resistant tree, *Eucalyptus camaldulensis*. Plant Physiol 164:683–693
Tamás L, Hutttová J, Mistrík I, Simonovicová B (2006) Aluminium-induced drought and oxidative stress in barley roots. J Plant Physiol 163:781–784
Taylor GJ, Basu A, Basu U, Slaski JJ, Zhang G, Good A (1997) Al-induced 51-kilodalton membrane-bound proteins are associated with resistance to Al in a segregating population of wheat. Plant Physiol 114:363–372
von Uexkull HR, Mutert E (1995) Global extent, development and economic impact of acid soils. In: Date RA, Grundon NJ, Rayment GE, Probert ME (eds) Plant–soil interactions at low pH: principles and management. Kluwer, Dordrecht, pp 5–19
Wagatsuma T, Yamasaka K (1985) Relationship between differential aluminum tolerance and plant-induced pH change of medium among barley cultivars. Soil Sci Plant Nutr 31:521–535
Wannaz ED, Rodriguez JH, Wolfsberger T, Carreras HA, Pignata ML, Fangmeier A, Franzaring J (2012) Accumulation of aluminium and physiological status of tree foliage in the vicinity of a large aluminium smelter. Sci World J 7. doi:10.1100/2012/865927
Watanabe T, Osaki M (2002) Mechanisms of adaptations to high aluminum conditions in native plant species growing in acid soil: a review. Commun Soil Sci Plant Anal 33:1247–1260
Watanabe T, Osaki M, Yoshihara T, Tadano T (1998) Distribution and chemical speciation of aluminum in the Al accumulator plant, *Melastoma malabathricum* L. Plant Soil 201:165–173
Wissemeier AH, Klotz F, Horst WJ (1987) Aluminum induced callose synthesis in roots of soybean. J Plant Physiol 129:487–492
Wissemeier AH, Hahn G, Marschner H (1998) Callose in roots of Norway spruce (*Picea abies* (L.) Karst.) is a sensitive parameter for aluminum supply at a forest site (Höglwald). Plant Soil 199:53–57
Yamaji N, Huang CF, Nagao S, Yano M, Sato Y, Nagamura Y, Ma JF (2009) A zinc finger transcription factor ART1 regulates multiple genes implicated in aluminum tolerance in rice. Plant Cell 21:3339–3349
Yamamoto Y, Kobayashi Y, Matsumoto H (2001) Lipid peroxidation is an early symptom triggered by aluminium, but not the primary cause of elongation inhibition in pea roots. Plant Physiol 125:199–208
Yamamoto Y, Kobayashi Y, Devi SR, Rikiishi S, Matsumoto H (2002) Aluminum toxicity is associated with mitochondrial dysfunction and the production of reactive oxygen species in plant cells. Plant Physiol 128:63–72

Yamane Y, Kashino Y, Koile H, Satoh K (1997) Increase in the fluorescence $F_0$ level reversible inhibition of Photosystem II reaction center by high-temperature treatments in higher plants. Photosynth Res 52:57–64
Yang LT, Jiang HX, Tang N, Chen LS (2011) Mechanisms of aluminum-tolerance in two species of citrus: secretion of organic acid anions and immobilization of aluminum by phosphorus in roots. Plant Sci 189:521–530
Yang LT, Jiang H, Qi YP, Chen LS (2012) Differential expression of genes involved in alternative glycolytic pathways, phosphorus scavenging and recycling in response to aluminum and phosphorus interactions in *Citrus* roots. Mol Biol Rep 39:6353–6366
Zúñiga R, Alberdi M, Reyes-Díaz M, Olivares E, Hess S, Bravo LA, Corcuera LJ (2006) Seasonal changes in the photosynthetic performance of two evergreen *Nothofagus* species in south central Chile. Rev Chil Hist Nat 79:489–504

# Diversity of Arbuscular Mycorrhizal Fungi in Acidic Soils and Their Contribution to Aluminum Phytotoxicity Alleviation

**Paula Aguilera, Jonathan Cumming, Fritz Oehl, Pablo Cornejo, and Fernando Borie**

**Abstract** Acidic conditions limit crop production on 40 % of the world's soils. These soils are characterized by a pH between 4.5 and 5.5, low phosphorus (P) availability, high contents of aluminum (Al) and manganese (Mn), and low soil basic cations. Edaphic conditions of acidic soils limit plant growth, mainly due to $Al^{3+}$ phytotoxicity, which reduces water and nutrient acquisition from soils and severely limits root growth of sensitive species. However, the association of symbiotic arbuscular mycorrhizal (AM) fungi with plant roots often modifies plant response to acid soil factors through enhanced P acquisition and reduced Al exposure. Several management practices are implemented to mitigate the negative effects of acidic soils on plant growth, among which are lime application, P fertilization, and the use of Al-tolerant plants. In this regard, the inoculation with AM fungi appears as another management alternative, because of the well-documented AM contribution to plants growing in acidic soils. Several reports have demonstrated that AM fungal structures and glomalin-related soil protein (GRSP) protect plants against stress produced by high levels of Al. However, there is broad functional diversity among AM fungal genera or species in their capacity to confer Al-resistance to host plants in acidic soils. Therefore, the aim of this review is to summarize AM fungal diversity present in acidic soils as well as relate their presence with the potential to alleviate Al phytotoxicity.

P. Aguilera • P. Cornejo • F. Borie (✉)
Center of Amelioration and Sustainability of Volcanic Soils, BIOREN-UFRO, Universidad de La Frontera, P.O. Box 54-D, Temuco, Chile
e-mail: fernando.borie@ufrontera.cl

J. Cumming
Department of Biology, West Virginia University, Morgantown, WV 26506, USA

F. Oehl
Agroscope, Institute for Sustainability Sciences, Plant-Soil-Interactions, Reckenholzstrasse 191, CH-8046 Zürich, Switzerland



S.K. Panda, F. Baluška (eds.), *Aluminum Stress Adaptation in Plants*, Signaling and Communication in Plants 24, DOI 10.1007/978-3-319-19968-9_11

# 1 Introduction

Crop production is limited by soil acidity on approximately 30–40 % of the world soils (von Uexküll and Mutert 1995). Soil acidification results from natural processes, such as high rainfall and intense biological activity, and from anthropogenic activities, including the use of ammonium fertilizers and deposition of strong mineral acids in precipitation, that induce soil base cation leaching and changes in aluminum (Al) solubility and speciation (Campbell 1998; Emmett 1999; Szott et al. 1999). Generally, acidic soils are characterized by high levels of solution and soil-bound Al, which are broadly toxic to plants (Kochian et al. 2005). Such soils have additional disadvantages to crop production, such as high P adsorption capacity, high $H^+$ levels, and low base soil cations (Mora et al. 1999, 2002). The most important diagnostic index of potential Al toxicity is Al saturation (%) of effective cation exchange capacity, which corresponds to the proportion of Al to the total amount of exchangeable Ca, Mg, K, Na, and Al (USDA 1996).

Aluminum is one of the most prevalent metals generating phytotoxicity due to its strong negative effects on plant root growth, directly reducing the capacity for water and nutrient acquisition (Kochian et al. 2005; Ma et al. 2001). In non-tolerant plants, Al exposure leads to broad disruptions in physiology and metabolism, including altered membrane transport processes, cell membrane structure, oxidative stress, and cell wall lesions (Kochian et al. 2005; Ma et al. 2001). To mitigate the negative effects of acid soils on plants, several management practices are typically implemented, such as lime application, P fertilization, and selection/use of Al-resistant plant genotypes. In addition, the inoculation/preservation of plants and soils with symbiotic arbuscular mycorrhizal (AM) fungi is another management alternative (Borie et al. 2010).

Arbuscular mycorrhizal fungi are obligated symbionts of about 80 % terrestrial plants, some of them growing in soils with serious constraints (Smith and Read 2008). Arbuscular mycorrhizal symbiosis is an association, which is established between soil fungi and most vascular plants, allowing a bidirectional interchange of nutrients and energy (Barea et al. 2013; Smith and Read 2008). Essentially, host plants improve their water and nutrient absorption capacity, and fungi receive carbon compounds (Barea et al. 2013).

Numerous studies have demonstrated that the AM symbiosis protects plants against stress produced by high levels of diverse toxic elements, including heavy metals (González-Guerrero et al. 2008; Bissonnette et al. 2010; Janoušková and Pavlíková 2010; Miransari 2010) and Al (Borie and Rubio 1999; Cumming and Ning 2003; Klugh and Cumming 2007; Klugh-Stewart and Cumming 2009). Importantly, there is wide functional diversity among AM fungal genera and species in their capacity to alter the rhizosphere (Clark and Zeto 2000; Kelly et al. 2005; Klugh and Cumming 2007), which can determine the outcome of AM fungi on Al phytotoxicity alleviation in acid soils.

## 2 Phylogeny and Taxonomic Identification of AM Fungi

The AM fungi have been identified by morphological and molecular analyses which have allowed their taxonomic organization into the phylum Glomeromycota, which contains three classes: *Glomeromycetes*, *Archaeosporomycetes*, and *Paraglomeromycetes* (Table 1). Within these classes, 5 orders and 14 families with 29 genera have been identified and grouped through concomitant morphological and molecular phylogenetic analyses (Oehl et al. 2011c). A more recent, slightly diverging classification was presented by Redecker et al. (2013). Following the most recent updates, 15 families and 38 genera are currently counted in the Glomeromycota (Błaszkowski et al. 2015; Marinho et al. 2014; Sieverding et al. 2014). They are listed and attributed to their orders and classes in Table 1.

Arbuscular mycorrhizal fungi have been evaluated by several methods based on their three forms present in soil, either as spores, hyphae, or colonized roots. Taxonomic classification requires the isolation of spores from soils/roots either from field-collected soils or from "trap cultures," in which fungi in collected soils are sustained to sporulation on a suitable host plant. Fungal spores are then collected by techniques such as wet sieving and decanting method (Gerdemann and Nicolson 1963; Sieverding 1991), differential water-sucrose centrifugation (Allen et al. 1979), sucrose gradient centrifugation (Ianson and Allen 1986), and adhesion and flotation-centrifugation (Horn et al. 1992). However, after spore isolation, it is necessary to complement these observations with morphological and molecular identification. Finally, appropriate results interpretation and AM fungi species identification are performed by accessing available databases such as INVAM (invam.caf.wvu.edu) and identification manuals (Błaszkowski 2012; Schenck and Pérez 1990).

In some genera, the morphological characters, which may be included in AM fungi identification, have been scarce; therefore, the identification by means of molecular techniques has made a substantial contribution to AM fungal ecology due to its simplicity because only a simple colonized root or spore is required (Redecker 2000, 2002). However, also morphological identification has made outstanding progresses lastly, and the combination of morphological and molecular tools has led to major advances in the taxonomy and classification of these fungi. The AM fungi diversity identification associated to the rhizosphere can be obtained by morphological studies, which allow the understanding of the interactions between plant species or cultivars and communities associated with AM fungi (Aguilera et al. 2014).

In AM fungal studies, the development and use of the monoxenic culture using transformed roots with one mycorrhizal fungus growing under in vitro conditions have become a common approach for molecular research (Bécard and Piché 1992; St-Arnaud et al. 1996). This approach generates adequate amounts of AM fungi under sterile conditions and has allowed the study of fungi under specific environmental conditions (Bago et al. 2004). In addition, several metabolic and molecular aspects of AM symbiosis have been reported using monoxenic culture, such as

**Table 1** Classification system of the phylum Glomeromycota after Oehl et al. (2011c), updated

| Class | Order | Family | Genus |
|---|---|---|---|
| *Glomeromycetes* | *Glomerales* | *Glomeraceae* | *Glomus* |
| | | | *Dominikia* |
| | | | *Funneliformis* |
| | | | *Kamienskia* |
| | | | *Rhizoglomus* |
| | | | *Sclerocystis* |
| | | | *Septoglomus* |
| | | | *Simiglomus* |
| | | *Entrophosporaceae* | *Claroideoglomus* |
| | | | *Albahypha* |
| | | | *Viscospora* |
| | | | *Entrophospora* |
| | *Diversisporales* | *Diversisporaceae* | *Diversispora* |
| | | | *Corymbiglomus* |
| | | | *Otospora* |
| | | | *Redeckera* |
| | | | *Tricispora* |
| | | *Sacculosporaceae* | *Sacculospora* |
| | | *Pacisporaceae* | *Pacispora* |
| | | *Acaulosporaceae* | *Acaulospora* |
| | | | *Kuklospora* |
| | *Gigasporales* | *Gigasporaceae* | *Gigaspora* |
| | | *Scutellosporaceae* | *Scutellospora* |
| | | | *Bulbospora* |
| | | | *Orbispora* |
| | | *Dentiscutataceae* | *Dentiscutata* |
| | | | *Fuscutata* |
| | | | *Quatunica* |
| | | *Intraornatosporaceae* | *Intraornatospora* |
| | | | *Paradentiscutata* |
| | | *Racocetraceae* | *Racocetra* |
| | | | *Cetraspora* |
| *Archaeosporomycetes* | *Archaeosporales* | *Ambisporaceae* | *Ambispora* |
| | | *Archaeosporaceae* | *Archaeospora* |
| | | | *Intraspora* |
| | | | *Palaeospora* |
| | | *Geosiphonaceae* | *Geosiphon* |
| *Paraglomeromycetes* | *Paraglomerales* | *Paraglomeraceae* | *Paraglomus* |

responses to oxidative stress produced by Cu (González-Guerrero et al. 2007) and Zn (González-Guerrero et al. 2005). In some genera, the morphological characters available for AM fungi identification are limited. In these cases, therefore, the use of molecular techniques aids in AM fungal identification due to the simplicity of their application and without the need for implementing trap cultures or transformed root systems (Clapp et al. 1995; Krüger et al. 2011; Oehl et al. 2005, 2006; Redecker 2000). Molecular methods can be directly applied to roots and rhizosphere soils allowing in situ assessment of AM fungal diversity under a variety of environmental conditions in the field or greenhouse (Daniell et al. 2001). Techniques such as temporal temperature gel electrophoresis (TTGE) analysis, using nested PCR on the NS31-Glo1 region of 18S rDNA, has been used in the identification of *Glomus* species colonizing roots of *Trifolium repens* and *Sorghum vulgare* (Cornejo et al. 2004). Denaturing gradient gel electrophoresis (DGGE) was used to assess AM fungi community structure of roots of *Zea mays* differing in P efficiency (Oliveira et al. 2009). Restriction fragment length polymorphism (RFLP) analyses have also been applied to characterize AM fungal community structure in agricultural ecosystems (Daniell et al. 2001).

In the AM fungi diversity studies, several molecular techniques have been used in the last years which have allowed many advances and they have been based principally on ribosomal DNA (rDNA) sequencing (Öpik et al. 2010, 2013). The rDNA has highly conserved regions as well as variable regions and can be used for phylogenetic identification from the level of phylum to species (Raab et al. 2005). The small subunit portion has been amplified by using PCR, where NS31, AM1, AML1, and AML2 primers have been commonly used for *Glomeraceae*, *Diversisporaceae*, *Gigasporaceae*, and *Acaulosporaceae* (Helgason et al. 1998; Lee et al. 2008). Several studies have shown that AM fungi cannot be identified reliably below the genera level using morphological characters, necessitating the use of molecular methods (Helgason et al. 1998, 1999).

## 3 Diversity of AM Fungi in Acidic Soils

Arbuscular mycorrhizal fungi are generally adapted to the edaphic environment from which they were isolated, and they can be found colonizing plants growing in soils with pH between 2.7 and 9.2 (Clark 1997; Siqueira et al. 1984). Although it has been reported that low pH and high Al concentrations in soil can negatively impact AM colonization (Göransson et al. 2008), local adaptation of AM fungi may play a central role in the functional success of mycorrhizae and their host plants in inhospitable soils (Jansa et al. 2008; Öpik et al. 2009). Indeed, functional diversity exists among AM fungal genera and species under diverse environmental stresses, which may reflect the operation of stress-specific adaptation mechanisms that modulate stresses associated with specific environments (Klugh and Cumming 2007; Klugh-Stewart and Cumming 2009; Medeiros et al. 1994a, b). Such functional diversity analyzes are focused on traits that the AM fungi provide to the host plant in regard to stress resistance, nutrient efficiency, and root system morphology resulting from changes in gene expression patterns in the AM symbiosis (Feddermann et al. 2010).

From acidic mountainous soils of the Sierra Nevada in Spain, two new *Acaulospora* species, *Ac. pustulata* and *Ac. tortuosa*, were described using also concomitant morphological and molecular analyses (Palenzuela et al. 2013) confirming that especially *Acaulospora* is an AM fungi genus with high diversity of species adapted to acidic soils (Oehl et al. 2011a, b).

Several species have been reported in acidic soils with high Al contents (Table 2). Morton (1986) described three new species from abandoned coal minesoils. Species were identified by using pot culture; subsequently, their taxonomic classification was made by morphological characterization. These species were: *Ac. dilatata*, *Ac. lacunosa* and *Ac. rugosa*. Whereas, other species have been found in acidic soils from upland Scotland. These species were established in pot culture, then they were analyzed by morphological and molecular methods, and finally, they were identified as *Ac. alpina* and *Ac. brasiliensis* (Krüger et al. 2011).

On the other hand, *Acaulosporaceae*, *Glomaceae*, and *Gigasporaceae* families have been found by using DGGE in maize rhizosphere from acidic soils of Brazil. The PCR amplification of 18S was made by means of nested PCR by using NS1, NS4 VANS1, and NS21 primers. Other primers were used for each specific AM fungus (Oliveira et al. 2009).

A large number of AM fungal genera and species are found in acidic soils with high Al contents (Table 2). Evaluation of these reports indicates that species richness declines slightly with soil pH over the range of 5.6–3.7 ($P = 0.044$ for the pH effect), but is independent of both method of isolation (i.e., field soil versus trap isolation) and method of identification (i.e., spore characteristics versus molecular fingerprint), although the data for identification based on molecular methods are limiting. The greatest diversity of species identified is in the *Glomus* (23) and *Acaulospora* (45) genera, with species within other genera being much less numerous (Table 2). Across all studies summarized in Table 2, *Ac. laevis*, *Ac. scrobiculata*, *Gi. margarita*, *Gl. diaphanum*, *Gl. fasciculatum*, *Rh. intraradices*, and *Fu. mosseae* were found in five or more of the studies.

Arbuscular mycorrhizal fungi can influence ecosystem sustainability, as they promote establishment of plants growing under unfavorable environmental conditions. In this regard, several of the reports on AM diversity in acidic soils have compared undisturbed and disturbed ecosystems. AM fungal diversity was studied in native and reforested *Araucaria angustifolia* forests with soils of pH 3.7 (Moreira-Souza et al. 2003). In this study, *Acaulospora* (eight species), *Entrophospora* (one species), *Gigaspora* (two species), *Glomus* (nine species), *Scutellospora* (four species) were found (Moreira-Souza et al. 2003). While these genera were generally shared between native forest soils and reforested sites, reforested sites had lower species richness (14 vs. 21 for field soils and 8 vs. 11 for trap cultures) (Moreira-Souza et al. 2003).

Land use also influences AM fungal diversity (Jansa et al. 2002, 2003; Oehl et al. 2003, 2009). In this regard, Oehl et al. (2010) reported AM fungal diversity present in temperate climate location of Europe obtained from AM fungi reproduced in trap culture. In this study, large differences in species richness were evident between grassland and arable systems on acidic soils, with an average

**Table 2** Arbuscular mycorrhizal fungal diversity associated to host plant growing in acidic soils. Classification updated

| Arbuscular mycorrhizal fungi | Host plants | Soil pH | Identification method | References |
|---|---|---|---|---|
| *Acaulospora dilatata*<br>*Ac. lacunosa*<br>*Ac. rugosa*<br>*Acaulospora* sp.<br>*Fuscutata heterogama*<br>*Glomus diaphanum*<br>*Paraglomus occultum*<br>*Entrophospora* sp. | Abandoned coal mine, USA<br>*Andropogon virginicus* (field)<br>*Sorghum sudanense* (trap) | pH $H_2O$ 3.6–4.2 | Morphology | Morton (1986) |
| *Gigaspora margarita*<br>*Rhizoglomus aggregatum*<br>*Funneliformis caledonius*<br>*Rh. clarum-like*<br>*Septoglomus constrictum*<br>*Claroideoglomus etunicatum*<br>*Rh. fasciculatum*<br>*Fu. geosporus*<br>*Rh. microaggregatum*<br>*Fu. mosseae*<br>*Sclerocystis rubiformis*<br>*Scutellospora aurogloba-like*<br>*S. callospora* | Converted meadow, Canada<br>*Hordeum vulgare* (field) | pH $H_2O$ 5.2 | Morphology | Hamel et al. (1994) |
| *Ac. bireticulata*<br>*Ac. mellea*<br>*Ac. trappei*<br>*Acaulospora* spp. (4)<br>*Gigaspora gigantea*<br>*Gi. rosea*<br>*Gi. ramisporophora*<br>*Rh. clarum*<br>*Septoglomus constrictum*<br>*Rh. fasciculatum*<br>*Ambispora leptoticha*<br>*Fu. mosseae*<br>*Glomus* spp.(4)<br>*S. calospora*<br>*Fuscutata hetrogama*<br>*Cetraspora pellucida*<br>*Dentiscutata reticulata* | Grassland, USA<br>*Anthoxanthum, Panicum, Plantago* (field)<br>*Sorghum vulgare* (trap) | pH 5.0 | Morphology | Bever et al. (1996) |

(continued)

**Table 2** (continued)

| Arbuscular mycorrhizal fungi | Host plants | Soil pH | Identification method | References |
|---|---|---|---|---|
| *Ac. bireticulata*<br>*Ac. foveata*<br>*Am. gerdemannii*<br>*Ac. laevis*<br>*Ac. scrobiculata*<br>*Ac. spinosa*<br>*Ac. rehmii*<br>*Acaulospora* spp. (2)<br>*Kuklospora colombiana*<br>*Gi. margarita*<br>*Gi. decipiens*<br>*Rh. agreggatum*<br>*Rh. clarum*<br>*Gl. diaphanum*<br>*Cl. etunicatum*<br>*Rh. fasciculatum*<br>*Gl. geosporum*<br>*Gl. macrocarpum*<br>*Gl. microcarpum*<br>*Gl. pansihalos*<br>*Cetraspora gilmorei*<br>*Dentiscutata nigra*<br>*Ce. pellucida*<br>*Scutellospora* sp.<br>*Unidentified* (1) | *Araucaria* forest, Brazil<br>*Araucaria* (field)<br>*Sorghum bicolor* (trap) | pH $CaCl_2$ 3.7 | Morphology | Moreira-Souza et al. (2003) |
| *Ac. dilatata*<br>*Ac. koskei*<br>*Ac. laevis*<br>*Acaulospora* sp. (2)<br>*Entrophospora schenckii*<br>*Archaeospora* sp. (1)<br>*Fu. coronatus*<br>*Gl. diaphanum*<br>*Cl. etunicatum*<br>*Rh. intraradices*<br>*Gl. macrocarpum*<br>*Sclerocystis rubiformis*<br>*Glomus* sp. (2)<br>*S. calospora*<br>*Scutellospora* sp. (3)<br>*Pacispora dominikii*<br>*Diversispora spurca* | Agricultural field, Chile<br>*Avena sativa* and *Triticum aestivum* rotation (field) | pH $H_2O$ 5.5 | Morphology | Castillo et al. (2006) |

(continued)

**Table 2** (continued)

| Arbuscular mycorrhizal fungi | Host plants | Soil pH | Identification method | References |
|---|---|---|---|---|
| *Ac. mellea*<br>*Ac. morrowiae*<br>*Ac. scrobiculata*<br>*Ac. spinosa*<br>*Gi. decipiens*<br>*Rh. clarum*<br>*Cl. etunicatum*<br>*Paraglomus occultum*<br>*Fuscutata heterogama*<br>*Ce. pellucida*<br>*Racocetra persica* | Apple orchard, Brazil<br>*Malus domestica* (field)<br>*Sorghum bicolor* (trap) | pH 4.0 | Morphology | Cavallazzy et al. (2007) |
| *Ac. mellea*<br>*Archaeospora myriocarpa*<br>*Ac. spinosa*<br>*Dentiscutata reticulata*<br>*Gi. margarita*<br>*Cl. claroideum*<br>*Sclerocystis coremioides*<br>*Rh. fasciculatum*<br>*Sclerocystis sinuosa*<br>*Racocetra verrucosa* | Papaya plantation, India<br>*Carica papaya* (field) | pH $H_2O$ 4.4–5.0 | Morphology | Khade et al. (2010) |
| *Fu. caledonius*<br>*Fu. mosseae*<br>*Septoglomus constrictum*<br>*Fu. geosporus*<br>*Cl. etunicatum*<br>*Paraglomus occultum* and *Paraglomus albidum*<br>*Gl. diaphanum*<br>*Gl. fasciculatum*<br>*Rh. clarum*<br>*Rh. intraradices*<br>*Diversispora versiformis*<br>*Dominikia aurea*<br>*Dominikia compressa* (*Gl.* sp. BR9)<br>*Rh. invermaium*<br>*Gl. macrocarpum*<br>*Gl. microcarpum*<br>*Gl. heterosporum*<br>*Gl. mortonii*<br>*Gl.* sp. BR17 | Grassland, pasture, maize-wheat rotation, Rhine Valley, Germany<br>*Lolium perenne*, *Trifolium pratense*, *Plantago lanceolata* (traps) | pH $H_2O$ 5.6 | Morphology | Oehl et al. (2010) |

(continued)

**Table 2** (continued)

| Arbuscular mycorrhizal fungi | Host plants | Soil pH | Identification method | References |
|---|---|---|---|---|
| *Gl.* sp. BR12<br>*Gl.* sp. BR18<br>*Gl.* sp. BR8<br>*Gl.* sp. BR6<br>*Simiglomus hoi*<br>*Archaeospora trappei*<br>*Ar.* sp. BR21<br>*Ambispora fennica*<br>*Am.* sp. BR10<br>*Entrophospora infrequens*<br>*Ac. paulinae*<br>*Ac. laevis*<br>*Ac. cavernata*<br>*Ac. longula*<br>*Ac. scrobiculata*<br>*Ac. thomii*<br>*Ac. sieverdingii* (*Ac.* sp. BR19)<br>*Ac. polonica*<br>*Ac.* sp. BR 20<br>*Ar. myriocarpa*<br>*Kuklospora colombiana*<br>*S. calospora*<br>*S. dipurpurescens*<br>*Ce. armeniaca*<br>*Ce. pellucida*<br>*Gi. margarita* | | | | |
| *Ac. laevis*<br>*Acaulospora* sp. (1)<br>*Ac. sieverdingii*<br>*Ac. longula*<br>*Pacispora dominikii*<br>*Cl. etunicatum*<br>*Cl. claroideum*<br>*Dominikia aurea*<br>*Gl. diaphanum*<br>*Glomus* sp. (2)<br>*Fu. mosseae*<br>*Rh. intraradices*<br>*Septoglomus constrictum*<br>*Simiglomus hoi*<br>*S. calospora*<br>*Ce. gilmorei*<br>*Cetraspora* sp (1)<br>*Am. gerdemannii*<br>*Ambispora* sp (1) | Agricultural field, Chile<br>*Triticum aestivum* (field) | pH $H_2O$ 4.7 | Morphology | Aguilera et al. (2014) |

(continued)

**Table 2** (continued)

| Arbuscular mycorrhizal fungi | Host plants | Soil pH | Identification method | References |
|---|---|---|---|---|
| *Ar. trappei*<br>*Ar. myriocarpa*<br>*Archaeospora* sp (1)<br>*Paraglomus occultum* | | | | |
| *Ac. longula*<br>*Ac. rugosa*<br>*Ac. scrobiculata*<br>*Ac. morrowiae*<br>*Archaeospora* sp.<br>*Gi. margarita*<br>*Fu. caledonius*<br>*Sclerocystis coremioides*<br>*Rh. manihotis*<br>*Fu. mosseae*<br>*Dentiscutata cerradensis* | Maize hybrid trials, Brazil<br>*Zea mays* (field) | pH $H_2O$ 5.2 | DNA (DGGE) | Oliveira et al. (2009) |
| *Ac. mellea*<br>*Ac. rugosa*<br>*Ac. spinosa*<br>*Rh. intraradices*<br>*Rh. fasciculatum*<br>*Glomus* spp. (4) | Legume pasture, Venezuela<br>*Centrosema macrocarpum* (field) | pH $H_2O$ 4.6–5.1 | SSUrDNA | Alguacil et al. (2010) |

of 32 species recovered from native soils compared with 18 from tilled systems across sandstone and granitic sedimentary parent materials. Agricultural management of these systems also affected reduced species-level diversity, with Shannon-Weaver Index being 1.8 compared to 2.5 for native acidic soil grasslands (Oehl et al. 2010).

Additionally, AM fungal diversity characterization has been carried out in forest and agricultural ecosystems of Southern Chile. In these soils about 39 AMF species were found, some of them reported for the first time. The genera found in the order of the highest to lowest abundance were: *Acaulospora, Glomus, Scutellospora* and *Archaeospora*. Whereas, in agricultural ecosystems, 22 AM fungal species were found belonging to the genera *Glomus* spp., *Acaulospora* spp. and *Scutellospora* spp. (Castillo 2005; Castillo et al. 2005).

Taken together, these reports suggest that the diversity of AM fungi in acidic soils is high, although estimates vary widely. The variation may be associated with the intensity of effort undertaken to assess the AM species in a given soil. For example, the highest species estimates in Table 2 (44) were from Oehl et al. (2010), and these were derived from both field and trap culture assessments, with the traps utilizing multiple species and multiple sample dates. Bever et al. (1996), Moreira-Souza et al. (2003), and Aguilera et al. (2014) also utilized intensive approaches to assessing AM fungal diversity (Table 2).

# 4 Contribution of AM Fungi to Aluminum Phytotoxicity Alleviation

Phytotoxicity caused by Al in acidic soils has been investigated in numerous studies and is due principally to the negative effects of the $Al^{3+}$ on root physiology and growth (Kochian et al. 2005). Root tips are the first specific sites exhibiting the negative effects of Al exposure, with rapid inhibition of cell elongation and division (Kinraide et al. 1992; Ryan et al. 1992). Aluminum resistance in plants primarily results from mechanisms that reduce the interactions of the toxic $Al^{3+}$ with root tip cells. Several researches carried out with genotypes of *Triticum aestivum* tolerant and sensitive to Al has shown that Al-tolerant plant root meristems accumulate less Al, suggesting that these plants have a mechanism of exclusion allowing to protect tissues against Al phytotoxicity (Rincón and Gonzales 1992; Silva et al. 2000). The AM symbiosis is involved in plant adaptation to stressful soil conditions (Seguel et al. 2013). Several studies have reported that AM fungal species promote growth and providing protection against acid soils factors, including limited P (Rubio et al. 2003), excess Al (Borie et al. 2010; Borie and Rubio 1999; Clark and Zeto 2000; Cumming and Ning 2003; Lux and Cumming 2001; Mendoza and Borie 1998; Klugh-Stewart and Cumming 2009), and elevated Fe and Mn (Rohyadi et al. 2004; Yano and Takaki 2005). Although often dependent on both host species/genotype investigated, the benefits of AM fungal for growth under acidic soil conditions have been shown for crop species, including *Hordeum vulgare* (Borie and Rubio 1999; Mendoza and Borie 1998), *Ipomoea batatas* (Yano and Takaki 2005), and *Vigna unguiculata* (Rohyadi et al. 2004), wild native species, such as *Andropogon virginicus* (Cumming and Ning 2003; Klugh-Stewart and Cumming 2009) and *Panicum virgatum* (Clark et al. 1999a, b), and tree species, including *Liriodendron tulipifera* (Klugh and Cumming 2007; Lux and Cumming 2001), *Malus prunifolia* (Cavallazzy et al. 2007), and *Musa acuminata* (Rufyikiri et al. 2000).

The association of AM fungi with roots improves plant performance on acid soils or under Al exposure due broad changes in root access to soil resources. In many cases, but not all, colonization allows plants to improve their ability to acquire water and nutrients (Borie et al. 2010; Borie and Rubio 1999; Rufyikiri et al. 2000). The acquisition of P, which is often found at low concentrations in the soil solution and bound to soil minerals, is a major limitation on acidic soils and AM fungi, through their extensive hyphal exploration of soils, may access distal P pools as well as solubilize pools otherwise not available to plants (Borie and Rubio 1999). This access to bound or insoluble P is facilitated by the production of metal-binding compounds by roots, such as the organic acids citrate and malate, which is maintained by acid-adapted AM fungi species under Al exposure (Klugh and Cumming 2007). In addition to fostering P uptake, the exudation of organic acids will improve the uptake of cations, notably Ca and Mg, which are often depressed under acid soil conditions (Borie and Rubio 1999).

Studies carried out with AM colonized plants frequently exhibit a decrease of Al or Mn binding to roots and translocation to leaves (Klugh and Cumming 2007; Rohyadi et al. 2004). The mechanisms underlying reduced metal accumulation are related to chelation of metals in the rhizosphere or sequestration in AM fungal structures, both of which reduce the direct availability of the metal ions, reducing their phytotoxicity. Chelation by organic acids (Cumming and Ning 2003; Klugh and Cumming 2007; Klugh-Stewart and Cumming 2009) is one mechanism purported to be active in the root zones of mycorrhizal plants. Additionally, novel AM fungal compounds, such as the glycoprotein glomalin, may play a role in sequestering and detoxifying Al in the soil solution (Aguilera et al. 2011). In addition, studies reported by Cuenca et al. (2001) have shown that Al is present in the mycelium of the AM fungi, principally in vesicles, suggesting that AM fungi may directly absorb and sequester Al and perhaps other metals in their hyphae.

In studies where multiple AM fungal species have been screened for attenuation of acidic soil stresses, there is a wide range of variation in effectiveness across AM fungal species investigated. In maize, three species, *Gl. diaphanum*, *Cl. etunicatum*, and *Rh. intraradices*, were equally effective at promoting growth on acid soils (Clark and Zeto 1996). Cavallazzy et al. (2007) noted that *Cl. etunicatum* and *S. pellucida* were most effective at conferring resistance to Al in *Malus prunifolia. Glomus clarum* was superior in conferring Al resistance to *Liriodendron tulipifera* (Klugh and Cumming 2007), and *Rh. clarum* and *S. heterogama* were most effective in protecting *Andropogon virginicus* (Klugh-Stewart and Cumming 2009). Interestingly, *S. heterogama* was not effective in protecting *M. prunfolia* (Cavallazzy et al. 2007). This variation in acidic soils resistance may be due to broad differences in resistance among AM fungal isolates within species or changes in resistance in isolates during maintenance culture (Kelly et al. 2005).

Mechanisms such as chelation or sequestration have been reported to influence stress attenuation of heavy metals, because it reduces the direct availability of the metal ion, reducing phytotoxicity in the rhizosphere (Cumming and Ning 2003; Rufyikiri et al. 2000, 2003). Studies reported by Cuenca et al. (2001) have shown that Al is present in the mycelium of the AM fungi, principally in vesicles. Recently, a study describing the effect of Al-tolerant wheat cultivars growing in an Andosol with phytotoxic Al levels on the AM fungi diversity was carried out and the authors have suggested the existence of a degree of co-adaptation among wheat cultivars and AM fungal communities that could have greater specialization. Additionally, there is a tolerance differential between the different plant species and plant cultivars with respect to the Al tolerance which confirm the fact that the AM fungal communities composition can be influenced by different management practices, tillage systems, crop species, and even plant cultivars within a plant species. In addition, The AM fungal species may have dominance when they are under stress conditions. In particular, species belonging to *Scutellospora* and *Acaulospora* genera were found in this study (Aguilera et al. 2014).

### 4.1 Species of AM Fungi Included in Technological Application Based on Acidic Condition Tolerance

Some AM fungal species have been recognized for their tolerance to acidic conditions (Table 3). Yano and Takaki (2005) studied beneficial aspects of *Gigaspora* on plant growth parameters at pH 4.2 and 5.2 (with lime application) and found that this fungus favored Al tolerance of *Ipomea batatas* growing in acidic soils with high Al levels. Whereas, Cavallazzy et al. (2007) selected AM fungi for improving the establishment of plants in acidic soils with high levels of Al and Mn and demonstrated that *Cl. etunicatum* and *S. pellucida* were the most efficient inocula. In these conditions, *Cl. etunicatum* and *S. pellucida* promoted plant growth and nutrient absorption.

Inoculation of *Rh.* clarum, *S. heterogama, and Cl. etunicatum* genera at high Al levels has demonstrated an increase in citrate and malate exudation from colonized plant roots. Additionally, inoculated plants presented the highest P amount and lowest Al concentration in their shoots. Both fungi and plant parameters were favored (Klugh-Stewart and Cumming 2009).

These technological applications are beneficial especially in tropical areas where Al toxicity reduces crop production. Rufyikiri et al. (2000) demonstrated that inoculated plants with *Rh. intraradices* showed great resistance under high Al concentrations.

Cuenca et al. (2001) described AM fungi tolerance to acidic conditions with high Al contents by using AM fungi inoculated plants from acidic and neutral soils growing at pH 3.0–5.0 and found that inocula from acidic conditions promoted a higher tolerance of *Clusia multiflora* to acidic soil. In this assay, *Gigaspora* and *Scutellospora* genera were found as dominant. On the other hand, Bartolome-Esteban and Schenck (1994) have studied AM fungal tolerance to Al by assessment of spore germination and hyphae growth at three different Al saturation percentage (6, 27, and 100 %) and observed that *Gigaspora* was the most efficient genus. However, Rohyadi (2005) evaluated the effect of Al increasing concentrations on germination of *Gi. margarita* spores and found that Al inhibited spores development; therefore, AM fungal colonization was reduced. However, some spores produced hyphae and subsequently colonized plant roots under excessive Al conditions.

Kelly et al. (2005) evaluated differential behavior of AM fungi testing five isolates of each *Rh. clarum*, *Ac. morrowiae*, and *S. heterogama* species using *Sorghum sudanense* as host plant. Plants were exposed at high Al levels, and they show different responses when they were exposed to fungal isolates from same

**Table 3** Arbuscular mycorrhizal fungal species and their efficiency in technological applications studies based on plant exposed to acidic conditions. Classification updated

| Acidic conditions | AM fungi | Host plant | Research highlights | References |
|---|---|---|---|---|
| 0–50–200 μM Al | *Acaulospora morrowiae* *Claroideoglomus claroideum* *Rhizoglomus clarum* (1) *Paraglomus brasilianum* | *Liriodendron tulipifera* L. | (1) Showed 88 % colonization rates, promoted Al plant resistance, colonized plants accumulated less Al in leaves, 2- to 10-fold higher P in leaves than other mycorrhizal plants and promoted citrate, malate, and oxalate exudation. | Klugh and Cumming (2007) |
| 0–100 μM Al | *Acaulospora morrowiae*, *Claroideoglomus claroideum*, *Rhizoglomus clarum* (1), *Cl. etunicatum* (2), *Paraglomus brasilianum*, *Fuscutata heterogama* (3) | *Andropogon virginicus* | (1) and (3) showed high hyphae lengths as well as citrate and malate exudation from colonized plant roots. (1) It favored the plant growth, while Al exposition affects all others colonized plants, and it presented the greatest colonization. (2) Accumulated greatest P amount in shoot under Al exposition. (1) Gave highest Al resistance to plants and accumulated lowest Al concentrations in plant shoots. | Klugh-Stewart and Cumming (2009) |
| pH 4.2 (originally) pH 3.0, 4.0, and 5.0 (experiment) | *Acaulospora scrobiculata* *Glomus* spp *Gigaspora* spp *Racocetra fulgida* | *Clusia multiflora* | Mycorrhizal plants produced higher dry matter of roots. Roots length was not affected by pH 3.0 and they showed low Al concentration. Inoculant plants showed higher AM colonization. | Cuenca et al. (2001) |
| pH 4 and 5 | *Rhizoglomus clarum* (1), *Glomus diaphanum* (2), | *Panicum virgatum* L. | Mycorrhizal plants increased 52- and 26-fold in dry matter | Clark et al. (1999a) |

(continued)

**Table 3** (continued)

| Acidic conditions | AM fungi | Host plant | Research highlights | References |
|---|---|---|---|---|
| | *Claroideoglomus etunicatum*, *Rh. intraradices*, *Gigaspora albida*, *Gi. margarita*, *Gi. rosea*, and *Acaulospora morrowiae* | | at pH 4 and 5, respectively. (1) and (2) showed the highest dry matter increase | |
| 4.48 $pH_W$ 88 % Al saturation | *Rhizoglomus clarum* (1), *Glomus diaphanum* (2), *Claroideoglomus etunicatum*, *Rhizoglomus intraradices*, *Gigaspora albida*, *Gi. margarita* (3), *Gi. rosea*, and *Acaulospora morrowiae* | *Panicum virgatum* L. | (1) and (2) increased 42- and 36-fold in dry matter (shoot and root), respectively, at pH 4 in relation to nonmycorrhizal plants. (1) increased 64- and 19-fold in root dry matter at pH 4 and 5, respectively. (3) showed the greatest root colonization at pH 4 and 5. | Clark et al. (1999b) |
| 0–400 μM Al | *Rhizoglomus clarum* (1) *Acaulospora morrowiae* *Fuscutata heterogama* | *Andropogon virginicus* | (1) The most efficient species that showed high colonization rates (78 %) and low Al translocation to shoots. | Kelly et al. (2005) |
| pH 4.2–5.2 | *Gigaspora margarita* | *Ipomoea batatas* | Mycorrhizal plants showed twofold in dry weight at pH 4.2, and these plants reduced their toxic symptoms. | Yano and Takaki (2005) |
| pH 4.0; 5.0 | *Claroideoglomus etunicatum* (1) *SCT110* *Cetraspora pellucida* (2) *SCT111* *Acaulospora scrobiculata SCT112* *Fuscutata heterogama SCT113* | *Malus prunifolia* | (2) showed high colonization levels; 68 % and 66 % at pH 4.0 and 5.0, respectively. (1) and (2) increased 132 % and 146 % the plant height respect to non-mycorrhizal plants. | Pereira et al. (2007) |

(continued)

**Table 3** (continued)

| Acidic conditions | AM fungi | Host plant | Research highlights | References |
|---|---|---|---|---|
| 180 μM Al | *Rhizoglomus intraradices* | *Musa acuminata* | Increase in biomass, water, and nutrient absorption. Al in roots and shoots was decreased and Al symptoms in plants were diminished. | Rufyikiri et al. (2000) |
| pH 4.7 | *Gigaspora margarita* | *Vigna unguiculata* L | Under 11.9 mg $kg^{-1}$ of available Al spores, germination decreased to about 40 % | Rohyadi (2005) |

species. Functional diversity observed in Al resistance can influence plant stability growing under phytotoxic Al levels. Species as *Rh. clarum, Cl. etunicatum, and Gi. margarita* have been reported like efficient isolates with high ability to withstand acidic environment and Al phytotoxicity (Clark et al. 1999a, b). Species as *Gi. margarita* and *Cl. etunicatum* are found commonly under acidic conditions, being *Cl. etunicatum* the most efficient (Rohyadi et al. 2004). Moreover, others species found in acidic soils are also *Acaulospora* sp., *Gigaspora* sp., and *Gl. manihotis*, which have been identified as particularly tolerant (Clark 1997).

Picone (2000) evaluated the species composition, dominance diversity curves, and Simpson diversity index in Ultisols with pH ranging from 3.9 to 5.6, reporting the presence of Acaulosporaceae (*Ac. foveata*, *Entrophospora* aff *colombiana*), Glomaceae (*Glomus* "small brown," *Gl. occultum*), and Gigasporaceae (*S. pellucida*, *Gigaspora* sp.) as the most frequent species.

## 4.2 Possible Role of GRSP on Al Phytotoxicity Alleviation in Acidic Soils

Glomalin is a glycoprotein released from AM fungi (Wright et al. 1996). It has been reported as glomalin related to soil protein (GRSP) in many soils in abundant amount (Wright et al. 2007; Wright and Upadhyaya 1996). Glomalin is known to be tightly bound in hyphae and spores (Driver et al. 2005). This glycoprotein has been recognized as heat-shock protein (Gadkar et al. 2006). Subsequently, glomalin was found in spore wall layers of in vitro culture of *Rh. intraradices* by using transmission electron microscopy (Purin and Rillig 2008). A monoclonal antibody has been used for identifying location of this protein in AM fungi (Purin and Rillig 2008; Rosier et al. 2008). Several reports have been focused on determining the ability of GRSP in sequestering diverse heavy metals (Cornejo et al. 2008a; González-Chávez et al. 2004; Vodnik et al. 2008) and Al (Aguilera et al. 2011).

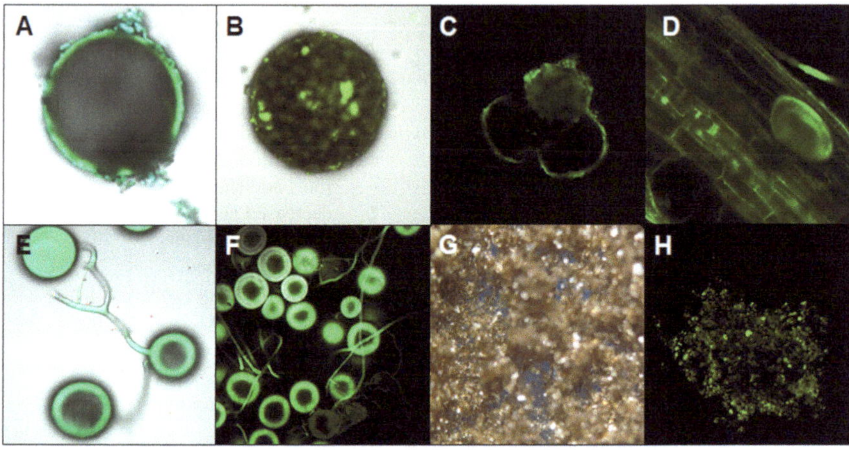


**Fig. 1** Visualization of fluorescence emission of Al in AM fungi structures by confocal laser scanning microscopy. (**a–c**) Spores from rhizosphere of *Triticum aestivum* plants growing in acidic soils with high Al saturation. (**d–f**) Fluorescence of spores and cell walls from colonized roots of in vitro culture. (**g**) Glomalin-related soil protein from acidic soil under dissection microscope. (**h**) Glomalin-related soil protein from acidic soil with $Al^{3+}$ addition (photography by P. Aguilera)

It has been shown in these studies that purified glomalin can sequester several heavy metals, especially Cu, Pb, and Cd (González-Chávez et al. 2004) presumably due to complex formation. In addition, AM fungi and GRSP are able to live under adverse environmental conditions with high levels of Cu, Pb, and Zn (Bedini et al. 2010). For improving glomalin analyses, it has been produced without the interference of other interfering compounds and soil constituents under a soilless system (Nichols 2010). Additionally, it has been demonstrated that agricultural practices can influence glomalin concentration in soils (Borie et al. 2006; Valarini et al. 2009). Cornejo et al. (2008a) reported high glomalin content in a high Cu polluted soil. This would suggest a possible role of GRSP in soil remediation, and the same role might be played in high Al content soils.

In vivo and in vitro assays were carried out for detecting $Al^{3+}$ in AM fungal propagules and GRSP extracted and purified from soils by using confocal laser scanning microscopy (CLSM) wherein fluorescence emission represents Al content inside the observed structures (Fig. 1). Although, there are reports that account for autofluorescence in AM fungi, this property is not yet attributed to specific compounds; however, glomalin is a structural AM compound. Therefore, we suggest that fluorescence is related to Al–glomalin interaction through stable complexes formation between this metal and AM fungal compound in immobilization sites (Aguilera et al. 2014; Cornejo et al. 2008a).

A role of GRSP on Al phytotoxicity alleviation has been evidenced based on their ability to sequester Al in the molecule, suggesting that this glycoprotein could form stable complexes with Al, explaining the benefits of some AM fungal strains in terms of increasing Al tolerance of crops growing in soils with a high phytotoxic Al levels (Aguilera et al. 2011).

## 5 Agricultural Practices Favoring AM Contribution

The interaction among AM fungi and different agronomic management practices has been studied in acidic soils, showing an increase in wheat production, P acquisition, and colonization rates when plants were inoculated with *Cl. etunicatum* and a no-tillage system (Rubio et al. 2003). The AM fungi increase nutrient absorption capacity of plants, mainly for P, N, and some microelements. However, with respect to N, the kind of fertilizer could influence the colonization development. Recent studies have shown that N-$NO_3^-$ fertilization in wheat plants inoculated with *Cl. etunicatum* favors mycorrhizal development and its function when they were compared with other plants fertilized with N-$NH_4^+$ (Cornejo et al. 2007). However, this response will be conditioned by the plant genotype. Other studies carried out in acidic soils have shown that N-$NO_3^-$ fertilization favors AM propagule development associated with wheat plants. In this regard, more than 4000 AMF spores per 100 g of soil were found at postharvest stage. High spore densities are often of special relevance to the amount of fungal propagules that remains in the soil and encourages the next establishing crop (Cornejo et al. 2008b).

In addition, another important factor that influences mycorrhizal behavior in acidic soils is the kind of tillage. In this sense, no-tillage promotes soil chemical properties by increasing P, C, N, and S concentrations after wheat harvest as well as mycorrhizal colonization indicators and GRSP amounts produced by AM fungi (Borie et al. 2006). Besides, it has been reported that native ecotype *Cl. etunicatum* favors Ca and Mg acquisition in Al-tolerant wheat plants. This effect is enhanced when liming applications are conducted (Borie and Rubio 1999).

According to previous reviews, agronomic management practices determine the role of mycorrhizal fungi associated with wheat plants growing in acidic soils. In this respect, AM can contribute to plant in alleviation of Al stress, through nutrient acquisition improving plant nutrition, by organic acidic exudation with chelating ability (Cumming and Ning 2003; Klugh and Cumming 2007; Klugh-Stewart and Cumming 2009) or by producing GRSP, which has a potential capacity of Al sequestration/immobilization. However, further research is needed for defining AM fungi contribution on stress alleviation produced by Al phytotoxicity. Several agronomic practices can influence on AM fungi diversity and functionality; by this way it is possible to favor the role of AM fungi associated to plants growing in acidic soil under phytotoxic Al levels especially in extensive agricultural systems.

## 6 Conclusions and Future Research Trends

This chapter performs a compilation of AM fungi species that have been studied in acidic soils with high levels of Al and analyzes the main contributions of AM fungi to plants growing in Andisols with high levels of Al phytotoxicity. Among the species found in acidic soils, *Glomus, Acaulospora, Gigaspora*, and *Scutellospora* prevail and have shown differences in the alleviation of Al phytotoxicity evaluated in terms of biomass production, organic acid exudation, and nutrient acquisition. Information provided in this review emphasizes the contribution of AM fungi to plants growing in acidic soils mainly based on increased nutrient acquisition, contributing to plant resistance, decrease in Al translocation into the shoot, and possible Al immobilization/sequestration by AM fungal structures and GRSP.

The AM fungal species and their efficiency under acidic conditions has been target of several studies focused mainly on technological applications; however, inoculant formulation for use in biofertilization requires a better understanding of physiological, molecular, and ecological bases regulating AM fungi diversity and their contribution to plants growing in soils with phytotoxic Al levels.

Biofertilization with AM fungi has often not been successfully developed mainly due to difficulties in culture establishment and maintenance. Specifically, it is necessary to figure out the optimum culturing procedures for different AM fungal species, to study the role of glomalin in these soils, and to select the most appropriate and efficient ecotypes. In this way, it would be possible to optimize plant performance in agricultural crops growing in acidic soils.

**Acknowledgments** Financial support of FONDECYT 3150175 (P. Aguilera), FONDECYT 1130541 (F. Borie) and FONDECYT 1120890 Grants from Comisión Nacional de Investigación Científica y Tecnológica, Chile.

## References

Aguilera P, Borie F, Seguel A, Cornejo P (2011) Fluorescence detection of aluminum in arbuscular mycorrhizal fungal structures and glomalin by using confocal laser scanning microscopy. Soil Biol Biochem 43:2417–2431

Aguilera P, Cornejo P, Borie F, Barea JM, von Baer E, Oehl F (2014) Diversity of arbuscular mycorrhizal fungi associated with *Triticum aestivum* L. plants growing in an Andosol with high aluminum level. Agr Ecosyst Environ 186:178–184

Alguacil MD, Lozano Z, Campoy M, Roldán A (2010) Phosphorus fertilisation management modifies the biodiversity of AM fungi in a tropical savanna forage system. Soil Biol Biochem 42:1114–1122

Allen MF, Moore TS, Christensen M, Stanton N (1979) Growth of vesicular-arbuscular mycorrhizal and non-mycorrhizal *Bouteloua gracilis* in a defined medium. Mycologia 71:666–669

Bago B, Cano C, Azcón-Aguilar C, Samson J, Coughlan AP, Piché I (2004) Differential morphogenesis of the extraradical mycelium of an arbuscular mycorrhizal fungus grown monoxenically on spatially heterogeneous culture media. Mycologia 96:452–460

Barea JM, Pozo M, López-Ráez JM, Aroca R, Ruíz-Lozano JM, Ferrol N, Azcón R, Azcón-Aguilar C (2013) Arbuscular Mycorrhizas and their significance in promoting soil-plant systems sustainability against environmental stresses. In: Rodelas B, Gonzalez-Lopez J (eds) Beneficial plant-microbial interactions: ecology and applications. CRC Press, Boca Raton, FL, pp 353–387

Bartolome-Esteban H, Schenck N (1994) Spore germination and hyphal growth of arbuscular mycorrhizal fungi in relation to soil aluminum saturation. Mycologia 86:217–226

Bécard G, Piché Y (1992) Establishment of vesicular-arbuscular mycorrhiza in root organ culture: review and proposed methodology. In: Norris J, Read D, Varma A (eds) Techniques for the study of mycorrhiza. Academic Press, New York, NY, pp 89–108

Bedini S, Turrini A, Rigo C, Argese E, Giovannetti M (2010) Molecular characterization and glomalin production of arbuscular mycorrhizal fungi colonizing a heavy metal polluted ash disposal island, downtown Venice. Soil Biol Biochem 42:758–765

Bever J, Morton J, Antonovics J, Schultz P (1996) Host-dependent sporulation and species diversity of arbuscular mycorrhizal fungi in a mown grassland. J Ecol 84:71–82

Bissonnette L, St-Arnaud M, Labrecque M (2010) Phytoextraction of heavy metals by two Salicaceae clones in symbiosis with arbuscular mycorrhizal fungi during the second year of a field trial. Plant Soil 332:55–67

Błaszkowski J (2012) Glomeromycota. W. Szafer Institute of Botany, Polish Academy of Sciences, Kraków, p 303

Błaszkowski J, Chwat G, Góralska A, Ryska P, Kovács GM (2015) Two new genera, *Dominikia* and *Kamienska*, and *D. disticha* sp. nov. in Glomeromycota. – Nova Hedwigia 100. Available online

Borie F, Rubio R (1999) Effects of arbuscular mycorrhizae and liming on growth and mineral acquisition of aluminum-tolerant and aluminum-sensitive barley cultivars. J Plant Nutr 22:121–137

Borie F, Rubio R, Rouanet JL, Morales A, Borie G, Rojas C (2006) Effects of tillage systems on soil characteristics, glomalin and mycorrhizal propagules in a Chilean Ultisol. Soil Till Res 88:253–261

Borie F, Rubio R, Morales A, Curaqueo G, Cornejo P (2010) Arbuscular mycorrhizae in agricultural and forest ecosystem in Chile. J Soil Sci Plant Nutr 10:185–206

Campbell LC (1998) Managing soil fertility decline. J Crop Prod 1:29–52

Castillo C (2005) Biodiversidad y efectividad de hongos micorrízicos arbusculares en ecosistemas agro-forestales del Centro-Sur de Chile. Doctoral Thesis in Sciences of Natural Resources, Universidad de La Frontera, Chile, p 124

Castillo C, Borie F, Godoy R, Rubio R, Sieverding E (2005) Diversity of mycorrhizal plant species and arbuscular mycorrhizal fungi in evergreen forest, deciduous forest and grassland ecosystems of Southern Chile. J Appl Bot Food Qual 80:40–47

Castillo C, Borie F, Godoy R, Rubio R, Sieverding E (2006) Diversity of mycorrhizal plant species and arbuscular mycorrhizal fungi in evergreen forest, deciduous forest and grassland ecosystems of Southern Chile. J Appl Bot Food Qual 80:40–47

Cavallazzy JRP, Klauberg Filho O, Stürmer SL, Rygiewicz PT, Mendonça MM (2007) Screening and selecting arbuscular mycorrhizal fungi for inoculating micropropagated apple rootstocks in acid soils. Plant Cell Tiss Organ Cult 90:117–129

Clapp JP, Young JPW, Merryweather JW, Fitter AH (1995) Diversity of fungal symbionts in arbuscular mycorrhizas from a natural community. New Phytol 130:259–265

Clark RB (1997) Arbuscular mycorrhizal adaptation, spore germination, root colonization, and host plant growth and mineral acquisition at low pH. Plant Soil 192:15–22

Clark RB, Zeto SK (1996) Mineral acquisition by mycorrhizal maize grown on acid and alkaline soil. Soil Biol Biochem 28:1495–1503

Clark RB, Zeto SK (2000) Mineral acquisition by arbuscular mycorrhizal plants. J Plant Nutr 23:867–902

Clark RB, Zeto SK, Zobel RW (1999a) Arbuscular mycorrhizal fungal isolate effectiveness on growth and root colonization of *Panicum virgatum* in acidic soil. Soil Biol Biochem 31:1757–1763

Clark RB, Zobel RW, Zeto SK (1999b) Effects of mycorrhizal fungus isolates on mineral acquisition by *Panicum virgatum* in acidic soil. Mycorrhiza 9:167–176

Cornejo P, Azcón-Aguilar C, Miguel Barea J, Ferrol N (2004) Temporal temperature gradient gel electrophoresis (TTGE) as a tool for the characterization of arbuscular mycorrhizal fungi. FEMS Microbiol Lett 241:265–270

Cornejo P, Borie F, Rubio R, Azcón R (2007) Influence of nitrogen source on the viability, functionality and persistence of *Glomus etunicatum* fungal propagules in an Andisol. Appl Soil Ecol 35:423–431
Cornejo P, Meier S, Borie G, Rillig MC, Borie F (2008a) Glomalin-related soil protein in a Mediterranean ecosystem affected by a copper smelter and its contribution to Cu and Zn sequestration. Sci Total Environ 406:154–160
Cornejo P, Rubio R, Castillo C, Azcon R, Borie F (2008b) Mycorrhizal effectiveness on wheat nutrient acquisition in an acidic soil from southern Chile as affected by nitrogen sources. J Plant Nutr 31:1555–1569
Cuenca G, De Andrade Z, Meneses E (2001) The presence of aluminum in arbuscular mycorrhizas of *Clusia multiflora* exposed to increased acidity. Plant Soil 231:233–241
Cumming J, Ning J (2003) Arbuscular mycorrhizal fungi enhance aluminium resistance of broomsedge (*Andropogon virginicus* L.). J Exp Bot 54:1447–1459
Daniell TJ, Husband R, Fitter AH, Young JPW (2001) Molecular diversity of arbuscular mycorrhizal fungi colonising arable crops. FEMS Microbiol Ecol 36:203–209
Driver JD, Holben WE, Rillig MC (2005) Characterization of glomalin as a hyphal wall component of arbuscular mycorrhizal fungi. Soil Biol Biochem 37:101–106
Emmett BA (1999) The impact of nitrogen on forest soils and feedbacks on tree growth. Water Air Soil Pollut 116:65–74
Feddermann N, Finlay R, Boller T, Elfstrand M (2010) Functional diversity in arbuscular mycorrhiza – the role of gene expression, phosphorous nutrition and symbiotic efficiency. Fungal Ecol 3:1–8
Gadkar V, Driver JD, Rillig MC (2006) A novel in vitro cultivation system to produce and isolate soluble factors released from hyphae of arbuscular mycorrhizal fungi. Biotechnol Lett 28:1071–1076
Gerdemann JW, Nicolson TH (1963) Spores of mycorrhizal Endogone species extracted from soil by wet sieving and decanting. T Brit Mycol Soc 46:235–244
González-Chávez MC, Carrillo-González R, Wright SF, Nichols KA (2004) The role of glomalin, a protein produced by arbuscular mycorrhizal fungi, in sequestering potentially toxic elements. Environ Pollut 130:317–323
González-Guerrero M, Azcón-Aguilar C, Mooney M, Valderas A, MacDiarmid CW, Eide DJ, Ferrol N (2005) Characterization of a *Glomus intraradices* gene encoding a putative Zn transporter of the cation diffusion facilitator family. Fungal Genet Biol 42:130–140
González-Guerrero M, Cano C, Azcón-Aguilar C, Ferrol N (2007) GintMT1 encodes a functional metallothionein in *Glomus intraradices* that responds to oxidative stress. Mycorrhiza 17:327–335
González-Guerrero M, Melville L, Ferrol N, Lott J, Azcón-Aguilar C, Peterson L (2008) Ultrastructural localization of heavy metals in the extraradical mycelium and spores of the arbuscular mycorrhizal fungus *Glomus intraradices*. Can J Microbiol 54:103–110
Göransson P, Olsson PA, Postma J, Falkengren-Grerup U (2008) Colonization by arbuscular mycorrhizal and fine endophytic fungi in four woodland grasses – variation in relation to pH and aluminium. Soil Biol Biochem 40:2260–2265
Hamel C, Dalpé Y, Lapierre C, Simard RR, Smith DL (1994) Composition of the vesicular-arbuscular mycorrhizal fungi population in an old meadow as affected by pH, phosphorus and soil disturbance. Agr Ecosyst Environ 49:223–231
Helgason T, Daniell TJ, Husband R, Fitter AH, Young JPW (1998) Ploughing up the wood-wide web? Nature 394:431
Helgason T, Fitter AH, Young JPW (1999) Molecular diversity of arbuscular mycorrhizal fungi colonising *Hyacinthoides* non-scripta (bluebell) in a seminatural woodland. Mol Ecol 8:659–666
Horn K, Hahn A, Pausch P, Hock B (1992) Isolation of pure spore and hyphal fractions from vesicular-arbuscular mycorrhizal fungi. J Plant Physiol 141:28–32

Ianson DC, Allen MF (1986) The effects of soil texture on extraction of vesicular-arbuscular mycorrhizal fungal spores from arid sites. Mycologia 78:164–168

Janoušková M, Pavlíková D (2010) Cadmium immobilization in the rhizosphere of arbuscular mycorrhizal plants by the fungal extraradical mycelium. Plant Soil 332:511–520

Jansa J, Mozafar A, Anken T, Ruh R, Sanders R, Frossard E (2002) Diversity and structure of AMF communities as affected by tillage in a temperate soil. Mycorrhiza 12:225–234

Jansa J, Mozafar A, Kuhn G, Anken T, Ruh R, Sanders R, Frossard E (2003) Soil tillage affects the community structure of mycorrhizal fungi in maize roots. Ecol Appl 13:1164–1176

Jansa J, Smith FA, Smith SE (2008) Are there benefits of simultaneous root colonization by different arbuscular mycorrhizal fungi? New Phytol 177:779–789

Kelly CN, Morton JB, Cumming JR (2005) Variation in aluminum resistance among arbuscular mycorrhizal fungi. Mycorrhiza 15:193–201

Khade SW, Rodrigues BF, Sharma PK (2010) Arbuscular mycorrhizal status and root phosphatase activities in vegetative Carica papaya L. varieties. Acta Physiol Plant 32:565–574

Kinraide TB, Ryan PR, Kochian LV (1992) Interactive effects of $Al^{3+}$, $H^{+}$, and other cations on root elongation considered in terms of cell-surface electrical potential. Plant Physiol 99:1461–1468

Klugh KR, Cumming JR (2007) Variations in organic acid exudation and aluminum resistance among arbuscular mycorrhizal species colonizing *Liriodendron tulipifera*. Tree Physiol 27:1103–1112

Klugh-Stewart K, Cumming JR (2009) Organic acid exudation by mycorrhizal *Andropogon virginicus* L. (broomsedge) roots in response to aluminum. Soil Biol Biochem 41:367–373

Kochian L, Piñeros M, Hoekenga O (2005) The physiology, genetics and molecular biology of plant aluminum resistance and toxicity. Plant Soil 274:175–195

Krüger M, Walker C, Schüßler A (2011) *Acaulospora brasiliensis* comb. nov. and *Acaulospora alpina* (Glomeromycota) from upland Scotland: morphology, molecular phylogeny and DNA-based detection in roots. Mycorrhiza 21:577–587

Lee J, Lee S, Young JPW (2008) Improved PCR primers for the detection and identification of arbuscular mycorrhizal fungi. FEMS Microbiol Ecol 65:339–349

Lux HB, Cumming JR (2001) Mycorrhizae confer aluminum resistance to tulip-poplar seedlings. Can J Forest Res 31:694–702

Ma JF, Ryan PR, Delhaize E (2001) Aluminium tolerance in plants and the complexing role of organic acids. Trends Plant Sci 6:273–278

Marinho F, Silva GA, Ferreira ACA, Veras JSN, Sousa NMF, Goto BT, Maia LC, Oehl F (2014) *Bulbospora minima*, new genus and new species in the Glomeromycetes from semi-arid Northeast Brazil. Sydowia 66:313–323

Medeiros CAB, Clark RB, Ellis JR (1994a) Effects of excess aluminum on mineral uptake in mycorrhizal sorghum. J Plant Nutr 17:1399–1416

Medeiros CAB, Clark RB, Ellis JR (1994b) Growth and nutrient uptake of sorghum cultivated with vesicular-arbuscular mycorrhiza isolates at varying pH. Mycorrhiza 4:185–191

Mendoza J, Borie F (1998) Effect of *Glomus etunicatum* inoculation on aluminum, phosphorus, calcium, and magnesium uptake of two barley genotypes with different aluminum tolerance. Commun Soil Sci Plant 29:681–695

Miransari M (2010) Contribution of arbuscular mycorrhizal symbiosis to plant growth under different types of soil stress. Plant Biol 12:563–569

Mora ML, Schnettler B, Demanet R (1999) Effect of liming and gypsum on soil chemistry, yield, and mineral composition of ryegrass grown in an acidic andisol. Commun Soil Sci Plant 30:1251–1266

Mora ML, Cartes P, Demanet R, Cornforth IS (2002) Effects of lime and gypsum on pasture growth and composition on an acid Andisol in Chile, South America. Commun Soil Sci Plant 33:2069–2081

Moreira-Souza M, Trufem SFB, Gomes-da-Costa SM, Cardoso EJBN (2003) Arbuscular mycorrhizal fungi associated with *Araucaria angustifolia* (Bert.) O. Ktze. Mycorrhiza 13:211–215

Morton J (1986) Three new species of *Acaulospora* (Endogonaceae) from high aluminum, low pH soils in West Virginia. Mycologia 78:641–648
Nichols KA (2010) Glomalin production and accumulation in soilless pot cultures. Can J Soil Sci 90:567–570
Oehl F, Sieverding E, Ineichen K, Mader P, Boller T, Wiemken A (2003) Impact of land use intensity on the species diversity of arbuscular mycorrhizal fungi in agroecosystems of Central Europe. Appl Environ Microbiol 69:2816–2824
Oehl F, Redecker D, Sieverding E (2005) *Glomus badium*, a new sporocarpic mycorrhizal fungal species from European grasslands with higher soil pH. J Appl Bot Food Qual 79:38–43
Oehl F, Sýkorová Z, Redecker D, Wiemken A, Sieverding E (2006) *Acaulospora alpina*, a new arbuscular mycorrhizal fungal species characteristic for high mountainous and alpine regions of the Swiss Alps. Mycologia 98:286–294
Oehl F, Sieverding E, Ineichen K, Mäder P, Wiemken A, Boller T (2009) Distinct sporulation dynamics of arbuscular mycorrhizal fungal communities from different agroecosystems in long-term microcosms. Agr Ecosyst Environ 134:257–268
Oehl F, Laczko E, Bogenrieder A, Stahr K, Bösch R, van der Heijden M, Sieverding E (2010) Soil type and land use intensity determine the composition of arbuscular mycorrhizal fungal communities. Soil Biol Biochem 42:724–738
Oehl F, Jansa J, Ineichen K, Mäder P, van der Heijden MGA (2011a) Arbuscular mycorrhizal fungi as bio-indicators in Swiss agricultural soils. Agrarforschung Schweiz 2:304–311
Oehl F, Schneider D, Sieverding E, Burga CA (2011b) Succession of arbuscular mycorrhizal communities in the foreland of the retreating Morteratsch glacier in the Central Alps. Pedobiologia 54:321–331
Oehl F, Sieverding E, Palenzuela J, Ineichen K, Silva GA (2011c) Advances in Glomeromycota taxonomy and classification. IMA Fungus 2:191–199
Oliveira CA, Sa NMH, Gomes EA, Marriel IE, Scotti MR, Guimaraes CT, Schaffert RE, Alves VMC (2009) Assessment of the mycorrhizal community in the rhizosphere of maize (*Zea mays* L.) genotypes contrasting for phosphorus efficiency in the acid savannas of Brazil using denaturing gradient gel electrophoresis (DGGE). Appl Soil Ecol 41:249–258
Öpik M, Metsis M, Daniell TJ, Zobel M, Moora M (2009) Large-scale parallel 454 sequencing reveals host ecological group specificity of arbuscular mycorrhizal fungi in a boreonemoral forest. New Phytol 184:424–437
Öpik M, Vanatoa A, Vanatoa E, Moora M, Davison J, Kalwij JM et al (2010) The online database MaarjAM reveals global and ecosystemic distribution patterns in arbuscular mycorrhizal fungi (Glomeromycota). New Phytol 188:223–241
Öpik M, Zobel M, Cantero JJ, Davison J, Facelli JM, Hiiesalu I et al (2013) Global sampling of plant roots expands the described molecular diversity of arbuscular mycorrhizal fungi. Mycorrhiza 23:411–430
Palenzuela J, Azcón-Aguilar C, Barea JM, Silva GA, Oehl F (2013) *Acaulospora pustulata* and *Acaulospora tortuosa*, two new species in the Glomeromycota from Sierra Nevada (southern Spain). Nova Hedwigia 97:305–319
Pereira J, Klauberg O, Strürmer S, Rygiewicz P, Mendoza M (2007) Screening and selecting arbuscular mycorrhizal fungi for inoculating micropropagated apple rootstocks in acid soils. Plant Cell Tiss Organ Cult 90:117–129
Picone C (2000) Diversity and abundance of arbuscular–mycorrhizal fungus spores in tropical forest and pasture. Biotropica 32:734–750
Purin S, Rillig MC (2008) Immuno-cytolocalization of glomalin in the mycelium of the arbuscular mycorrhizal fungus *Glomus intraradices*. Soil Biol Biochem 40:1000–1003
Raab PA, Brennwald A, Redecker D (2005) Mitochondrial large ribosomal subunit sequences are homogeneous within isolates of *Glomus* (arbuscular mycorrhizal fungi, Glomeromycota). Mycol Res 109:1315–1322
Redecker D (2000) Specific PCR primers to identify arbuscular mycorrhizal fungi within colonized roots. Mycorrhiza 10:73–80

Redecker D (2002) Molecular identification and phylogeny of arbuscular mycorrhizal fungi. Plant Soil 244:67–73

Redecker D, Schüssler A, Stockinger H, Stürmer SL, Morton JB, Walker C (2013) An evidence-based consensus for the classification of arbuscular mycorrhizal fungi (Glomeromycota). Mycorrhiza 23:515–531

Rincón M, Gonzales RA (1992) Aluminum partitioning in intact roots of aluminum-tolerant and aluminum-sensitive wheat (*Triticum aestivum* L.) cultivars. Plant Physiol 99:1021–1028

Rohyadi A (2005) Spore germination and colonization of *Gigaspora margarita* as influenced by aluminium concentration. J Microbiol Indones 10:71–74

Rohyadi A, Smith FA, Murray RS, Smith SE (2004) Effects of pH on mycorrhizal colonization and nutrient uptake in cowpea under conditions that minimise confounding effects of elevated available aluminium. Plant Soil 260:283–290

Rosier CL, Piotrowski JS, Hoye AT, Rillig MC (2008) Intraradical protein and glomalin as a tool for quantifying arbuscular mycorrhizal root colonization. Pedobiologia 52:41–50

Rubio R, Borie F, Schalchli C, Castillo C, Azcón R (2003) Occurrence and effect of arbuscular mycorrhizal propagules in wheat as affected by the source and amount of phosphorus fertilizer and fungal inoculation. Appl Soil Ecol 23:245–255

Rufyikiri G, Declerck S, Dufey JE, Delvaux B (2000) Arbuscular mycorrhizal fungi might alleviate aluminium toxicity in banana plants. New Phytol 148:343–352

Rufyikiri G, Thiry Y, Declerck S (2003) Contribution of hyphae and roots to uranium uptake and translocation by arbuscular mycorrhizal carrot roots under root-organ culture conditions. New Phytol 158:391–399

Ryan PR, Shaff JE, Kochian LV (1992) Aluminum toxicity in roots: correlation among ionic currents, ion fluxes, and root elongation in aluminum-sensitive and aluminum-tolerant wheat cultivars. Plant Physiol 99:1193–1200

Schenck NC, Pérez Y (1990) Manual for the identification of VA mycorrhizal fungi. Synergistic Publications, Gainesville, FL

Seguel A, Cumming J, Klugh-Stewart K, Cornejo P, Borie F (2013) The role of arbuscular mycorrhizas in decreasing aluminium phytotoxicity in acidic soils: a review. Mycorrhiza 23:167–183

Sieverding E (1991) Vesicular arbuscular mycorrhiza management in tropical agrosystems in Deutsche Gesellschaft für Technische Zusammenarbeiit (GTZ). GmbH, Eschborn, p 371

Sieverding E, Silva GA, Berndt R, Oehl F (2014) *Rhizoglomus*, a new genus in the Glomeraceae. Mycotaxon 129:373–386

Silva IR, Smyth TJ, Moxley DF, Carter TE, Allen NS, Rufty TW (2000) Aluminum accumulation at nuclei of cells in the root tip. Fluorescence detection using lumogallion and confocal laser scanning microscopy. Plant Physiol 123:543–552

Siqueira JO, Hubbell DH, Mahmud AW (1984) Effect of liming on spore germination, germ tube growth and root colonization by vesicular-arbuscular mycorrhizal fungi. Plant Soil 76:115–124

Smith S, Read D (2008) Mycorrhizal symbiosis. Elsevier, New York, NY

St-Arnaud M, Hamel C, Vimard B, Caron M, Fortin JA (1996) Enhanced hyphal growth and spore production of the arbuscular mycorrhizal fungus *Glomus intraradices* in an *in vitro* system in the absence of host roots. Mycol Res 100:328–332

Szott LT, Buresh RJ, Palm CA (1999) Ecosystem fertility and fallow function in the humid and subhumid tropics. Agroforest Syst 47:163–196

USDA (1996) Soil survey laboratory methods manual. Soil Survey Investigations Report N° 42, Washington, DC, USA, p 693

Valarini PJ, Curaqueo G, Seguel A, Manzano K, Rubio R, Cornejo P, Borie F (2009) Effect of compost application on some properties of a volcanic soil from Central South Chile. Chil J Agr Res 69:416–425

Vodnik D, Grčman H, Maček I, van Elterenb J, Kovačevič M (2008) The contribution of glomalin-related soil protein to Pb and Zn sequestration in polluted soil. Sci Total Environ 392:130–136 von Uexküll HR, Mutert E (1995) Global extent, development and economic impact of acid soils. Plant Soil 171:1–15

Wright S, Upadhyaya A (1996) Extraction of an abundant an unusual protein from soil and comparison with hyphal protein of arbuscular mycorrhizal fungi. Soil Sci 161:575–586

Wright SF, Franke-Snyder M, Morton JB, Upadhyaya A (1996) Time-course study and partial characterization of a protein on hyphae of arbuscular mycorrhizal fungi during active colonization of roots. Plant Soil 181:193–203

Wright SF, Green VS, Cavigelli MA (2007) Glomalin in aggregate size classes from three different farming systems. Soil Till Res 94:546–549

Yano K, Takaki M (2005) Mycorrhizal alleviation of acid soil stress in the sweet potato (*Ipomoea batatas*). Soil Biol Biochem 37:1569–1572

# Specificity of Ion Uptake and Homeostasis Maintenance During Acid and Aluminium Stresses

**Jayakumar Bose, Olga Babourina, Yanling Ma, Meixue Zhou, Sergey Shabala, and Zed Rengel**

**Abstract** Low pH (proton toxicity) and aluminium toxicity coexist in acid soils, affecting plant growth worldwide. Decades of research concluded that proton and aluminium toxicity mechanisms are complex and remain unclear. Among the Al tolerance mechanisms, exudation of organic acid anions received considerable attention, leading to the identification of novel genes involved in organic acid anion metabolism and transport. As a downside, the major focus on exudation of organic acid anions has overshadowed research on other potential Al tolerance mechanisms (e.g. reduced cell wall binding, rhizosphere alkalisation, phosphate exudation, enhanced uptake of essential nutrients) that may be operating. In this work, the current knowledge on how proton and aluminium toxicity and tolerance mechanisms are operating when plants are exposed to acid soils is reviewed. Special emphasis has been given to the question of how uptake and homeostasis of hydrogen, potassium, phosphorus, calcium, and magnesium ions in plants are affected and regulated during low-pH and aluminium stresses. There is enough evidence to suggest that low-pH and combined low-pH/aluminium stresses differentially affect root tissues and, consequently, the rhizosphere. Less disturbed phosphorus, calcium, and magnesium uptake and homeostasis maintenance help plants to cope with low-pH and combined low-pH/aluminium stresses.

---

J. Bose (✉)
School of Earth and Environment, University of Western Australia, Crawley, WA 6009, Australia

School of Land and Food, University of Tasmania, Private Bag 54, Hobart, TAS 7001, Australia
e-mail: Jay.Bose@utas.edu.au

O. Babourina • Z. Rengel
School of Earth and Environment, University of Western Australia, Crawley, WA 6009, Australia

Y. Ma • M. Zhou • S. Shabala
School of Land and Food, University of Tasmania, Private Bag 54, Hobart, TAS 7001, Australia



S.K. Panda, F. Baluška (eds.), *Aluminum Stress Adaptation in Plants*, Signaling and Communication in Plants 24, DOI 10.1007/978-3-319-19968-9_12

# 1 Introduction

Acidic soils are formed mainly due to the weathering of acidic parent material and the leaching of basic cations by soil water. As a result, soils in high rainfall areas and older soils exhibit greater acidity. A number of other factors also contribute to soil acidification, including imbalances in the nitrogen, carbon, and sulphur cycles (Goulding et al. 1998; Mannion 1998); use of $NH_4^+$-forming fertilisers (Rowell and Wild 1985; Tang et al. 2000); atmospheric acidification (Vries and Breeuwsma 1987; Galloway 1989); nitrogen fixation by legumes (Bolan et al. 1991; Shen et al. 2004); and excessive uptake of cations by plants (Shen et al. 2004). Thus, soil acidification is a continuous process, which means the problem of acid soils is exacerbated over time in severity and extent (Rengel 2004).

In acidic soils, plant growth may be limited by various toxicities (H, Al, Mn) and deficiencies ($NH_4$-N, P, Ca, Mg, and $MoO_4$) (for references, see Kidd and Proctor 2001). Among these complex factors, aluminium (Al) toxicity received considerable attention because Al becomes increasingly soluble when the $pH_{(water)}$ decreases below 5 (Kochian 1995). In particular, activity of trivalent cationic $Al(H_2O)_6^{3+}$ (hereafter $Al^{3+}$ for convenience) often peaks at around pH 4.2–4.3, severely affecting root growth in acid soils (Kinraide 1990, 1991, 1993; Matsumoto 2000; Taylor et al. 2000; Poschenrieder et al. 2008). Interestingly, low pH ($H^+$ toxicity) alone can affect growth in diverse plant species (Bose et al. 2010b). There are some low-pH soils (e.g. organic soils) where $Al^{3+}$ ions are present in low concentration; thereby $H^+$ ions dominate the composition of the soil solution (Kidd and Proctor 2001). These $H^+$-ion-dominated soils account for a high proportion of acid soils around the globe. For instance, histosols occupy 200 million ha worldwide (Brady and Weil 1990). Hence, low-pH and combined low-pH/$Al^{3+}$ stresses need to be separated in order to understand stress-specific toxicity and tolerance mechanisms in plants. In this chapter, the existing knowledge on how $Al^{3+}$ and $H^+$ toxicity mechanisms are operating when plants are exposed to acid soils is reviewed. Special emphasis is placed on how $Al^{3+}$ and $H^+$ toxicities specifically affect ion uptake and homeostasis regulation in plants.

# 2 $Al^{3+}$ Toxicity in Plants

## *2.1 Inhibition of Root Growth by $Al^{3+}$ Toxicity*

An early symptom of $Al^{3+}$ toxicity to plants is inhibition of root growth that becomes measurable within minutes of exposure to micromolar concentrations of $Al^{3+}$ (see Delhaize and Ryan 1995; Rengel 2004 for references). Thus, roots have been the focus of research to decipher the mechanisms of $Al^{3+}$ toxicity and tolerance in plants.

Root growth is a complex and dynamic phenomenon that involves a series of biochemical and physiological processes differing in various root tissues (Street 1966; Wang et al. 2006). Detailed investigations of the spatial sensitivity to $Al^{3+}$ in different root zones revealed that the root apex (Ryan et al. 1993), particularly the distal elongation zone within the root apex (Sivaguru and Horst 1998; Kollmeier et al. 2000; Illes et al. 2006), is the primary site of $Al^{3+}$ toxicity. The distal elongation zone (due to its specific architecture) has extraordinary capability to sense various environmental stimuli and act as a "plant command centre" to integrate sensory inputs into adaptive responses (Baluška et al. 2004). Accordingly, the distal elongation zone needs to be studied in detail for a greater understanding of the primary mechanisms of $Al^{3+}$ toxicity and tolerance. However, $Al^{3+}$ toxicity and tolerance studies on distal elongation zone are relatively rare. Some studies have shown that $Al^{3+}$ also affects physiological and biochemical processes in other root zones, such as the root cap, meristem, elongation zone, and mature zone (Brady et al. 1993; Olivetti et al. 1995; Rengel 1996; Bose et al. 2010a, b, 2013).

The mature root zone is the longest, accounts for more than 90 % of root biomass, and is the principal area for nutrient absorption (Gahoonia and Nielsen 1998; Parker et al. 2000; Bibikova and Gilroy 2002). Taking $K^+$ as an example, 10 out of 15 $K^+$ transporters (KT/KUPs) are expressed in the mature zone (Ahn et al. 2004). Further, $H^+$, $K^+$ $Ca^{2+}$, and $Mg^{2+}$ uptake at the mature zone is different to that at the root apex (Ferguson and Clarkson 1975, 1976; Kiegle et al. 2000; Newman 2001; Demidchik et al. 2002; Bose et al. 2010b; Guo et al. 2010; Bose et al. 2013). Interestingly, $Al^{3+}$ concentration in the internal tissues of the mature zone is higher than in the cortex (Babourina and Rengel 2009) and cytosolic Ca signals propagate from mature zone to root cap during $Al^{3+}$ stress (Rincon-Zachary et al. 2010). Because of this, the response of the mature zone to $Al^{3+}$ might be different from that of the root apex. Indeed, $H^+$, $K^+$, and $Mg^{2+}$ fluxes in response to $Al^{3+}$ stress differ between mature root zone and elongation zone in Arabidopsis (Bose et al. 2010a, b, 2013). However, how these ion fluxes modulate the root growth during $Al^{3+}$ stress remains to be elucidated.

## 2.2 *Inhibition of Cell Division and Cell Elongation*

Early work by Clarkson (1965) revealed that $Al^{3+}$ toxicity strongly altered root development and pointed at the hampering of cell division by $Al^{3+}$ ions as a primary cause of root growth inhibition. Indeed, (1) binding of Al to nucleic acids in root tips along with inhibition of cell division (Matsumoto et al. 1976; Morimura et al. 1978) and (2) reduction in the mitotic index along with different abnormalities such as chromosome bridges, breaks, sticky metaphases, nuclear dissolution, cell death, and in some cells chromosome duplication under $Al^{3+}$ stress have been observed in maize and onion roots (De Campos and Viccini 2003). In contrast, $Al^{3+}$-induced stimulation of cell division was also reported under low concentrations of $Al^{3+}$, mainly in cell culture experiments. For example, cell cycle activity was enhanced in the Al-tolerant cell culture line of Coffea arabica, whereas inhibition was observed in the Al-sensitive cell line (Valadez-Gonzalez et al. 2007).

The $Al^{3+}$-induced alterations of the cell cycle received considerable attention because (1) it has been well established that Al can enter the symplasm quite rapidly (Silva et al. 2000; Taylor et al. 2000; Babourina and Rengel 2009), and (2) $Al^{3+}$ could alter the cell cycle through a signalling cascade without the need for Al to reach nuclei of meristematic cells (Poschenrieder et al. 2009). Further, $Al^{3+}$ toxicity is not restricted to inhibition of root length. More detailed temporal and spatial study on the maize root cell patterning under $Al^{3+}$ stress revealed that 5-min $Al^{3+}$ exposure was sufficient to inhibit cell division in the proximal meristem zone and stimulate cell division in the distal elongation zone. Protrusion of an incipient lateral root was observed in the distal elongation zone after 180 min. These observations suggest a rapid change in the cell patterning events along the root axis upon a short-time $Al^{3+}$ exposure (Doncheva et al. 2005).

Stiffening of cell walls and a consequent inhibition of root growth have been observed in response to $Al^{3+}$ stress under different experimental conditions (Tabuchi and Matsumoto 2001; Ma et al. 2004; Jones et al. 2006). Indeed, large amounts of Al accumulate in the cell walls and intercellular spaces of root tips. For example, 85–99.9 % of Al was found in the apoplasm of root cells (Taylor et al. 2000; Ma 2007). Apart from precipitation of Al on the root surface and in intercellular spaces, binding of exchangeable Al to the negative charges of the pectin substances in the cell wall was also observed (Blamey 2001). In an in vitro study, Al treatment did not cause cell wall stiffening in dead root tips of maize (Ma et al. 2004), indicating that it is a biochemical process and not purely physical cross-linking between pectin material and $Al^{3+}$. This leads to the conclusion that Al binds to the newly formed cell wall material, which is required for cell elongation growth, thereby altering mechanical properties of cell wall and hampering cell elongation (Ma et al. 2004; Ma 2007). The cross-linking of other polar cell wall constituents, such as hydroxyproline-rich glycoproteins (HRGPs) by reactive oxygen species in combination with callose deposition, has been shown to inhibit cell elongation in Arabidopsis thaliana (De Cnodder et al. 2005).

## 2.3 Production of Reactive Oxygen Species

Reactive oxygen species (ROS) are natural by-products of aerobic respiration formed when oxygen is partially reduced. ROS can be toxic to plant cells or can act as signalling molecules depending on the circumstances (Scholz-Starke et al. 2005). ROS are essential for (1) root elongation because quenching of root ROS resulted in inhibition of root elongation in *Arabidopsis thaliana* (Demidchik et al. 2003), (2) regulation of hyperpolarisation-activated cation channels (HACC) present in the epidermis of the root elongation zone (Demidchik et al. 2003; Foreman et al. 2003), and (3) activation of a potassium outward-rectifying channel (KORC) and a non-selective cation channel (NSCC), which mediate, respectively,

$K^+$ efflux and $Ca^{2+}$ influx in root hair tips of C3 and C4 plants (Demidchik et al. 2003).

The formation of ROS in response to $Al^{3+}$ has been observed in many studies (Darko et al. 2004; Tamás et al. 2004; Babourina et al. 2006; Jones et al. 2006; Tahara et al. 2008), even though $Al^{3+}$ is not a transition metal and therefore cannot catalyse redox reactions. However, $Al^{3+}$ in combination with iron caused peroxidation of lipids in the plasma membrane of soybean (Cakmak and Horst 1991) and rice roots (Meriga et al. 2004) and cultured tobacco cells (Ono et al. 1995; Yamamoto et al. 1997). Further, $Al^{3+}$ induced the expression of several genes encoding antioxidant enzymes such as glutathione S-transferase, peroxidase, and superoxide dismutase (SOD) in Arabidopsis thaliana (Richards et al. 1998; Ezaki et al. 2000), which established the significance of ROS production under $Al^{3+}$ toxicity.

A number of hypotheses have been proposed for $Al^{3+}$-induced rapid production of ROS, including dysfunction of mitochondria (Yamamoto et al. 2002), formation of aluminium superoxide semi-reduced radicals (Exley 2004), and activation of oxidising enzymes (Šimonovicová et al. 2004a, b). However, time-dependent studies demonstrated that cell death and protein oxidation occurred several hours after the cessation of root growth (Boscolo et al. 2003; Šimonovicová et al. 2004b). For example, ROS production and loss of growth were observed after 12 h of Al exposure in tobacco (Yamamoto et al. 2002). Considering the time taken to produce ROS, it appears ROS production may not be the primary mechanism of $Al^{3+}$ toxicity. Yamamoto et al. (2002) suggested that ROS production is not important for root growth inhibition, but rather important for callose biosynthesis. Indeed, cross-linking of ROS with hydroxyproline-rich glycoproteins (HRGPs) was accompanied by callose deposition and was shown to be an important mechanism for inhibition of cell elongation induced by the ethylene precursor 1-aminocyclopropane-1-carboxylic acid (ACC) in *Arabidopsis thaliana* (De Cnodder et al. 2005).

## *2.4 Disturbance of Cytoskeleton*

Cytoskeletal structures (microtubules and microfilaments) are pivotal for cell divisions and the elongation of growing roots (cf. Sivaguru et al. 1999, 2000; Kochian et al. 2005). Al-induced disturbance to organisation of microtubules and microfilaments in the root cells was well documented (e.g. Sivaguru et al. 1999, 2003; Amenos et al. 2009). Such Al-induced structural changes in the root cells might underlie morphological changes and structural malformations observed in Al-stressed roots (Kochian et al. 2005).

## 2.5 Changes in the Plasma Membrane Properties

As Al can enter the symplasm rather rapidly (Silva et al. 2000; Taylor et al. 2000; Babourina and Rengel 2009), $Al^{3+}$ stress is likely to occur at the plasma membrane (Ahn and Matsumoto 2006). $Al^{3+}$ has a strong affinity for the plasma membrane surface (560-fold stronger than $Ca^{2+}$) (Akeson et al. 1989). Yermiyahu et al. (1997) demonstrated that the surface charge of the plasma membrane vesicles isolated from the Al-sensitive wheat cv. Scout was 26 % lower than that of vesicles from the Al-tolerant cv. Atlas, allowing more Al to bind to the Scout vesicles, thereby causing greater Al toxicity compared with Atlas. Moreover, Ahn et al. (2001) reported that 50 μM Al neutralised the surface charge of the plasma membrane and caused a surface potential shift from −20 to +1 mV in squash roots. These results indicated that membrane surface charge regulated the accessibility of Al ions to cells. Indeed, strong correlation was observed between $Al^{3+}$ toxicity and the concentration of adsorbed Al on the membrane surface, as calculated by Gouy–Chapman–Stern model (Kinraide et al. 1992; Kinraide 1994). Such binding of Al to the plasma membrane (1) alters its fluidity and structure (Chen et al. 1991) in addition to the surface potential (Kinraide 2001), (2) induces organic anion release (Ryan et al. 1995; Osawa and Matsumoto 2002), (3) blocks $Ca^{2+}$ transport (Ding et al. 1993; Pineros and Tester 1993), and/or (4) inhibits $H^+$-ATPase activity (Ahn et al. 2002). These changes would alter the plasma membrane potential. However, there are contrasting results reported in the literature about $Al^{3+}$ stress effects on the plasma membrane potential. For example, $Al^{3+}$ stress induced depolarisation in intact roots of Al-sensitive wheat genotype Scout but not in Al-resistant genotype Atlas (Miyasaka et al. 1989). In some studies, $Al^{3+}$ induced hyperpolarisation in Al-sensitive but not in Al-tolerant genotypes (Kinraide 1993; Lindberg and Strid 1997; Johnson et al. 2005; Wherrett et al. 2005). The reason for contradicting results may be plants either growing (for a few days) or just being conditioned (for a few hours) in the low-pH (≈4.5) medium before root cells were impaled with a measuring electrode (longer time in the low-pH medium may allow plants to recover from low-pH-induced depolarisation) (Kinraide 1993). Further research, especially on low-pH stress studied separately from $Al^{3+}$ stress, is thus needed to understand $Al^{3+}$-specific changes in the plasma membrane potential.

Our recent study involving Arabidopsis wild type (Col-0) clearly separated low-pH stress from combined low-pH/$Al^{3+}$ stresses. The low-pH treatment induced plasma membrane depolarisation, which was significantly diminished ($P \leq 0.05$) when combined stresses (low-pH/$Al^{3+}$) were imposed (Bose et al. 2010b). Further, Al-tolerant alr104 and the wild type had depolarised plasma membranes for the entire 30-min measurement period under combined low-pH/$Al^{3+}$ treatment, whereas in the Al-sensitive mutants (als3 and als5), initial depolarisation to around −60 mV became hyperpolarisation at −110 mV after 20 min (Bose et al. 2010a). Thus, the ability of plants to maintain plasma membrane depolarisation during Al stress is critical for Al tolerance.

## 2.6 *Inhibition of Nutrient Uptake*

Long-term exposure to $Al^{3+}$ (from hours to days) results in a deficiency of one or more nutrients, such as Ca, Mg, $NH_4$-N, P, K, and B (cf. Mugwira et al. 1980; Grimme 1983; Foy 1988; Keltjens 1988; Rengel and Robinson 1989; Rengel 1990; Rengel and Elliott 1992; Keltjens and Tan 1993; Lenoble et al. 1996; Mariano and Keltjens 2005). These deficiencies may be due to (1) direct inhibition of uptake system and/or (2) $Al^{3+}$-induced impairment of root growth and a consequent decrease in the nutrient-absorbing surface area (Clarkson 1985). The latter cause of deficiency is common after prolonged exposure to $Al^{3+}$ (hours to days), whereby root growth reduction is associated with decreased nutrient accumulation (see Rengel 1992 for references). Therefore, long-term $Al^{3+}$ exposure studies may not provide information about specific $Al^{3+}$ effects on nutrient uptake. Further complication with long-term studies is that $Al^{3+}$ may inhibit root growth without reducing nutrient uptake. For example, root growth inhibition under $Al^{3+}$ was observed in Norway spruce, small birch, and wheat without reduction in $Ca^{2+}$ and/or $Mg^{2+}$ uptake (Göransson and Eldhuset 1995; Ryan et al. 1997; Godbold and Jentschke 1998). Hence, short-term $Al^{3+}$ exposure studies involving direct measurements of ion fluxes are essential for understanding immediate $Al^{3+}$ effects on nutrient uptake.

### 2.6.1 Calcium Uptake

The interaction between $Al^{3+}$ toxicity and $Ca^{2+}$ uptake received considerable attention because symptoms of severe $Al^{3+}$ toxicity resemble $Ca^{2+}$ deficiency in plants (see Foy 1988; Rengel and Elliott 1992 for references), and exogenous application of relatively high (millimolar) concentrations of $Ca^{2+}$ alleviated $Al^{3+}$ toxicity in many plant species (Brady et al. 1993; Keltjens and Tan 1993; Kinraide et al. 2004). Thus, the capacity of genotypes to maintain $Ca^{2+}$ influx from low-pH environments may contribute to low-pH tolerance. Indeed, low-pH-tolerant Arabidopsis mutants (alr104 and als5) (Bose et al. 2010a) recorded higher $Ca^{2+}$ influx in the distal elongation zone than the wild type and als3 mutant in the low-pH treatment (Bose et al., unpublished results). However, the combined low-pH/50 μM Al stress caused $Ca^{2+}$ efflux from both distal elongation and mature root zones within a minute in all four genotypes (Col-0, als3, als5, and alr104). Such an initial Al-induced $Ca^{2+}$ efflux is likely to have been due to extensive displacement of apoplastic $Ca^{2+}$ by Al ions.

$Al^{3+}$ might inhibit $Ca^{2+}$ influx into intact root cells (Huang et al. 1992; Ryan and Kochian 1993), protoplasts (Rengel and Elliott 1992; Rengel 1994), and the membrane vesicles (Huang et al. 1996; White 1998) through binding of $Al^{3+}$ on the plasma membrane surface (Akeson et al. 1989). Such binding of $Al^{3+}$ to the plasma membrane surface may block $Ca^{2+}$-permeable channels in the plasma membrane. Indeed, both the hyperpolarisation-activated $Ca^{2+}$-permeable channels (Ding et al. 1993; Kiegle et al. 2000; Very and Davies 2000) and depolarisation-activated

$Ca^{2+}$ channels (Rengel et al. 1995; Pineros and Tester 1997) are sensitive to Al, but $Ca^{2+}$ influx inhibition was higher in the former (87 ± 7 %) (Kiegle et al. 2000) than the latter (only 44 %) (Rengel and Zhang 2003). During Al stress, $Ca^{2+}$ fluxes in the distal root elongation of Al-tolerant genotypes (wild type and alr104) recovered to show net influx after the initial Al-induced $Ca^{2+}$ efflux, but $Ca^{2+}$ influx in Al-sensitive genotypes (als3 and als5) remained inhibited. Given that combined low-pH/50 μM Al stress caused less depolarisation and eventual hyperpolarisation of $E_m$ in the Al-sensitive mutants (als3 and als5), it may be suggested that $Al^{3+}$ stress inhibited hyperpolarisation-activated $Ca^{2+}$-permeable channels in Al-sensitive mutants (Bose et al. unpublished results).

As the above $Ca^{2+}$ influx inhibition following $Al^{3+}$ exposure precedes root growth inhibition (Huang et al. 1992; Ryan and Kochian 1993), it could be one of potential primary causes of $Al^{3+}$ phytotoxicity (Rengel 1992; Rengel and Zhang 2003). However, further studies revealed that low concentration of $Al^{3+}$ can inhibit root growth without affecting $Ca^{2+}$ influx, and addition of ameliorating cations ($Mg^{2+}$ and $Na^{+}$) improved root growth, even though the net $Ca^{2+}$ influx remained inhibited (Ryan and Kochian 1993; Ryan et al. 1997). Similarly, $Al^{3+}$ caused root hair growth inhibition without affecting $Ca^{2+}$ influx in *Limnobium stoloniferum* (Jones et al. 1995). Poor correlation between Al-induced $Ca^{2+}$ influx inhibition and elongation growth of Chara (Reid et al. 1995) indicated that Al-induced inhibition of $Ca^{2+}$ influx alone cannot be a critical factor in triggering Al toxicity in plants. However, prolonged inhibition of $Ca^{2+}$ influx into Al-treated root cells disrupts Ca nutrition, which in turn exacerbates Al toxicity in plants (Rengel and Zhang 2003).

### 2.6.2 Magnesium Uptake

$Mg^{2+}$ is unique among the major biological cations due to the largest hydrated radius (0.428 nm), the smallest ionic radius (0.072 nm), and the highest charge density. Because it binds water molecules 3–4 orders of magnitude more tightly than do other cations, $Mg^{2+}$ often interacts with other molecules while maintaining its hydration sphere (Maguire and Cowan 2002). As a result, $Mg^{2+}$ binds quite weakly to the negatively charged groups in the root cell wall, so the excess cations like $H^{+}$ and $Al^{3+}$ present in acid soils can inhibit $Mg^{2+}$ loading in the apoplasm and uptake across the plasma membrane (Marschner 1991, 1995).

$Al^{3+}$-induced inhibition of $Mg^{2+}$ uptake has been observed in diverse plant species (Grimme 1983; Keltjens 1988; Rengel and Robinson 1989; Rengel 1990). $Al^{3+}$ might cause $Mg^{2+}$ uptake inhibition through competitive interactions between $Al^{3+}$ and plasma membrane transporters for $Mg^{2+}$ (Rengel and Robinson 1989; Rengel 1990) because (1) Al and Mg have similar hydrated ionic radii, and (2) plants preferentially take up heavy isotope $^{26}Mg$ (the daughter nuclei of $^{27}Al$) from a mix of $Mg^{2+}$ isotopes in nutrient solutions and store it in tissues (reviewed in Bose et al. 2011a). This might be true because *Arabidopsis thaliana* magnesium transporters (AtMGT1 and AtMGT10) are highly sensitive to $Al^{3+}$, providing potential molecular targets for $Al^{3+}$ toxicity in plants (Li et al. 2001). On the contrary, overexpression of $Mg^{2+}$ transporter genes in yeast (MacDiarmid and Gardner 1998), *Nicotiana benthamiana* (Deng et al. 2006), and rice (Chen et al. 2012) conferred Al tolerance by potentially alleviating Al-induced magnesium deficiency (Chen and Ma 2013), but these studies did not provide sufficient evidence of enhanced magnesium uptake and an increase in intracellular $Mg^{2+}$ concentration in the presence of $Al^{3+}$ ions. This issue has been addressed by measuring $Mg^{2+}$ uptake using $Mg^{2+}$-selective microelectrodes and fluorescent dye in Arabidopsis roots during short-term (0–60 min) exposure to $Al^{3+}$ stress (Bose et al. 2013). The results showed that enhanced $Mg^{2+}$ uptake and increased intracellular free $Mg^{2+}$ concentration correlated with an improved capacity of Arabidopsis genotypes to cope with low-pH and combined low-pH/Al stresses (Bose et al. 2013).

### 2.6.3 Potassium Uptake

$K^+$ is essential for cell division through polymerisation of actin (Alberts et al. 1994) and turgor-dependent cell elongation caused by accumulation of $K^+$ in the vacuole (Frensch 1997; Dolan and Davies 2004; Sano et al. 2007). However, there is no causal relationship between $Al^{3+}$ toxicity and $K^+$ nutrition in plants because $Al^{3+}$ induced either inhibition (Matsumoto and Yamaya 1986; Nichol et al. 1993) or an increase in $K^+$ uptake (Lee and Pritchard 1984; Lindberg 1990; Tanoi et al. 2005). The reason for increased $K^+$ uptake under $Al^{3+}$ stress may be a decrease in net $K^+$ efflux rather than an increase in uptake (Horst et al. 1992; Olivetti et al. 1995; Sasaki et al. 1995). Several patch clamp studies demonstrated that Al ions decrease the open probability of $K^+$ inward-rectifying channels through internal blocking (Gassmann and Schroeder 1994; Liu and Luan 2001). In contrast, Al induced or maintained $K^+$ efflux in Al-tolerant wheat genotypes together with enhanced malate release (Ryan et al. 1995; Osawa and Matsumoto 2002; Wherrett et al. 2005), probably to balance charges created by exudation of weak organic acid anions (Ryan et al. 1995; Matsumoto 2000; Ma et al. 2001; Osawa and Matsumoto 2002; Wherrett et al. 2005). This notion is also confirmed in Arabidopsis thaliana wherein Al-tolerant genotypes (alr104 and Col-0) showed greater $K^+$ efflux than Al-sensitive genotypes (*als3* and *als5*) during $Al^{3+}$ stress (Bose et al. 2010a).

### 2.6.4 Phosphorus Uptake

Apart from causing direct toxicity to roots, $Al^{3+}$ ions also exacerbate P deficiency by binding with P to form sparingly soluble Al–P complexes that are not plant available (Haynes and Mokolobate 2001). Hence, even in acidic soils that have relatively high total concentration of P, availability of P is limiting (Kochian et al. 2004; Fukuda et al. 2007). Selection of genotypes for either P efficiency or Al tolerance independently may be unsuccessful because these two soil constraints occur jointly in acidic soils (Foy 1988; Yan et al. 1995). For example, Al-tolerant soybean genotype 416937 selected under controlled conditions was found to be sensitive to acid soils in the field (Ritchey and Carter 1993; Ferrufino et al. 2000). In contrast, an Al-sensitive soybean genotype was found to be relatively tolerant to acid soils (e.g. Foy et al. 1992). These discrepancies might be due to failure in the selection process to account for interactions between Al and P that normally occur in acid soils. Thus, a thorough understanding of the Al–P interactions is essential for improving the productivity of crops in acid soils.

Generally, plants may respond to both Al toxicity and P deficiency by exuding organic acid anions (Lopez-Bucio et al. 2000; Haynes and Mokolobate 2001; Shen et al. 2002). Exudation of low-molecular-weight organic acid anions (e.g. citrate, malate) in the rooting media is advantageous because organic acid anions can (1) protect plants from Al toxicity by forming non-phytotoxic Al-organic anion complexes and (2) enhance P availability and thus improve plant P uptake by chelating Al from the Al–P complexes, thus liberating P for plant uptake (Subbarao et al. 1997; Ishikawa et al. 2002).

The signal perception of Al toxicity or P deficiency and translocation of this signal into activation of organic acid synthesis and exudation are pivotal for P nutrition and Al tolerance in acid soils. Proteomic (Fukuda et al. 2007) and transcriptomic (Wasaki et al. 2003) analysis of rice roots grown in Al-toxic and P-deficient low-pH solution revealed that (1) modifications of root protein expression were similar under Al toxicity and P deficiency, and (2) carbon supply to the tri-carboxylic acid (TCA) cycle to produce organic acids was maintained by enhancing glycolysis. Indeed, P-efficient genotypes were able to enhance Al tolerance in acid soils by stimulated exudation of different Al-chelating organic acid anions in soybean (Liao et al. 2006), cowpea (Jemo et al. 2007), and barley (Delhaize et al. 2009). Interestingly, Liao et al. (2006) found that Al toxicity induced citrate exudation, P deficiency triggered oxalate exudation, and malate release was induced by either Al toxicity or P deficiency in soybean. In contrast, Ligaba et al. (2004) reported that citrate exudation was enhanced by P deficiency but not by Al toxicity in purple lupin. These controversial results clearly suggest that there are important differences in how Al toxicity and P deficiency may effect organic anion exudation, which is of huge importance when these two environmental stresses occur together as they regularly do in acid soils.

## 3 Disturbance of Ion Homeostasis

The maintenance of optimal concentrations of inorganic ions such as $H^+$, $K^+$, $Ca^{2+}$, and $Mg^{2+}$ (ionic homeostasis) inside plant cells and organelles is pivotal for the functioning of biopolymers (Andreev 2001). Ion homeostasis in plants is regulated by controlled flux of ions across the plasma membrane and the endomembranes in addition to storage in organelles (Bose et al. 2011a). Entry of Al ions into the cytoplasm (Silva et al. 2000; Babourina and Rengel 2009) may affect homeostasis of various ions inside the cell.

### 3.1 $H^+$ Homeostasis

The change of pH (ΔpH) between the cytoplasm and the apoplasm is the major driving force for the translocation of ions in plant cells. Under no stress, the pH is 7.3–7.6 in the cytoplasm, 4.5–5.9 in vacuoles, ~7 in mitochondria, 7.2–7.8 in chloroplasts, and ~5.5 in the apoplast (Kurkdjian and Guern 1989; Bose et al. 2011a, b). Thus, cytoplasm is less acidic when compared to vacuoles and the apoplast. This pH difference is regulated by proton pumps ($H^+$-ATPase and $H^+$-PPase) located at the plasma membrane and the tonoplast, driving $H^+$ from the cytoplasm to either the apoplast or the vacuole (Marty 1999). Hence, disturbance in $H^+$-ATPase activity by environmental stresses would affect cytoplasmic pH regulation. Indeed, transient changes in cytoplasmic pH are pivotal for the signal cascades to elicit defence mechanisms or developmental processes in response to a variety of environmental stimuli (Roos et al. 2006). The low-pH treatment caused net $H^+$ influx into the root tissue and caused intracellular acidification (Gerendas et al. 1990; Plieth et al. 1999; Babourina and Rengel 2009; Bose et al. 2010b). Further, genes controlling cytoplasmic pH were downregulated in the low-pH-hypersensitive Arabidopsis stop1 mutant (Iuchi et al. 2007; Sawaki et al. 2009). Thus, cytoplasmic acidification may be responsible for poor root growth in the treatments with low pH only. Interestingly, low-pH tolerance of als5 and alr104 mutants coincided with high $H^+$ influx, suggesting that aforementioned mutants may possess effective mechanisms to prevent cytoplasmic acidification despite high $H^+$ influx from the external environment; in contrast, such mechanisms would be absent/ineffective in the low-pH-sensitive mutant (als3) (Bose et al. 2010a).

Modulation of cytosolic pH by combined low-pH/$Al^{3+}$ stress can act as a secondary messenger to activate/inactivate transporters and enzymes and, in turn, regulate synthesis of organic acid anions and their subsequent release. An increase in intracellular pH (from ≈5.7 under control pH $5.5_{[water]}$) towards pH ≈ 6.5 following a combined low-pH/100 μM Al stress was observed in Arabidopsis wild type (Bose et al. 2010b). This rise in intracellular pH would favour deprotonation of organic acids inside the cytoplasm (Davies 1986) and potentially their anion exudation into the rhizosphere. Though $Al^{3+}$ decreased the $H^+$-ATPase activity in the plasma membrane vesicles prepared from Al-treated seedlings of barley (Matsumoto 1988; Matsumoto et al. 1992), wheat (Sasaki et al. 1995), and squash (Ahn et al. 2001, 2002), inhibition of $H^+$-ATPase activity appears to be dependent on $Al^{3+}$ concentration. For example, $Al^{3+}$ concentrations lower than the threshold $Al^{3+}$ phytotoxicity caused upregulation of $H^+$-ATPase, whereas phytoxic $Al^{3+}$ concentrations resulted in $H^+$-ATPase inhibition in maize (Facanha and Okorokova-Facanha 2002) and soybean roots (Shen et al. 2005). In addition, cytoplasmic pH may also vary depending on the $Al^{3+}$ concentrations used. Thus, more work is needed to understand $Al^{3+}$ concentration's influence on the cytoplasmic pH homeostasis.

## 3.2 $Ca^{2+}$ Homeostasis

Being the secondary messenger, free cytosolic $Ca^{2+}$ activities are pivotal for transduction of hormonal and environmental signals to the responsive elements of cellular metabolism (see Rengel and Zhang 2003; Bose et al. 2011b for references). Free cytosolic $Ca^{2+}$ activities in plant cells are usually maintained in the 100–200 nM concentration range (Bush 1995; Webb et al. 1996). However, $Ca^{2+}$ activities in the cell wall (apoplasm) and other internal organelles (e.g. vacuoles and endoplasmic reticulum) are higher than the cytosolic $Ca^{2+}$ by 3–4 orders of magnitude (Clarkson 1984; DuPont et al. 1990; Evans et al. 1991; Bose et al. 2011b). Low concentrations of cytosolic $Ca^{2+}$ are maintained by ATP-dependent $Ca^{2+}$ pumps and $Ca^{2+}$ exchangers (CaX) in the plasma membrane and the endomembranes via (1) sequestration into different organelles and (2) pumping $Ca^{2+}$ into the apoplasm (Evans et al. 1991; Hirschi 2001; Miedema et al. 2001; Bose et al. 2011b).

$Al^{3+}$ affects the $Ca^{2+}$ homeostasis maintenance in three ways. Firstly, $Ca^{2+}$ is essential for cross-linking the pectic materials in the cell wall. Aluminium displaces pectin-bound $Ca^{2+}$ because Al has a higher affinity for pectic material than $Ca^{2+}$ (Blamey 2001), and overexpression of pectin methylesterase enzyme in *Solanum tuberosum* resulted in severe Al toxicity (Schmohl et al. 2000). In fact, between 90 % (Reid et al. 1995) and 99.99 % (Taylor et al. 2000) of cell-wall-bound $Ca^{2+}$ is displaced by $Al^{3+}$ in Chara internodal cells. In *Arabidopsis thaliana* roots, initial Al-induced $Ca^{2+}$ efflux was higher in the Al-sensitive genotypes (*als3* and *als5*) than in the wild type and Al-tolerant alr104 mutant, suggesting extensive displacement of apoplastic $Ca^{2+}$ by Al ions in the Al-sensitive mutants (Bose et al. unpublished results). Hence, displacement of $Ca^{2+}$ by $Al^{3+}$ would severely alter the physical properties of the cell wall, including extensibility, rigidity, and permeability (Reid et al. 1995; Tabuchi and Matsumoto 2001; Jones et al. 2006; Horst et al. 2007), thereby detrimentally affecting cell division and elongation. However, contradicting results were observed in onion root tips where the particle-induced X-ray emission technique indicated that $Ca^{2+}$ in the root tips was not displaced by Al (Schofield et al. 1998). These discrepancies might be due to different experimental systems and environmental conditions. Secondly, $Al^{3+}$ inhibits the $Ca^{2+}$ influx (reviewed in Sect. 6.1). Thirdly, Al disturbs cytosolic $Ca^{2+}$ activity, thereby affecting the signal transduction pathways involved in root growth. However, a disagreement exists in the literature about Al effects on cytosolic $Ca^{2+}$ homeostasis and its involvement in Al toxicity.

In many plant species, $Al^{3+}$ toxicity caused elevation of cytoplasmic $Ca^{2+}$ activity, with such elevation being higher in Al-sensitive than Al-tolerant genotypes of the same species (Jones et al. 1998a; Zhang and Rengel 1999; Ma et al. 2002; Rengel and Zhang 2003). This cytosolic $Ca^{2+}$ rise would play a major role in the expression of $Al^{3+}$ toxicity because the cell-responsive elements may stop responding to transient rises in cytosolic $Ca^{2+}$ caused by a variety of signals (Rengel and Zhang 2003). For example, an increase in cytosolic $Ca^{2+}$ caused closure of plasmodesmata (Holdaway-Clarke et al. 2000) and inhibited plasmodesmata-mediated cell-to-cell transport in Al-sensitive wheat roots (Sivaguru et al. 2000). A good correlation was observed between Al-induced cytosolic $Ca^{2+}$ rise (within 30 min) and root growth inhibition in wheat genotypes (Zhang and Rengel 1999), leading to the hypothesis that disruption of $Ca^{2+}$ homeostasis may be the primary cause of $Al^{3+}$ toxicity (Rengel and Zhang 2003). However, in a recent study on Arabidopsis, an Al-induced cytoplasmic $Ca^{2+}$ rise started in the mature (least Al-sensitive) root zone (in 48 s) and proceeded towards the root cap (in 100 s). Moreover, a $Ca^{2+}$ rise did not differ among Al-resistant, Al-sensitive, and the wild-type Arabidopsis roots (Rincon-Zachary et al. 2010). Similarly, a lack of correlation between Al-induced growth inhibition and alteration in cytosolic $Ca^{2+}$ in the root hairs of Arabidopsis thaliana wild-type, sensitive, and tolerant genotypes (Jones et al. 1998a) indicated that alteration in cytosolic $Ca^{2+}$ may not be responsible for growth inhibition. In some studies, such as in tobacco cell cultures, Al decreased the cytosolic $Ca^{2+}$ concentration along with growth inhibition (Jones et al. 1998b). More detailed comparison (Plieth et al. 1999) of low-pH and combined low-pH/$Al^{3+}$ effects on cytosolic $Ca^{2+}$ dynamics using *Arabidopsis thaliana* indicated that intact roots responded to low pH by a sustained elevation of cytosolic $Ca^{2+}$. However, this low-pH-mediated elevation in cytosolic $Ca^{2+}$ activity was abolished in the presence of Al, suggesting that $Ca^{2+}$-mediated protection mechanism against low pH is irreversibly inhibited by Al (Plieth et al. 1999). More information, especially during the first few seconds of low-pH and $Al^{3+}$ stress, is clearly needed to resolve many discrepancies in the literature.

## 3.3 *$Mg^{2+}$ Homeostasis*

$Al^{3+}$ and $Mg^{2+}$ ions have similar hydrated radii; hence, $Al^{3+}$ ions compete with $Mg^{2+}$ ions for apoplastic binding, uptake via $Mg^{2+}$-permeable cation channels and transporters, and binding with enzymes, ATP, and anions (reviewed in Bose et al. 2011a). As a result, $Mg^{2+}$ transport and metabolism under $Al^{3+}$ stress might be impaired in all the compartments of the cell (Bose et al. 2011a). However, little information is available on how $Al^{3+}$ stress modulates the cytosolic free $Mg^{2+}$ concentration. To shed light on this issue, intracellular free $Mg^{2+}$ concentrations were measured in the epidermal root cells of Arabidopsis genotypes using an $Mg^{2+}$-selective fluorescence dye (Magnesium Green$^{TM}$). Under control conditions (pH 5.45), free cytosolic $Mg^{2+}$ concentrations were in the range of 0.8–1.4 mM. The Al-resistant mutant alr104 recorded the highest intracellular $Mg^{2+}$ concentration followed by als5 ~ wild type > als3. The low-pH (4.2) stress did not alter the free cytosolic $Mg^{2+}$ concentration, whereas combined low-pH/50 mM Al stress raised the intracellular $Mg^{2+}$ concentration in all genotypes tested but to a different extent. The Al-tolerant genotypes (wild-type Col-0 and *alr104* mutant) recorded a higher intracellular $Mg^{2+}$ concentration than the Al-sensitive mutants (*als3* and *als5*) (Bose et al. 2013). The ability of Al-tolerant genotypes (Col-0 and alr104) to maintain the influx of $Mg^{2+}$ ions into the root tissue from the external medium is the primary reason for enhanced intracellular $Mg^{2+}$ concentration in these genotypes.

Elevated intracellular $Mg^{2+}$ might play a pivotal role in the maintenance of $H^+$-ATPase activity, acid phosphatase activity, organic acid synthesis and metabolism, cytosolic $Ca^{2+}$ dynamics, and reactive oxygen species homeostasis during $Al^{3+}$ stress (Bose et al. 2011a; Chen and Ma 2013). Interestingly, exposure of Arabidopsis wild-type (Col-0) roots to Al concentrations higher than 50 μM (i.e. 100 and 500 μM $AlCl_3$ treatments, pH 4.2) decreased the intracellular $Mg^{2+}$ concentration in a dose-dependent manner. This decline is explained by the decreased $Mg^{2+}$ influx, or increased efflux, at 500 μM $Al^{3+}$, caused by Al inhibition of the plasma membrane cation channels (Bose et al. 2013). Above observations suggest that the efficacy of phytotoxic Al to block $Mg^{2+}$ transport through cation channels is concentration and genotype dependent. More work is needed to identify threshold $Al^{3+}$ concentration for different crop species.

# 4 Conclusions

Low-pH and combined low-pH/$Al^{3+}$ stresses differentially affect uptake and homeostasis of hydrogen, phosphorus, potassium, calcium, and magnesium. Plants with a superior capacity to take up hydrogen at the same time preventing cytoplasmic acidification along with enhanced uptake of phosphorus, calcium, and magnesium ions perform well under low-pH and combined low-pH/$Al^{3+}$ stresses. In the case of potassium, enhanced uptake may help plants resist low-pH stress, whereas an enhanced potassium loss to balance charges with the organic acid anion exudation is the preferred strategy to combat the combined low-pH/$Al^{3+}$ stress. Breeding for enhanced nutrition of phosphorus, calcium, and magnesium under $Al^{3+}$ stress may be both possible and desirable approach to improve crop growth in acid soils.

**Acknowledgement** Jayakumar Bose was a recipient of Endeavour International Postgraduate Research Scholarship and University of Western Australia Postgraduate Award. Additional support was provided by the Australian Research Council. This work was supported by the Australian Research Council grants to Zed Rengel, Meixue Zhou, and Sergey Shabala.

# References

Ahn SJ, Matsumoto H (2006) The role of the plasma membrane in the response of plant roots to aluminum toxicity. Plant Signal Behav 1:37–45

Ahn SJ, Sivaguru M, Osawa H, Chung GC, Matsumoto H (2001) Aluminum inhibits the $H^+$-ATPase activity by permanently altering the plasma membrane surface potentials in squash roots. Plant Physiol 126:1381–1390

Ahn SJ, Sivaguru M, Chung GC, Rengel Z, Matsumoto H (2002) Aluminium-induced growth inhibition is associated with impaired efflux and influx of $H^+$ across the plasma membrane in root apices of squash (*Cucurbita pepo*). J Exp Bot 53:1959–1966

Ahn SJ, Shin R, Schachtman DP (2004) Expression of KT/KUP genes in Arabidopsis and the role of root hairs in $K^+$ uptake. Plant Physiol 134:1135–1145

Akeson MA, Munns DN, Burau RG (1989) Adsorption of $Al^{3+}$ to phosphatidylcholine vesicles. Biochim Biophys Acta 986:33–40

Alberts B, Bray D, Lewis J, Raff M, Roberts K, Watson JD (1994) Molecular biology of the cell. Garland Publishing, New York, NY

Amenos M, Corrales I, Poschenrieder C, Illes P, Baluska F, Barcelo J (2009) Different effects of aluminum on the actin cytoskeleton and brefeldin A-sensitive vesicle recycling in root apex cells of two maize varieties differing in root elongation rate and aluminum tolerance. Plant Cell Physiol 50:528–540

Andreev IM (2001) Functions of the vacuole in higher plant cells. Russ J Plant Physiol 48:672–680

Babourina O, Rengel Z (2009) Uptake of aluminium into Arabidopsis root cells measured by fluorescent lifetime imaging. Ann Bot 104:189–195

Babourina O, Ozturk L, Cakmak I, Rengel Z (2006) Reactive oxygen species production in wheat roots is not linked with changes in $H^+$ fluxes during acidic and aluminium stresses. Plant Signal Behav 1:71–76

Baluška F, Mancuso S, Volkmann D, Barlow P (2004) Root apices as plant command centres: the unique 'brain-like' status of the root apex transition zone. Biologia (Bratislava) 59:7–19

Bibikova T, Gilroy S (2002) Root hair development. J Plant Growth Regul 21:383–415

Blamey FPC (2001) The role of the root cell wall in aluminum toxicity. In: Ae N, Arihara J, Okada K, Srinivasan A (eds) Plant nutrient acquisition: new perspectives. Springer, Tokyo, pp 201–226

Bolan NS, Hedley MJ, White RE (1991) Processes of soil acidification during nitrogen cycling with emphasis on legume based pastures. Plant Soil 134:53–63

Boscolo PRS, Menossi M, Jorge RA (2003) Aluminum-induced oxidative stress in maize. Phytochemistry 62:181–189

Bose J, Babourina O, Shabala S, Rengel Z (2010a) Aluminium-induced ion transport in Arabidopsis: the relationship between Al tolerance and root ion flux. J Exp Bot 61:3163–3175

Bose J, Babourina O, Shabala S, Rengel Z (2010b) Aluminum dependent dynamics of ion transport in Arabidopsis: specificity of low pH and aluminum responses. Physiol Plant 139:401–412

Bose J, Babourina O, Rengel Z (2011a) Role of magnesium in alleviation of aluminium toxicity in plants. J Exp Bot 62:2251–2264

Bose J, Pottosin I, Shabala SS, Palmgren MG, Shabala S (2011b) Calcium efflux systems in stress signalling and adaptation in plants. Front Plant Sci 2:85

Bose J, Babourina O, Shabala S, Rengel Z (2013) Low-pH and aluminium resistance in Arabidopsis correlates with high cytosolic magnesium content and increased magnesium uptake by plant roots. Plant Cell Physiol 54:1093–1104

Brady NC, Weil RR (1990) Nature and properties of soils. Macmillan, New York, NY

Brady DJ, Edwards DG, Asher CJ, Blamey FPC (1993) Calcium amelioration of aluminum toxicity effects on root hair development in soybean [*Glycine max* (L) Merr]. New Phytol 123:531–538

Bush DS (1995) Calcium regulation in plant cells and its role in signaling. Annu Rev Plant Biol 46:95–122

Cakmak I, Horst WJ (1991) Effect of aluminium on lipid peroxidation, superoxide dismutase, catalase, and peroxidase activities in root tips of soybean (*Glycine max*). Physiol Plant 83:463–468

Chen ZC, Ma JF (2013) Magnesium transporters and their role in Al tolerance in plants. Plant Soil 368:51–56

Chen JP, Sucoff EI, Stadelmann EJ (1991) Aluminum and temperature alteration of cell membrane permeability of *Quercus rubra*. Plant Physiol 96:644–649
Chen ZC, Yamaji N, Motoyama R, Nagamura Y, Ma JF (2012) Up-regulation of a magnesium transporter gene OsMGT1 is required for conferring aluminum tolerance in rice. Plant Physiol 159:1624–1633
Clarkson DT (1965) The effect of aluminium and some other trivalent metal cations on cell division in the root apices of *Allium cepa*. Ann Bot 29:309–315
Clarkson DT (1984) Calcium transport between tissues and its distribution in the plant. Plant Cell Environ 7:449–456
Clarkson DT (1985) Factors affecting mineral nutrient acquisition by plants. Annu Rev Plant Biol 36:77–115
Darko E, Ambrus H, Stefanovits-Banyai E, Fodor J, Bakos F, Barnaba B (2004) Aluminium toxicity, Al tolerance and oxidative stress in an Al-sensitive wheat genotype and in Al-tolerant lines developed by in vitro microspore selection. Plant Sci 166:583–591
Davies DD (1986) The fine control of cytosolic pH. Physiol Plant 67:702–706
De Campos JMS, Viccini LF (2003) Cytotoxicity of aluminum on meristematic cells of *Zea mays* and *Allium cepa*. Caryologia 56:65–73
De Cnodder T, Vissenberg K, Van Der Straeten D, Verbelen JP (2005) Regulation of cell length in the *Arabidopsis thaliana* root by the ethylene precursor 1-aminocyclopropane-1-carboxylic acid: a matter of apoplastic reactions. New Phytol 168:541–550
Delhaize E, Ryan PR (1995) Aluminum toxicity and tolerance in plants. Plant Physiol 107:315–321
Delhaize E, Taylor P, Hocking PJ, Simpson RJ, Ryan PR, Richardson AE (2009) Transgenic barley (*Hordeum vulgare* L.) expressing the wheat aluminium resistance gene (TaALMT1) shows enhanced phosphorus nutrition and grain production when grown on an acid soil. Plant Biotech J 7:391–400
Demidchik V, Bowen HC, Maathuis FJM, Shabala SN, Tester MA, White PJ, Davies JM (2002) Arabidopsis thaliana root non-selective cation channels mediate calcium uptake and are involved in growth. Plant J 32:799–808
Demidchik V, Shabala SN, Coutts KB, Tester MA, Davies JM (2003) Free oxygen radicals regulate plasma membrane $Ca^{2+}$- and $K^{+}$-permeable channels in plant root cells. J Cell Sci 116:81–88
Deng W, Luo K, Li D, Zheng X, Wei X, Smith W, Thammina C, Lu L, Li Y, Pei Y (2006) Overexpression of an Arabidopsis magnesium transport gene, AtMGT1, in *Nicotiana benthamiana* confers Al tolerance. J Exp Bot 57:4235–4243
Ding JP, Badot PM, Pickard BG (1993) Aluminium and hydrogen ions inhibit a mechanosensory calcium-selective cation channel. Aust J Plant Physiol 20:771–778
Dolan L, Davies J (2004) Cell expansion in roots. Curr Opin Plant Biol 7:33–39
Doncheva S, Amenos M, Poschenrieder C, Barcelo J (2005) Root cell patterning: a primary target for aluminium toxicity in maize. J Exp Bot 56:1213–1220
DuPont FM, Bush DS, Windle JJ, Jones RL (1990) Calcium and proton transport in membrane vesicles from barley roots. Plant Physiol 94:179–188
Evans DE, Briars SA, Williams LE (1991) Active calcium transport by plant cell membranes. J Exp Bot 42:285–303
Exley C (2004) The pro-oxidant activity of aluminum. Free Rad Biol Med 36:380–387
Ezaki B, Gardner RC, Ezaki Y, Matsumoto H (2000) Expression of aluminum-induced genes in transgenic Arabidopsis plants can ameliorate aluminum stress and/or oxidative stress. Plant Physiol 122:657–666
Facanha AR, Okorokova-Facanha AL (2002) Inhibition of phosphate uptake in corn roots by aluminum-fluoride complexes. Plant Physiol 129:1763–1772
Ferguson IB, Clarkson DT (1975) Ion transport and endodermal suberization in the roots of *Zea mays*. New Phytol 75:69–79

Ferguson IB, Clarkson DT (1976) Simultaneous uptake and translocation of magnesium and calcium in barley (*Hordeum vulgare* L.) roots. Planta 128:267–269

Ferrufino A, Smyth TJ, Israel DW, Carter TE Jr (2000) Root elongation of soybean genotypes in response to acidity constraints in a subsurface solution compartment. Crop Sci 40:413

Foreman J, Demidchik V, Bothwell JHF, Mylona P, Miedema H, Torres MA, Linstead P, Costa S, Brownlee C, Jones JDG (2003) Reactive oxygen species produced by NADPH oxidase regulate plant cell growth. Nature 422:442–446

Foy CD (1988) Plant adaptation to acid, aluminum-toxic soils. Commun Soil Sci Plant Anal 19:959–987

Foy CD, Duke JA, Devine TE (1992) Tolerance of soybean germplasm to an acid Tatum subsoil. J Plant Nutr 15:527–547

Frensch J (1997) Primary responses of root and leaf elongation to water deficits in the atmosphere and soil solution. J Exp Bot 48:985–999

Fukuda T, Saito A, Wasaki J, Shinano T, Osaki M (2007) Metabolic alterations proposed by proteome in rice roots grown under low P and high Al concentration under low pH. Plant Sci 172:1157–1165

Gahoonia TS, Nielsen NE (1998) Direct evidence on participation of root hairs in phosphorus (P-32) uptake from soil. Plant Soil 198:147–152

Galloway JN (1989) Atmospheric acidification: projections for the future. Ambio 18:161–166

Gassmann W, Schroeder JI (1994) Inward-rectifying $K^+$ channels in root hairs of wheat – a mechanism for aluminum-sensitive low-affinity $K^+$ uptake and membrane-potential control. Plant Physiol 105:1399–1408

Gerendas J, Ratcliffe RG, Sattelmacher B (1990) $^{31}P$ nuclear magnetic resonance evidence for differences in intracellular pH in the roots of maize seedlings grown with nitrate or ammonium. J Plant Physiol 137:125–128

Godbold DL, Jentschke G (1998) Aluminium accumulation in root cell walls coincides with inhibition of root growth but not with inhibition of magnesium uptake in Norway spruce. Physiol Plant 102:553–560

Göransson A, Eldhuset TD (1995) Effects of aluminium ions on uptake of calcium, magnesium and nitrogen in *Betula pendula* seedlings growing at high and low nutrient supply rates. Water Air Soil Pollut 83:351–361

Goulding KWT, Bailey NJ, Bradbury NJ, Hargreaves P, Howe M, Murphy DV, Poulton PR, Willison TW (1998) Nitrogen deposition and its contribution to nitrogen cycling and associated soil processes. New Phytol 139:49–58

Grimme H (1983) Aluminium induced magnesium deficiency in oats. Zeitsch Pflanzenern Bodenk 146:666–676

Guo KM, Babourina O, Christopher DA, Borsic T, Rengel Z (2010) The cyclic nucleotide-gated channel AtCNGC10 transports $Ca^{2+}$ and $Mg^{2+}$ in Arabidopsis. Physiol Plant 139:303–312

Haynes R, Mokolobate M (2001) Amelioration of Al toxicity and P deficiency in acid soils by additions of organic residues: a critical review of the phenomenon and the mechanisms involved. Nutr Cycl Agroecosyst 59:47–63

Hirschi K (2001) Vacuolar $H^+/Ca^{2+}$ transport: who's directing the traffic? Trends Plant Sci 6:100–104

Holdaway-Clarke TL, Walker NA, Hepler PK, Overall RL (2000) Physiological elevations in cytoplasmic free calcium by cold or ion injection result in transient closure of higher plant plasmodesmata. Planta 210:329–335

Horst WJ, Asher CJ, Cakmak I, Szulkiewicz P, Wissemeier AH (1992) Short-term responses of soybean roots to aluminum. J Plant Physiol 140:174–178

Horst WJ, Kollmeier M, Schmohl N, Sivaguru M, Wang Y, Felle HH, Hedrich R, Schröder W, Staß A (2007) Significance of the root apoplast for aluminium toxicity and resistance of maize. In: Sattelmacher B, Horst WJ (eds) The apoplast of higher plants: compartment of storage, transport and reactions. Springer, Dordrecht, pp 49–66

Huang JW, Shaff JE, Grunes DL, Kochian LV (1992) Aluminum effects on calcium fluxes at the root apex of aluminum-tolerant and aluminum-sensitive wheat cultivars. Plant Physiol 98:230–237

Huang JW, Pellet DM, Papernik LA, Kochian LV (1996) Aluminum interactions with voltage-dependent calcium transport in plasma membrane vesicles isolated from roots of aluminum-sensitive and-resistant wheat cultivars. Plant Physiol 110:561–569

Illes P, Schlicht M, Pavlovkin J, Lichtscheidl I, Baluska F, Ovecka M (2006) Aluminium toxicity in plants: internalization of aluminium into cells of the transition zone in Arabidopsis root apices related to changes in plasma membrane potential, endosomal behaviour, and nitric oxide production. J Exp Bot 57:4201–4213

Ishikawa S, Adu-Gyamfi J, Nakamura T, Yoshihara T, Watanabe T, Wagatsuma T (2002) Genotypic variability in phosphorus solubilizing activity of root exudates by pigeonpea grown in low-nutrient environments. In: Adu-Gyamfi JJ (ed) Food security in nutrient-stressed environments: exploiting plants' genetic capabilities. Kluwer Academic, Dordrecht, pp 111–121. doi:10.1007/978-94-017-1570-6_13

Iuchi S, Koyama H, Iuchi A, Kobayashi Y, Kitabayashi S, Kobayashi Y, Ikka T, Hirayama T, Shinozaki K, Kobayashi M (2007) Zinc finger protein STOP1 is critical for proton tolerance in Arabidopsis and coregulates a key gene in aluminum tolerance. Proc Natl Acad Sci U S A 104:9900–9905

Jemo M, Abaidoo RC, Nolte C, Horst WJ (2007) Aluminum resistance of cowpea as affected by phosphorus-deficiency stress. J Plant Physiol 164:442–451

Johnson VJ, Tsunoda M, Murray TF, Sharma RP (2005) Decreased membrane fluidity and hyperpolarization in aluminum-treated PC-12 cells correlates with increased production of cellular oxidants. Environ Toxicol Pharmacol 19:221–230

Jones DL, Shaff JE, Kochian LV (1995) Role of calcium and other ions in directing root hair tip growth in *Limnobium stoloniferum*. Planta 197:672–680

Jones DL, Gilroy S, Larsen PB, Howell SH, Kochian LV (1998a) Effect of aluminum on cytoplasmic $Ca^{2+}$ homeostasis in root hairs of *Arabidopsis thaliana* (L.). Planta 206:378–387

Jones DL, Kochian LV, Gilroy S (1998b) Aluminum induces a decrease in cytosolic calcium concentration in BY-2 tobacco cell cultures. Plant Physiol 116:81–89

Jones DL, Blancaflor EB, Kochian LV, Gilroy S (2006) Spatial coordination of aluminium uptake, production of reactive oxygen species, callose production and wall rigidification in maize roots. Plant Cell Environ 29:1309–1318

Keltjens WG (1988) Short-term effects of Al on nutrient uptake, $H^+$ efflux, root respiration and nitrate reductase activity of two sorghum genotypes differing in Al-susceptibility. Commun Soil Sci Plant Anal 19:1155–1163

Keltjens WG, Tan K (1993) Interactions between aluminium, magnesium and calcium with different monocotyledonous and dicotyledonous plant species. Plant Soil 155–156:485–488

Kidd PS, Proctor J (2001) Why plants grow poorly on very acid soils: are ecologists missing the obvious? J Exp Bot 52:791–799

Kiegle E, Gilliham M, Haseloff J, Tester M (2000) Hyperpolarisation-activated calcium currents found only in cells from the elongation zone of Arabidopsis thaliana roots. Plant J 21:225–229

Kinraide TB (1990) Assessing the rhizotoxicity of the aluminate ion, $Al(OH)_4^-$. Plant Physiol 93:1620–1625

Kinraide TB (1991) Identity of the rhizotoxic aluminium species. Plant Soil 134:167–178

Kinraide TB (1993) Aluminum enhancement of plant growth in acid rooting media – a case of reciprocal alleviation of toxicity by 2 toxic cations. Physiol Plant 88:619–625

Kinraide TB (1994) Use of a Gouy-Chapman-Stern model for membrane-surface electrical potential to interpret some features of mineral rhizotoxicity. Plant Physiol 106:1583–1592

Kinraide TB (2001) Ion fluxes considered in terms of membrane-surface electrical potentials. Aust J Plant Physiol 28:605–616

Kinraide TB, Ryan PR, Kochian LV (1992) Interactive effects of $Al^{3+}$, $H^{+}$, and other cations on root elongation considered in terms of cell-surface electrical potential. Plant Physiol 99:1461–1468
Kinraide TB, Pedler JF, Parker DR (2004) Relative effectiveness of calcium and magnesium in the alleviation of rhizotoxicity in wheat induced by copper, zinc, aluminum, sodium, and low pH. Plant Soil 259:201–208
Kochian LV (1995) Cellular mechanisms of aluminum toxicity and resistance in plants. Annu Rev Plant Physiol Plant Mol Biol 46:237–260
Kochian LV, Hoekenga OA, Piñeros MA (2004) How do crop plants tolerate acid soils? Mechanisms of aluminum tolerance and phosphorous efficiency. Annu Rev Plant Biol 55:459–493
Kochian LV, Pineros MA, Hoekenga OA (2005) The physiology, genetics and molecular biology of plant aluminum resistance and toxicity. Plant Soil 274:175–195
Kollmeier M, Felle HH, Horst WJ (2000) Genotypical differences in aluminum resistance of maize are expressed in the distal part of the transition zone. Is reduced basipetal auxin flow involved in inhibition of root elongation by aluminum? Plant Physiol 122:945–956
Kurkdjian A, Guern J (1989) Intracellular pH: measurement and importance in cell activity. Annu Rev Plant Biol 40:271–303
Lee J, Pritchard M (1984) Aluminium toxicity expression nutrient uptake, growth and root morphology of *Trifolium repens* L. cv. 'Grasslands Huia'. Plant Soil 82:101–116
Lenoble ME, Blevins DG, Sharp RE, Cumbie BG (1996) Prevention of aluminium toxicity with supplemental boron. I. Maintenance of root elongation and cellular structure. Plant Cell Environ 19:1132–1142
Li L, Tutone AF, Drummond RSM, Gardner RC, Luan S (2001) A novel family of magnesium transport genes in Arabidopsis. Plant Cell 13:2761–2775
Liao H, Wan H, Shaff J, Wang X, Yan X, Kochian LV (2006) Phosphorus and aluminum interactions in soybean in relation to aluminum tolerance. Exudation of specific organic acids from different regions of the intact root system. Plant Physiol 141:674–684
Ligaba A, Yamaguchi M, Shen H, Sasaki T, Yamamoto Y, Matsumoto H (2004) Phosphorus deficiency enhances plasma membrane $H^{+}$-ATPase activity and citrate exudation in greater purple lupin(*Lupinus pilosus*). Funct Plant Biol 31:1075–1083
Lindberg S (1990) Aluminium interactions with $K^{+}$ ($^{86}Rb^{+}$) and $^{45}Ca^{2+}$ fluxes in three cultivars of sugar beet (*Beta vulgaris*). Physiol Plant 79:275–282
Lindberg S, Strid H (1997) Aluminium induces rapid changes in cytosolic pH and free calcium and potassium concentrations in root protoplasts of wheat (*Triticum aestivum*). Physiol Plant 99:405–414
Liu K, Luan S (2001) Internal aluminum block of plant inward $K^{+}$ channels. Plant Cell 13:1453–1465
Lopez-Bucio J, Nieto-Jacobo MF, Ramırez-Rodrıguez V, Herrera-Estrella L (2000) Organic acid metabolism in plants: from adaptive physiology to transgenic varieties for cultivation in extreme soils. Plant Sci 160:1–13
Ma JF (2007) Syndrome of aluminum toxicity and diversity of aluminum resistance in higher plants. Int Rev Cytol 264:225–253
Ma JF, Ryan PR, Delhaize E (2001) Aluminium tolerance in plants and the complexing role of organic acids. Trends Plant Sci 6:273–278
Ma Q, Rengel Z, Kuo J (2002) Aluminium toxicity in rye (*Secale cereale*): root growth and dynamics of cytoplasmic $Ca^{2+}$ in intact root tips. Ann Bot 89:241–244
Ma JF, Shen R, Nagao S, Tanimoto E (2004) Aluminum targets elongating cells by reducing cell wall extensibility in wheat roots. Plant Cell Physiol 45:583–589
MacDiarmid CW, Gardner RC (1998) Overexpression of the *Saccharomyces cerevisiae* magnesium transport system confers resistance to aluminum ion. J Biol Chem 273:1727–1732
Maguire ME, Cowan JA (2002) Magnesium chemistry and biochemistry. Biometals 15:203–210
Mannion AM (1998) Global environmental change: the causes and consequences of disruption to biogeochemical cycles. Geogr J 164:168–182

Mariano ED, Keltjens WG (2005) Long-term effects of aluminum exposure on nutrient uptake by maize genotypes differing in aluminum resistance. J Plant Nutr 28:323–333

Marschner H (1991) Mechanisms of adaptation of plants to acid soils. Plant Soil 134:1–20

Marschner H (1995) Mineral nutrition of higher plants. Academic Press, London

Marty F (1999) Plant vacuoles. Plant Cell 11:587–600

Matsumoto H (1988) Inhibition of proton transport activity of microsomal membrane vesicles of barley roots by aluminum. Soil Sci Plant Nutr 34:499–506

Matsumoto H (2000) Cell biology of aluminum toxicity and tolerance in higher plants. Int Rev Cytol 200:1–46

Matsumoto H, Yamaya T (1986) Inhibition of potassium uptake and regulation of membrane-associated $Mg^{2+}$-ATPase activity of pea roots by aluminum. Soil Sci Plant Nutr 32:179–188

Matsumoto H, Hirasawa E, Torikai H, Takahashi E (1976) Localization of absorbed aluminium in pea root and its binding to nucleic acids. Plant Cell Physiol 17:127–137

Matsumoto H, Yamamoto Y, Kasai M (1992) Changes of some properties of the plasma membrane-enriched fraction of barley roots related to aluminum stress: membrane-associated ATPase, aluminum and calcium. Soil Sci Plant Nutr 38:411–419

Meriga B, Krishna Reddy B, Rajender Rao K, Ananda Reddy L, Kavi Kishor PB (2004) Aluminium-induced production of oxygen radicals, lipid peroxidation and DNA damage in seedlings of rice (*Oryza sativa*). J Plant Physiol 161:63–68

Miedema H, Bothwell JHF, Brownlee C, Davies JM (2001) Calcium uptake by plant cells–channels and pumps acting in concert. Trends Plant Sci 6:514–519

Miyasaka SC, Kochian LV, Shaff JE, Foy CD (1989) Mechanisms of aluminum tolerance in wheat – an investigation of genotypic differences in rhizosphere pH, $K^+$, and $H^+$ transport, and root cell membrane potentials. Plant Physiol 91:1188–1196

Morimura E, Takahashi E, Matsumoto H (1978) Association of aluminium with nuclei and inhibition of cell division in onion (*Allium cepa*) roots. Zeitschr Pflanzenphysiol 88:395–408

Mugwira LM, Patel SU, Fleming AL (1980) Aluminium effects on growth and Al, Ca, Mg, K and P levels in triticale, wheat and rye. Plant Soil 57:467–470

Newman IA (2001) Ion transport in roots: measurement of fluxes using ion-selective microelectrodes to characterize transporter function. Plant Cell Environ 24:1–14

Nichol BE, Oliveira LA, Glass ADM, Siddiqi MY (1993) The effects of aluminum on the influx of calcium, potassium, ammonium, nitrate, and phosphate in an aluminum-sensitive cultivar of barley (*Hordeum vulgare* L.). Plant Physiol 101:1263–1266

Olivetti GP, Cumming JR, Etherton B (1995) Membrane-potential depolarization of root cap cells precedes aluminum tolerance in snapbean. Plant Physiol 109:123–129

Ono K, Yamamoto Y, Hachiya A, Matsumoto H (1995) Synergistic inhibition of growth by aluminum and iron of tobacco (*Nicotiana tabacum* L.) cells in suspension culture. Plant Cell Physiol 36:115–125

Osawa H, Matsumoto H (2002) Aluminium triggers malate-independent potassium release via ion channels from the root apex in wheat. Planta 215:405–412

Parker JS, Cavell AC, Dolan L, Roberts K, Grierson CS (2000) Genetic interactions during root hair morphogenesis in Arabidopsis. Plant Cell 12:1961–1974

Pineros M, Tester M (1993) Plasma-membrane $Ca^{2+}$ channels in roots of higher plants and their role in aluminum toxicity. Plant Soil 156:119–122

Pineros M, Tester M (1997) Calcium channels in higher plant cells: selectivity, regulation and pharmacology. J Exp Bot 48:551–577

Plieth C, Sattelmacher B, Hansen UP, Knight MR (1999) Low-pH-mediated elevations in cytosolic calcium are inhibited by aluminium: a potential mechanism for aluminium toxicity. Plant J 18:643–650

Poschenrieder C, Gunsé B, Corrales I, Barceló J (2008) A glance into aluminum toxicity and resistance in plants. Sci Total Environ 400:356–368

Poschenrieder C, Amenos M, Corrales I, Doncheva S, Barcelo J (2009) Root behavior in response to aluminum toxicity. In: Baluska F (ed) Plant-environment interactions. Springer, Berlin, pp 21–43

Reid RJ, Tester MA, Smith FA (1995) Calcium/aluminium interactions in the cell wall and plasma membrane of Chara. Planta 195:362–368

Rengel Z (1990) Competitive $Al^{3+}$ inhibition of net $Mg^{2+}$ uptake by *intact Lolium multiflorum* roots: II. Plant age effects. Plant Physiol 93:1261–1267

Rengel Z (1992) Role of calcium in aluminium toxicity. New Phytol 121:499–513

Rengel Z (1994) Effects of Al, rare earth elements, and other metals on net $^{45}Ca^{2+}$ uptake by Amaranthus protoplasts. J Plant Physiol 143:47–51

Rengel Z (1996) Uptake of aluminium by plant cells. New Phytol 134:389–406

Rengel Z (2004) Aluminium cycling in the soil-plant-animal-human continuum. Biometals 17:669–689

Rengel Z, Elliott DC (1992) Mechanism of aluminum inhibition of net $^{45}Ca^{2+}$ uptake by Amaranthus protoplasts. Plant Physiol 98:632–638

Rengel Z, Robinson DL (1989) Competitive $Al^{3+}$ inhibition of net $Mg^{2+}$ uptake by intact *Lolium multiflorum* roots: I. Kinetics. Plant Physiol 91:1407–1413

Rengel Z, Zhang WH (2003) Role of dynamics of intracellular calcium in aluminium toxicity syndrome. New Phytol 159:295–314

Rengel Z, Pineros M, Tester M (1995) Transmembrane calcium fluxes during Al stress. Plant Soil 171:125–130

Richards KD, Schott EJ, Sharma YK, Davis KR, Gardner RC (1998) Aluminum induces oxidative stress genes in *Arabidopsis thaliana*. Plant Physiol 116:409–418

Rincon-Zachary M, Teaster ND, Sparks JA, Valster AH, Motes CM, Blancaflor EB (2010) Fluorescence resonance energy transfer-sensitized emission of Yellow Cameleon 3.60 reveals root zone-specific calcium signatures in Arabidopsis in response to aluminum and other trivalent cations. Plant Physiol 152:1442–1458

Ritchey KD, Carter TE (1993) Emergence and growth of two non-nodulated soybean genotypes (*Glycine max* (L.) Merr.) in response to soil acidity. Plant Soil 151:175–183

Roos W, Viehweger K, Dordschbal B, Schumann B, Evers S, Steighardt J, Schwartze W (2006) Intracellular pH signals in the induction of secondary pathways – the case of *Eschscholzia californica*. J Plant Physiol 163:369–381

Rowell DL, Wild A (1985) Causes of soil acidification: a summary. Soil Use Manag 1:32–33

Ryan PR, Kochian LV (1993) Interaction between aluminum toxicity and calcium uptake at the root apex in near-isogenic lines of wheat (*Triticum aestivum* L.) differing in aluminum tolerance. Plant Physiol 102:975–982

Ryan PR, Ditomaso JM, Kochian LV (1993) Aluminium toxicity in roots: an investigation of spatial sensitivity and the role of the root cap. J Exp Bot 44:437–446

Ryan PR, Delhaize E, Randall PJ (1995) Characterization of Al-stimulated efflux of malate from the apices of Al-tolerant wheat roots. Planta 196:103–110

Ryan PR, Reid RJ, Smith FA (1997) Direct evaluation of the $Ca^{2+}$-displacement hypothesis for Al toxicity. Plant Physiol 113:1351–1357

Sano T, Becker D, Ivashikina N, Wegner LH, Zimmermann U, Roelfsema MRG, Nagata T, Hedrich R (2007) Plant cells must pass a $K^+$ threshold to re-enter the cell cycle. Plant J 50:401–413

Sasaki M, Kasai M, Yamamoto Y, Matsumoto H (1995) Involvement of plasma membrane potential in the tolerance mechanism of plant roots to aluminum toxicity. Plant Soil 171:119–124

Sawaki Y, Iuchi S, Kobayashi Y, Kobayashi Y, Ikka T, Sakurai N, Fujita M, Shinozaki K, Shibata D, Kobayashi M, Koyama H (2009) STOP1 regulates multiple genes that protect Arabidopsis from proton and aluminum toxicities. Plant Physiol 150:281–294

Schmohl N, Pilling J, Fisahn J, Horst WJ (2000) Pectin methylesterase modulates aluminium sensitivity in *Zea mays* and *Solanum tuberosum*. Physiol Plant 109:419–427

Schofield RMS, Pallon J, Fiskesjö G, Karlsson G, Malmqvist KG (1998) Aluminum and calcium distribution patterns in aluminum-intoxicated roots of *Allium cepa* do not support the calcium-displacement hypothesis and indicate signal-mediated inhibition of root growth. Planta 205:175–180

Scholz-Starke J, Gambale F, Carpaneto A (2005) Modulation of plant ion channels by oxidizing and reducing agents. Arch Biochem Biophys 434:43–50

Shen H, Yan X, Zhao M, Zheng S, Wang X (2002) Exudation of organic acids in common bean as related to mobilization of aluminum-and iron-bound phosphates. Environ Exp Bot 48:1–9

Shen J, Tang C, Rengel Z, Zhang F (2004) Root-induced acidification and excess cation uptake by $N_2$-fixing Lupinus albus grown in phosphorus-deficient soil. Plant Soil 260:69–77

Shen H, He LF, Sasaki T, Yamamoto Y, Zheng SJ, Ligaba A, Yan XL, Ahn SJ, Yamaguchi M, Sasakawa H, Matsumoto H (2005) Citrate secretion coupled with the modulation of soybean root tip under aluminum stress. Up-regulation of transcription, translation, and threonine-oriented phosphorylation of plasma membrane $H^+$-ATPase. Plant Physiol 138:287–296

Silva IR, Smyth TJ, Moxley DF, Carter TE, Allen NS, Rufty TW (2000) Aluminum accumulation at nuclei of cells in the root tip. Fluorescence detection using lumogallion and confocal laser scanning microscopy. Plant Physiol 123:543–552

Šimonovicová M, Huttova J, Mistrik I, Široká B, Tamas L (2004a) Peroxidase mediated hydrogen peroxide production in barley roots grown under stress conditions. Plant Growth Regul 44:267–275

Šimonovicová M, Huttová J, Mistrik I, Široká B, Tamas L (2004b) Root growth inhibition by aluminum is probably caused by cell death due to peroxidase-mediated hydrogen peroxide production. Protoplasma 224:91–98

Sivaguru M, Horst WJ (1998) The distal part of the transition zone is the most aluminum-sensitive apical root zone of maize. Plant Physiol 116:155–163

Sivaguru M, Baluska F, Volkmann D, Felle HH, Horst WJ (1999) Impacts of aluminum on the cytoskeleton of the maize root apex. Short-term effects on the distal part of the transition zone. Plant Physiol 119:1073–1082

Sivaguru M, Fujiwara T, Samaj J, Baluska F, Yang Z, Osawa H, Maeda T, Mori T, Volkmann D, Matsumoto H (2000) Aluminum-induced 1-3-β-D-glucan inhibits cell-to-cell trafficking of molecules through plasmodesmata. A new mechanism of aluminum toxicity in plants. Plant Physiol 124:991–1006

Sivaguru M, Pike S, Gassmann W, Baskin TI (2003) Aluminum rapidly depolymerizes cortical microtubules and depolarizes the plasma membrane: evidence that these responses are mediated by a glutamate receptor. Plant Cell Physiol 44:667–675

Street HE (1966) The physiology of root growth. Annu Rev Plant Physiol 17:315–344

Subbarao GV, Ae N, Otani T (1997) Genotypic variation in iron-, and aluminum-phosphate solubilizing activity of pigeonpea root exudates under P deficient conditions. Soil Sci Plant Nutr 43:295–305

Tabuchi A, Matsumoto H (2001) Changes in cell-wall properties of wheat (*Triticum aestivum*) roots during aluminum-induced growth inhibition. Physiol Plant 112:353–358

Tahara K, Yamanoshita T, Norisada M, Hasegawa I, Kashima H, Sasaki S, Kojima K (2008) Aluminum distribution and reactive oxygen species accumulation in root tips of two Melaleuca trees differing in aluminum resistance. Plant Soil 307:167–178

Tamás L, Šimonoviová M, Huttová J, Mistrik I (2004) Aluminium stimulated hydrogen peroxide production of germinating barley seeds. Environ Exp Bot 51:281–288

Tang C, Raphael C, Rengel Z, Bowden JW (2000) Understanding subsoil acidification: effect of nitrogen transformation and nitrate leaching. Austr J Soil Res 38:837–850

Tanoi K, Junko H, Kazutoshi S, Yoshitake H, Hiroki N, Tomoko MN (2005) Analysis of potassium uptake by rice roots treated with aluminum using a positron emitting nuclide, $^{38}$K. Soil Sci Plant Nutr 51:715–717

Taylor GJ, McDonald-Stephens JL, Hunter DB, Bertsch PM, Elmore D, Rengel Z, Reid RJ (2000) Direct measurement of aluminum uptake and distribution in single cells of *Chara corallina*. Plant Physiol 123:987–996
Valadez-Gonzalez N, Colli-Mull JG, Brito-Argaez L, Muñoz-Sánchez JA, Aguilar JJZ, Castano E, Hernández-Sotomayor SMT (2007) Differential effect of aluminum on DNA synthesis and CDKA activity in two *Coffea arabica* cell lines. J Plant Growth Regul 26:69–77
Very A-A, Davies JM (2000) Hyperpolarization-activated calcium channels at the tip of Arabidopsis root hairs. Proc Natl Acad Sci U S A 97:9801–9806
Vries W, Breeuwsma A (1987) The relation between soil acidification and element cycling. Water Air Soil Poll 35:293–310
Wang H, Inukai Y, Yamauchi A (2006) Root development and nutrient uptake. Crit Rev Plant Sci 25:279–301
Wasaki J, Yonetani R, Kuroda S, Shinano T, Yazaki J, Fujii F, Shimbo K, Yamamoto K, Sakata K, Sasaki T (2003) Transcriptomic analysis of metabolic changes by phosphorus stress in rice plant roots. Plant Cell Environ 26:1515–1523
Webb AAR, McAinsh MR, Taylor JE, Hetherington AM (1996) Calcium ions as intracellular second messengers in higher plants. Adv Bot Res 22:45–96
Wherrett T, Ryan PR, Delhaize E, Shabala S (2005) Effect of aluminium on membrane potential and ion fluxes at the apices of wheat roots. Funct Plant Biol 32:199–208
White PJ (1998) Calcium channels in the plasma membrane of root cells. Ann Bot 81:173–183
Yamamoto Y, Hachiya A, Matsumoto H (1997) Oxidative damage to membranes by a combination of aluminum and iron in suspension-cultured tobacco cells. Plant Cell Physiol 38:1333–1339
Yamamoto Y, Kobayashi Y, Devi SR, Rikiishi S, Matsumoto H (2002) Aluminum toxicity is associated with mitochondrial dysfunction and the production of reactive oxygen species in plant cells. Plant Physiol 128:63–72
Yan X, Lynch JP, Beebe SE (1995) Genetic variation for phosphorus efficiency of common bean in contrasting soil types: I. Vegetative response. Crop Sci 35:1086
Yermiyahu U, Brauer DK, Kinraide TB (1997) Sorption of aluminum to plasma membrane vesicles isolated from roots of Scout 66 and Atlas 66 cultivars of wheat. Plant Physiol 115:1119–1125
Zhang WH, Rengel Z (1999) Aluminium induces an increase in cytoplasmic calcium in intact wheat root apical cells. Austr J Plant Physiol 26:401–410

# Aluminum-Induced Inhibition of Root Growth: Roles of Cell Wall Assembly, Structure, and Function

**Zhong-Bao Yang and Walter J. Horst**

**Abstract** Aluminum (Al) toxicity is the most important soil constraint for plant growth and development in acid soils. It is a matter of debate whether the primary lesions of Al toxicity are apoplastic or symplastic, while there is increasing physiological, biochemical, and molecular evidence showing that the modification of cell wall properties contributes to the Al-induced inhibition of root growth. The rapid binding of Al in the root cell wall particularly to the pectin matrix and hemicellulose can affect cell wall properties. Most recent studies have revealed that the local accumulation of auxin in the most Al-sensitive root zone of the root apex is a major factor leading to Al-induced root-growth inhibition. Evidence suggests that the auxin effect is mediated mainly via modification of cell wall structural properties. A further in-depth characterization of the Al-induced apoplastic reactions in the most Al-sensitive zone of the root apex is urgently required to better understand the phytohormone-mediated signaling network leading to Al-induced inhibition of root growth.

## 1 Introduction

Soil acidity with pH ≤ 5.5 is one of the most important factors limiting crop production worldwide on approximately 30 % of the world's total land area and as much as 50 % of the world's potentially arable lands. The tropics and subtropics account for 60 % of the acid soils in the world. In tropical areas, about 43 % of soils are acidic comprising about 68 % of tropical America, 38 % of tropical Asia, and 27 % of tropical Africa (von Uexküll and Mutert 1995). When the soil pH drops below 5, $Al^{3+}$ is solubilized into the soil solution and becomes a major constraint for

Z.-B. Yang (✉)
The Key Laboratory of Plant Cell Engineering and Germplasm Innovation, Ministry of Education, College of Life Science, Shandong University, Jinan 250100, People's Republic of China
e-mail: zbyang@sdu.edu.cn

W.J. Horst
Institute of Plant Nutrition, Leibniz Universität Hannover, Herrenhaeuser Str. 2, 30419 Hannover, Germany

 
S.K. Panda, F. Baluška (eds.), *Aluminum Stress Adaptation in Plants*, Signaling and Communication in Plants 24, DOI 10.1007/978-3-319-19968-9_13

plant growth and development in acidic soils (Kinraide et al. 1992). The easily observable symptom of Al toxicity is a rapid (minutes to few hours) inhibition of root growth (Horst et al. 1992; Delhaize and Ryan 1995), resulting in a reduced and damaged root system that limits mineral nutrient and water uptake (Kochian et al. 2004). The rapidity of this response indicates that Al first inhibits root cell expansion and elongation and consequently cell division over the longer term (Delhaize and Ryan 1995; Kochian 1995). Another most sensitive indicator of Al injury on roots is the induction of callose synthesis (Wissemeier et al. 1987), particularly in the root apex (Wissemeier and Horst 1995).

Although much progress has been made during recent years in the understanding of Al resistance, the molecular and physiological mechanisms leading to Al-induced inhibition of root elongation are still not well understood. There are a number of excellent reviews in recent years summarizing the state of knowledge and addressing knowledge gaps (Horst 1995; Kochian et al. 2004; Panda and Matsumoto 2007; Panda et al. 2009; Horst et al. 2010; Delhaize et al. 2012; Liu et al. 2014; Kochian et al. 2015). Particularly, the relative importance of symplastic versus apoplastic lesions of Al toxicity remains a matter of debate. The studies by Zheng et al. (Yang et al. 2008, 2011; Zhu et al. 2012) and especially by Horst et al. (2010) focused on the attention on the role of the apoplast in Al toxicity regarding short-term inhibition of root elongation by Al. Here we summarize the current understanding of the role of root cell wall structure and assembly in Al-induced inhibition of root growth of plants.

## 2 Al Toxicity Is Targeted Primarily to the Root-Apex Transition Zone Whereas the Elongation Zone Is Not, or Less, Affected Depending on Plant Species

The inhibition of root growth has been established as the main symptom of Al toxicity in barley (*Hordeum vulgare*) and rye (*Secale cereale*) early in 1918 (Hartwell and Pember 1918). Seventy-five years after this finding, Ryan et al. (1993) confirmed that the root apex is the major perception site of Al toxicity in maize (*Zea mays*). The root apex of higher plants is quite sensitive to environmental stimuli. As the most prominent plant organ, the root cap can sense diverse physical and chemical stimuli such as gravity, light, humidity, oxygen, and mineral elements. Subsequently, the motoric responses to these stimuli are transmitted to the elongation zone (Baluška and Mancuso 2013). The transition zone, which is located between the apical meristem and basal elongation zone of the root, has a unique role as the determiner of cell fate and root growth (Baluška et al. 2010). In Lixis, an Al-sensitive genotype of maize, Sivaguru and Horst (1998) specified that the distal part of the transition zone (DTZ, 1–2 mm) is the most Al-sensitive zone of the root apex. Application of Al to the DTZ but not the elongation zone (EZ) reduced cell elongation in the EZ to the same extent as application to the entire 10 mm root apex (Fig. 1). The transition zone (TZ) of the root apex as the

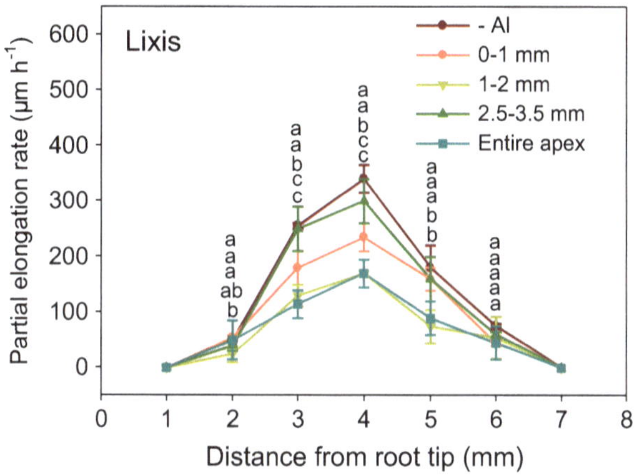


**Fig. 1** Effect of Al supply (90 μM, 1 h) to the entire root apex or specific 1-mm apical root zones on partial elongation rates of apical 1-mm root segments of the primary root of the maize cv Lixis (Al-sensitive). Values are means of five independent measurements ±SD. Different letters indicate significant differences at $p < 0.05$ (Tukey test). From Kollmeier et al. (2000)

primary Al-toxic site of plants has been also proposed in other species, such as *Arabidopsis thaliana* (Illéš et al. 2006; Yang et al. 2014) and *Sorghum bicolor* (Sivaguru et al. 2013). While in common bean (*Phaseolus vulgaris*), a leguminous plant, Rangel et al. (2007) showed that in addition to the TZ also the EZ of the root apex responds to Al exposure. This difference might be due to the fact that dicotyledons and grasses (*Poales*) are different in the composition of their cell walls. Particularly, the pectin content of the cell walls is higher in dicotyledons (Carpita and Gibeaut 1993) which explains their enhanced Al-binding capacity. The important role of the cell wall pectin content for Al accumulation and Al sensitivity has been demonstrated repeatedly (Horst et al. 2010).

Cells in the TZ are very active in cytoskeletal rearrangements, in endocytosis and endocytic vesicle recycling, as well as in electric activities (Baluška and Mancuso 2013). In Arabidopsis, Illéš et al. (2006) found that Al can rapidly depolarize the plasma membrane, while the full recovery of the membrane potential was slower in the cells of the DTZ than in the proximal transition zone (PTZ) after the removal of external Al stress, implying that the DTZ is the most Al-sensitive site in the root apex. Further combination of morin staining to detect Al accumulation in cells and FM4-64 for endosomal/vacuolar membrane observation was used; these authors found that the Al internalization mainly occurred in the cells of the DTZ rather than the proximal transition zone and EZ. Using monoclonal tubulin and actin antibodies, Sivaguru et al. (1999) found that the more sensitive response to Al of root elongation in the DTZ results from a higher Al accumulation in this zone accompanied by Al-mediated alterations to microtubules and actin microfilaments, Al-induced depolarization of the plasma membrane, and callose formation particularly in the outer cortex cells of the DTZ (Sivaguru and Horst 1998). It is possible that the rapid Al-induced changes to cytosolic $Ca^{2+}$ ($[Ca^{2+}]_{cyt}$) may mediate cytoskeletal disorders, since Rincón-Zachary et al. (2010) using fluorescence resonance energy transfer (FRET)-sensitized emission to image *Arabidopsis thaliana* roots expressing the yellow cameleon 3.60 $Ca^{2+}$ reporter demonstrated that Al evoked an increase of $[Ca^{2+}]_{cyt}$ within seconds primarily in the TZ of the root apex. The elevated $[Ca^{2+}]_{cyt}$ and modification of the plasma membrane are known to be crucial for callose deposition (Kauss et al. 1990) and the effect of Al on actin microfilaments via involvement of $Ca^{2+}$-mediated kinases and phosphatases (Grabski et al. 1998).

Exclusion of Al from the apoplast through exudation of organic acid such as citrate and malate has been identified as one of the major Al resistance mechanisms in plants (Ma et al. 2001; Kochian et al. 2004; Ryan et al. 2011). Al-activated citrate transporters (MATE) and Al-activated malate transporters (ALMT) have been identified to be crucial for citrate and malate exudation from root tips conferring Al exclusion and Al resistance (Delhaize et al. 2012). Using the laser-capture microdissection (LCM) technique, Sivaguru et al. (2013) found that Al-induced *SbMATE* gene expression and protein synthesis were specifically localized to the epidermal and outer cortical cell layers of the root-apex DTZ in the Al-resistant near-isogenic sorghum line. In this root zone, Al induced the greatest cell damage and generation of reactive oxygen species (ROS). The specific Al-induced ROS induction in the DTZ may play a signaling role in the induction of *SbMATE* gene expression. The $H_2O_2$-induced *ALMT1* gene expression and malate exudation from the root apex in Arabidopsis (Kobayashi et al. 2013) may support this hypothesis.

# 3 Auxin Is Involved in Al-Induced Inhibition of Root Elongation

## *3.1 Auxin Accumulation in the Transition Zone and Inhibition of Auxin Transport into the EZ Are Involved in Inhibition of Cell Elongation in the EZ*

In the root-apex TZ, the meristematic cells exit the cell division phase and prepare for filamentous actin (F-actin)-dependent rapid cell elongation (Baluška et al. 1992, 2001; Verbelen et al. 2006). Cell division in the MZ and cell elongation in the EZ are inhibited as primary effects of Al occurring in the adjacent transition zone, in which these processes are less active. This observation suggests that putative primary Al signals are transduced from the transition zone to both the root meristem and the fast elongation zone. Auxin is an important regulator of root cell division, elongation, and differentiation. Hence, auxin controls overall root growth. High concentrations of auxin, however, inhibit the elongation of certain cell types (Teale et al. 2005). Auxin signaling within the root-apex TZ is sensitive not only to developmental signals but also to environmental cues (Baluška et al. 2010), including Al (Sivaguru and Horst 1998). Several studies have demonstrated that aluminum may interact with auxin signaling pathways, leading to alterations of auxin accumulation and distribution in roots (Kollmeier et al. 2000; Doncheva et al. 2005; Shen et al. 2008). In maize, local application of Al to the DTZ promoted auxin accumulation in the MZ and DTZ, while reduced auxin level in the EZ. External

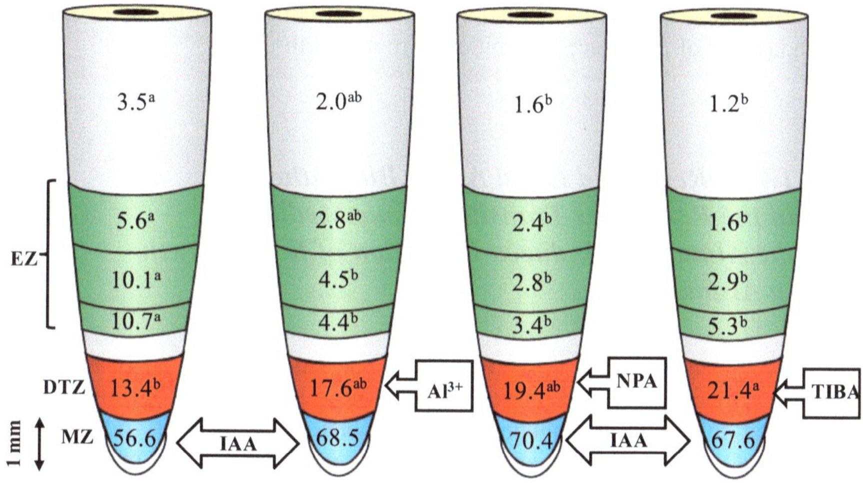


**Fig. 2** Effect of application of Al and IAA transport inhibitors to the DTZ of primary roots of the maize cv Lixis (Al-sensitive) on the relative distribution of [$^3$H]IAA applied to the MZ (total uptake = 100 %). Application of 90 mM mononuclear Al, 10 mM NPA, or 10 mM TIBA in nutrient solution, pH 4.3, in 0.6 % (w/v) agarose gel to the DTZ for 30 min. Control roots were treated only with nutrient solution in agarose blocks, pH 4.3. [$^3$H]IAA (0.1 mM in 1.2 % [w/v] agarose blocks containing nutrient solution) was applied to the MZ for 30 min. Values are means of five independent replicates ±SD. Different letters indicate significant differences at $p < 0.05$ (Tukey test). From Kollmeier et al. (2000)

supply of auxin to the EZ was able to partially overcome the inhibition of root growth imposed by the application of Al to the DTZ (Kollmeier et al. 2000). The coincidence of the response to Al of auxin distribution in the root apex with the local supply of the auxin polar transport inhibitor N-1-naphthylphthalamic acid (NPA) and 2,3,5-triiodobenzoic acid (TIBA) (Fig. 2) suggested that the blockage of auxin polar transport and thus auxin signaling from the DTZ to the EZ could be the primary cause of Al-induced inhibition of root growth. Also, Doncheva et al. (2005) found that after the local application of the auxin polar transport inhibitor NPA to the DTZ of the root apex, abundant S-phase nuclei were observed in the distal elongation zone (DEZ) at 2.5–3 mm from the root tip, suggesting that the inhibition of auxin transport plays a role in Al-induced alteration of root cell patterning.

The cells of the DTZ are unique from all other root cells from the perspective of endocytosis, vesicle recycling, polar auxin transport, and as the first region of the root apex which is not covered with mucilage (Baluška et al. 2010). A relationship between Al toxicity, endocytosis, endosome, and vesicle recycling in the TZ cells of Arabidopsis roots has been demonstrated by Illéš et al. (2006). In addition, in Arabidopsis, Shen et al. (2008) have found that Al inhibited root to shoot auxin transport and thus root growth mainly through the blockage of the transport of PIN2 vesicles from plasma membrane to endosomes. However, the recent study by Yang et al. (2014) suggests that the auxin transporter PIN1 rather than PIN2 is involved in the modulation of Al-induced inhibition of root growth.

### *3.2 TAA1-Regulated Local Auxin Biosynthesis in the Root-Apex Transition Zone Mediates the Aluminum-Induced Inhibition of Root Growth in Arabidopsis*

Generally, auxin is transported to roots via a polar transport system from auxin-synthesizing shoot tissues (Petrášek and Friml 2009). However, in addition to being synthesized in the shoots, auxin is also generated in the roots (Overvoorde et al. 2010). In fact, the gradients of auxin in root apices depend on both local biosynthesis and directional intercellular auxin transport (Petersson et al. 2009; Petrášek and Friml 2009). It is becoming clear that the root-generated auxin contributes to the maintenance of the gradients and maxima required for normal root development, and many auxin biosynthesis genes have been identified in the root apex (Ljung et al. 2005; Petersson et al. 2009; Overvoorde et al. 2010). High-resolution auxin-measurement and gene-expression analysis in specific cell types after fluorescence-activated cell sorting revealed that the root apex is the site of auxin biosynthesis, and a substantial contribution of local biosynthesis to auxin homeostasis in the root tip was proposed (Petersson et al. 2009). Only recently the complete auxin biosynthesis pathway has been established: a two-step pathway concerts tryptophan (Trp) to indol-3-acetic acid (IAA) in plants, in which Trp is first converted to indole-3-pyruvate (IPA) by the TAA family of amino acid transferases and subsequently IAA is produced from IPA by the YUC family of flavin monooxygenases (Zhao 2012). The transcriptional analysis of maize roots growing in acid soil provided indirect evidence of enhanced auxin biosynthesis in Al-induced inhibition of root growth (Mattiello et al. 2010). Genes encoding enzymes involved in auxin biosynthesis such as *IAA amidohydrolase* (Zm.3056.1.A1_at) and *anthranilate phosphoribosyltransferase* (Zm.1556.1.A1_at) were upregulated in the root apex of the soil acidity-sensitive line S1587-17, while the auxin-degrading enzyme *indole-3-acetate beta-glucosyltransferase* (Zm.18805.1.A1_at) was downregulated after 3 days exposure to soil acidity. Recently, we clearly showed that the *TAA1*-regulated local auxin biosynthesis in the root-apex TZ mediates Al-induced inhibition of root growth (Yang et al. 2014 ), and this local induction of auxin biosynthesis depended on the ethylene signaling pathway (Fig. 3). However, we cannot exclude that other genes belonging to the YUC family could also contribute to Al-induced local auxin biosynthesis, since the mutant line of *YUC1D* also showed reduced Al toxicity (Yang et al. 2014).

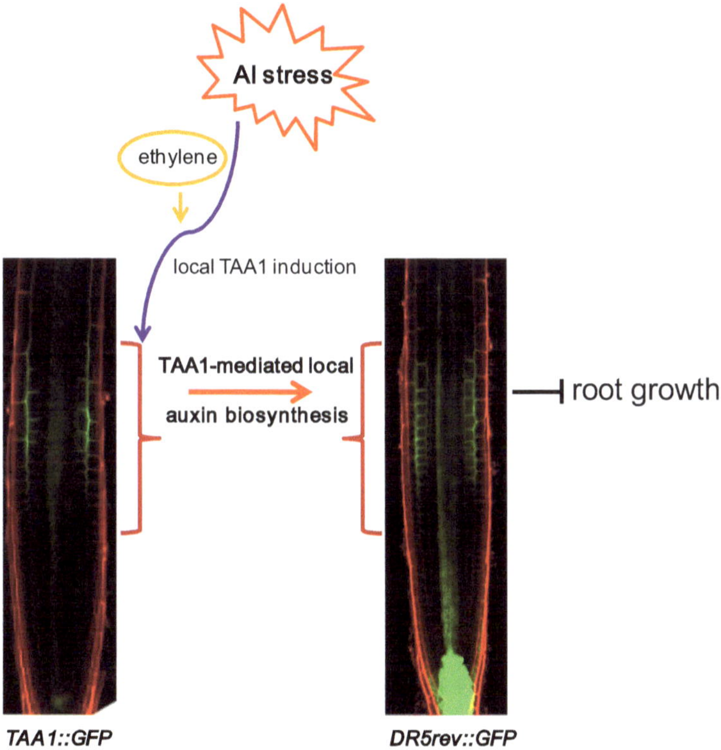


**Fig. 3** Schematic representation of *TAA1*-regulated local auxin biosynthesis in the transition zone (TZ) mediating root-growth inhibition in response to Al stress. Exposure of Arabidopsis roots to Al induces a localized enhancement of auxin signaling in the root-apex TZ that is dependent on *TAA1*, which encodes a Trp aminotransferase and regulates auxin biosynthesis. *TAA1* is specifically upregulated in the root apex TZ in response to Al stress, thus mediating local auxin biosynthesis and inhibition of root growth. The *TAA1*-regulated local auxin biosynthesis in the root apex TZ in response to Al stress is dependent on ethylene. From Yang et al. (2014)

## *3.3 Cell Wall Modification Is a Downstream Response to Al Stress and Contributes to the Auxin-Mediated Root-Growth Inhibition*

Cell walls are dynamic structures and the primary walls of plants consist of a cellulose–hemicellulose interlinked network embedded in a matrix of pectins and cell wall structure proteins (Carpita and Gibeaut 1993; Cosgrove 1997, 2005). The rapid enlargement of cells requires wall loosening, which involves modification of the molecular interactions within the CW network, resulting in the relaxation of wall tension (Perrot-Rechenmann 2010). Auxin was shown to induce rapid cell elongation in stem, coleoptile, or hypocotyl segments within minutes after auxin treatment (Rayle and Cleland 1992; Cleland 1995). According to the acid-growth theory, this rapid effect is believed to result from the activation of a plasma-membrane $H^+$-ATPase, inducing extrusion of $H^+$ for apoplastic acidification, activation of expansins, and subsequent wall loosening (Hager 2003). Takahashi et al. (2012) showed that auxin activates the plasma-membrane $H^+$-ATPase by phosphorylation and regulates hypocotyl elongation in Arabidopsis. However, it appears that Al-induced inhibition of root growth can hardly be explained by reduced auxin activation of the plasma-membrane $H^+$-ATPase, since several studies have shown that the Al-induced plasma-membrane $H^+$-ATPase activity rather contributed to Al resistance (Shen et al. 2004; Yang et al. 2007; Chen et al. 2013). However, the possibility that auxin-induced excess acidification in the cell wall leading to the inhibition of root growth under Al stress cannot be ruled out.

Compared to the hypothesis of cell wall acidification, it is more probable that auxin mediates Al-induced inhibition of root growth directly by interaction with CW proteins through auxin-responsive factors (ARFs) (Fig. 4). The transcriptomic analysis presented by Yang et al. (2014) revealed that many of the differentially transcribed genes associated with cell wall modification were regulated by the transcription factors ARF10 and ARF16, suggesting that the auxin-regulated Al-induced inhibition of root growth arises from auxin signaling regulated modification of cell wall structure and/or structural components. Pitaksaringkarn et al. (2014) found that auxin regulates *XTH19* and *XTH20* expression, which are involved in cell proliferation in incised Arabidopsis inflorescence stems. However, the study by Zhu et al. (2013) revealed that auxin enhances Al toxicity via an alteration of *ALUMINUM-SENSITIVE1*-mediated Al distribution in the symplast. In spite of this, the recent study by Wu et al. (2014) in rice indicates that overexpression of *OsPIN2* alleviates the Al-induced cell rigidity in the root apex by modulating PIN2-based auxin transport, IAA efflux, and CW acidification. The reduction of Al accumulation mainly in the CW of the *OsPIN2* overexpression line further supports the hypothesis that the CW modification is probably a downstream response to Al exposure and contributes to the auxin-mediated root-growth inhibition by Al stress.

### *3.4 Al Toxicity Requires Al Accumulation/Binding in the Cell Wall*

The accumulation of Al in root tips is characterized by a rapid initial phase and a low rate at later stages (Zhang and Taylor 1989, 1990). The rapid initial phase reflects the binding of Al in the apoplast (Taylor et al. 2000; Wang et al. 2004; Horst et al. 2007; Rangel et al. 2009) in which the negatively charged carboxylic groups of pectin provide the $Al^{3+}$-binding sites (Blamey et al. 1990; Chang et al. 1999). In fact, the involvement of pectin in Al resistance mainly depends on its degree of methylation (DM), since the DM is responsible for the negativity of the CW (Eticha et al. 2005), which is controlled by pectin methylesterase (PME) (Bordenave 1996;

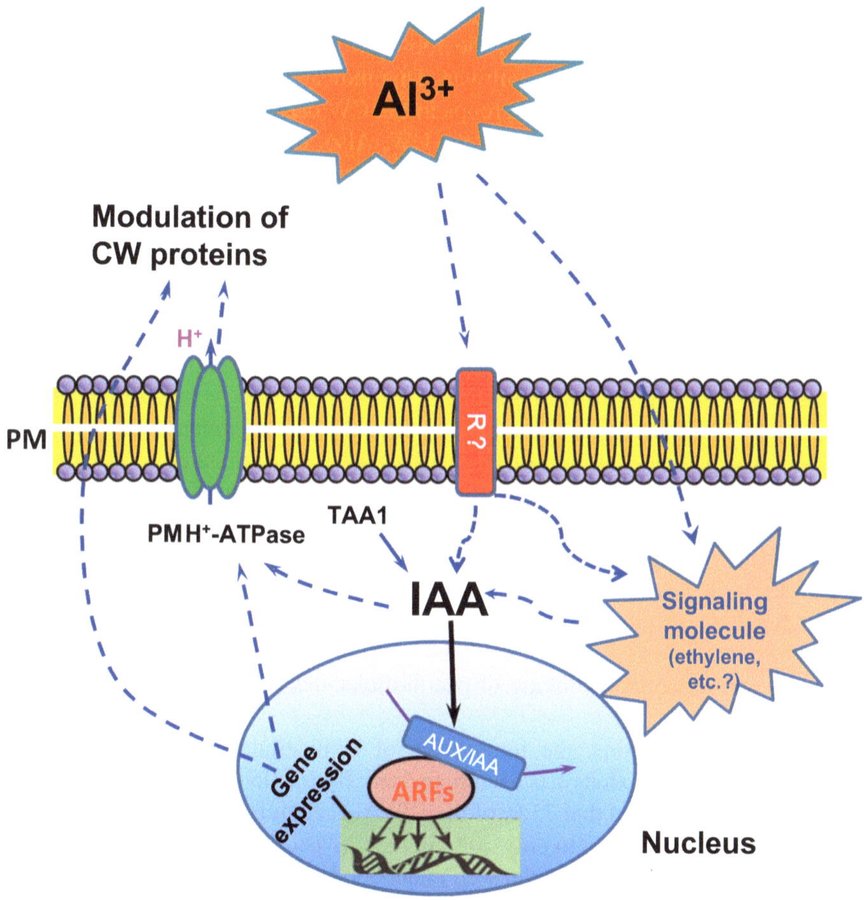


**Fig. 4** Hypothetical scheme of the regulatory role of IAA in Al-induced inhibition of root growth via modification of cell wall (CW) properties. Al rapidly triggers a signaling pathway through acting on an unknown receptor (R) localized at the plasma membrane (PM), or through a signaling molecule such as ethylene, etc. Consequently, the downstream signaling of TAA1-regulated synthesized auxin is activated, in which auxin-responsive factors (ARFs) directly or indirectly regulate CW proteins or activate gene expression or activity of the PM $H^+$-ATPase modulating the activities of CW proteins through changes of the apoplastic pH

Gerendás 2007). In potato (*Solanum tuberosum* L.), higher Al accumulation and callose production in the roots and more severe inhibition of root growth were found in transgenic plants with higher PME expression than the wild type when exposed to Al (Schmohl et al. 2000). Short-term PME treatment of intact maize roots enhanced Al accumulation and Al-induced inhibition of root elongation (Horst et al. 2007). In two differential Al-resistant cultivars of maize, Eticha et al. (2005) observed that the Al-sensitive cultivar had lower DM and greater Al accumulation, and thus were more severely injured by Al compared with the Al-resistant cultivar, while no difference was found in pectin content. Similarly, in rice (*Oryza sativa*), Yang et al. (2008) found that CW PME activity and related content of demethylated pectin in the root tips were higher in the Al-sensitive cultivar than in the Al-resistant cultivar. This indicates that the higher density of polygalacturonic acid carboxylic groups in the CW causes a corresponding higher Al accumulation in the root tips and the CW. Also, transcriptional analysis of Al resistance in maize by Maron et al. (2008) revealed that Al upregulated the expression of the *PME* gene in both Al-resistant and Al-sensitive genotypes, while the level of upregulation of *PME* was higher in Al-sensitive genotypes. Furthermore, Horst et al. (1999) reported that short-term Al accumulation of roots was closely related to the pectin content in apical root sections of maize and faba bean (*Vicia faba*), and the binding of Al to the pectic matrix was closely positively correlated with Al-induced callose formation and thus Al sensitivity. Therefore, it appears that the binding of Al to pectins is closely related to Al sensitivity, since it was also reported that the Al-induced increase in pectin content of Al-sensitive cultivars was greater than that of Al-resistant cultivars (Eticha et al. 2005; Yang et al. 2008). Also in common bean, Rangel et al. (2009) found that the Al-induced root-growth inhibition was closely negatively related particularly to strongly bound CW Al. This suggests that the strong binding of Al to the pectic matrix of the CW is a main factor in Al toxicity rather than a resistance mechanism in common bean, although in earlier studies by Van et al. (1994), it was suggested that Al is detoxified by binding to pectins since the free carboxyl groups of pectin can bind or chelate $Al^{3+}$ ions and cause cross-linking of pectin molecules (Klimashevskii and Dedov 1975).

In addition to pectins, the importance of hemicellulose in CW Al-binding capacity and thus Al toxicity has been suggested. By fractionating CW components, Yang et al. (2011) found that 75 % of the CW Al accumulated in the hemicellulose fraction 1 (HC1) compared to only 20 % in the CW pectin fraction. The interaction of Al with hemicellulose is not yet well understood, since according to the analysis of uronic acids in the different CW component fractions, the percentage of uronic acids in the pectin, HC1, and HC2 fractions were 72 %, 15 %, and 13 %, respectively. Xyloglucan is the most abundant hemicellulosic polysaccharide primary cell walls of dicotyledons. It functions by forming load-bearing cross-links among microfibrils, where they play a central role in modulating the mechanical properties of CWs (reviewed by Nishitani 1997). In Arabidopsis, Zhu et al. (2012) provided evidence that Al interacts specifically with xyloglucans. They postulated that the formation of an Al-xyloglucan complex inhibits cell wall loosening in the elongation zone of roots and thus contributes to inhibition of root elongation by Al.

The xyloglucan endotransglucosylase/hydrolases (XTHs) are enzymes that specifically use xyloglucan as a substrate and catalyze xyloglucan endotransglucosylase (XET) and/or xyloglucan endohydrolase (XEH) activities. They play a key role in the modification of CW structure and extensibility through the cleavage and re-formation of bonds between xyloglucan chains (Rose et al. 2002; Bray 2004). The XTH proteins are a large family of CW proteins which have 33 members known in the Arabidopsis genome (Rose et al. 2002; Bray 2004). In Arabidopsis, the expression of *XTH31* was suppressed by Al stress and thus has been suggested to play a crucial role in the modulation of Al resistance through the regulation of the CW xyloglucan content and thus Al accumulation in roots (Zhu et al. 2012). Combination of the yeast two-hybrid assay and coimmunoprecipitation analysis revealed that XTH17 can interact with XTH31 in vitro (Zhu et al. 2014). These authors conclude that XTH17 and XTH31 may exist as a dimer at the plasma membrane conferring in vivo XET action, thus modulating CW Al-binding capacity and Al sensitivity of Arabidopsis. Further studies indicated that the O-acetylation of xyloglucan by the putative O-acetyltransferase TRICHOME BIREFRINGENCE-LIKE27 (TBL27 [AXY4]) affects Al sensitivity by modulation of Al-binding capacity in the hemicellulose xyloglucan (Zhu et al. 2014).

A decisive role of Al binding in the CW for Al toxicity is further supported by the studies of Xia et al. (2010) and Li et al. (2014) in rice. They characterized the plasma membrane-localized transporter, Nrat1 (Nramp aluminum transporter 1), belonging to the Nramp (natural resistance-associated macrophage protein) family specifically transporting the trivalent Al ion through the plasma membrane. The effective transport of Al from the apoplast to the symplast where it is ultimately sequestered in the vacuole plays an important role in the remarkably high Al tolerance of rice by reducing the level of toxic Al in the root CW.

# 4 Al Affects Cell Wall Properties

## *4.1 Cell Wall Extensibility*

It has been demonstrated that Al treatment reduces root CW extensibility (Tabuchi and Matsumoto 2001; Ma et al. 2004). Through the analysis of CW components in root tips of both Al-resistant and Al-sensitive cultivars of wheat (*Triticum aestivum*), Tabuchi and Matsumoto (2001) and Zakir Hossain et al. (2006) showed that Al increased both the molecular mass of hemicellulosic polysaccharides and the amount of wall-bound ferulic acids particularly in the Al-sensitive cultivar. They speculated that phenolic acids may cross-link with other CW components such as hemicellulosic polysaccharides and thus induce the mechanical rigidity of the CW leading to the decrease in CW extensibility and inhibition of root elongation. Irrespective of whether pectin or hemicellulose is the primary binding site of Al in the CW, the binding of Al to either CW component will affect CW extension either directly physically or indirectly. Figure 5 schematically depicts the possible pathways how Al may affect the CW extensibility. For the indirect effect of CW Al, the replacement from the CW pectic matrix of $Ca^{2+}$, which plays a key role in controlling CW extensibility by the formation and cleavage of Ca bonds during cell elongation (Boyer 2009), or decreasing the effectiveness of CW-loosening enzymes, such as XTHs (Tabuchi and Matsumoto 2001; Ma et al. 2004; Wehr et al. 2004), and of cell wall structural proteins such as expansin (Cosgrove 1989) may be responsible. The studies in Arabidopsis (Yang et al. 2011) and bean

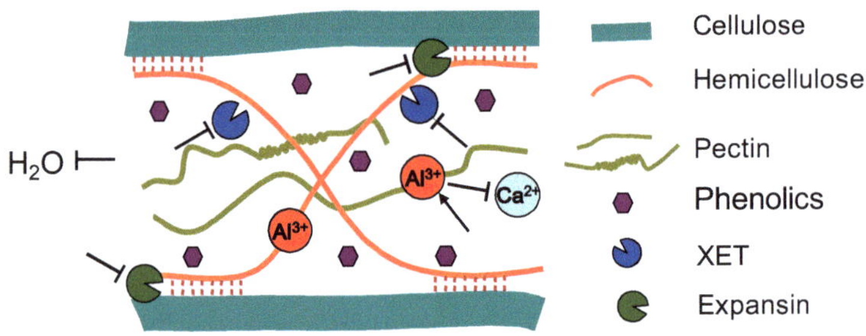


**Fig. 5** A simplified model representing the effect of Al on cell wall (CW) extension. Under Al stress, $Al^{3+}$ strongly binds to pectins and hemicellulose affecting the chemical and mechanical properties of the CW either directly and/or indirectly: (1) The Al cross-linked pectic and hemocellulosic cell wall matrix may lose its extensibility physically and/or physiologically by decreasing the activities of CW-loosening enzymes such as xyloglucan endotransglucosylase (XET). (2) Al rapidly and irreversibly displaces $Ca^{2+}$ at the site of $Ca^{2+}$-pectate cross-linkages, which play a key role in controlling CW extensibility and thus cell elongation and development. Direct interaction of Al with CW structural proteins (expansin) may also affect CW extensibility. In addition, the Al-induced accumulation of phenolics involved in the cross-linking of structural CW components thus strengthening the CW wall may affect CWl porosity and limit water flow. Based on Yang et al. (2013)

(*Phaseolus vulgaris*) (Yang et al. 2012) have demonstrated that Al stress resulted in the inhibition of the expression of XTH genes and the activity of the CW-loosening enzyme XET in roots, which was related to the inhibition of root elongation by Al. In Arabidopsis roots, the reduction of the activity of this enzyme was accompanied with the deposition of callose an indicator of Al sensitivity (Yang et al. 2011). However, a direct interaction of $Al^{3+}$ with structural CW proteins thus directly affecting CW extensibility cannot be ruled out.

The interaction of Al with CW components may also indirectly affect CW extension through the cell wall–plasma membrane–cytoskeleton continuum (Horst et al. 1999; Sivaguru et al. 2000). The interaction between CW and plasma membrane may be mediated by a cell wall-associated pectin receptor kinase (Kohorn and Kohorn 2012) which has been implicated in Al-induced root-growth inhibition in Arabidopsis (Sivaguru et al. 2003).

## 4.2 Cell Wall Porosity

The plant cell wall is a composite structure consisting of a cellulose–hemicellulose framework embedded within a matrix of pectins and proteins as mentioned above. The pores of the CW are the first barrier for mobile solutes such as ions, proteins, and water penetrating the wall (Brett and Waldron 1996), and plant cells interact with their environment through the porous network of the CW (Carpita et al. 1979). Generally, the pore diameter of the plant CW is in the range of 3.5–5.5 nm, which mainly depends on structure, hydrophobicity, chemical composition, and physical properties of the CW (Carpita et al. 1979; Chesson et al. 1997). This porous structure of the matrix permits low-molecular-weight solutes to diffuse across the CW and interact with the plasma membrane, while for high-molecular-weight solutes the pore size impedes transport (Sattelmacher 2001). According to Baron-Epel et al. (1988), the pore size of the CW is mainly controlled by the pectic matrix. The Al-induced enhancement of the CW pectin content in the root tips and its cross-linking by Al through binding to the negatively charged sites (Horst et al. 2010) may affect the porosity of the wall. Whether binding of Al to hemicellulose through Al–xyloglucan interaction also affects root porosity needs further clarification.

Any change in the factors affecting the pectic matrix may change the porosity. For example, it was reported that low temperature decreased the pore size of the CW by modifying CW composition (Bauchot et al. 1999; Rajashekar and Lafta 1996). In general, dicots display a stronger response to B supply than monocots, which may be due to the higher B requirement of dicots. Enhanced Al toxicity in B-deficient dicot plant species (Stass et al. 2007) could also be related to the pore size of the cell wall which is affected by borate ester cross-linking of the pectic polysaccharide RG II (Fleischer et al. 1999). In a study on the interaction of Al toxicity and drought stress in common bean, Yang et al. (2010, 2011, 2013) presented circumstantial evidence that polyethylene glycol 6000 induces a rearrangement of the wall polymers and thus affects CW porosity. This restricted the penetration of $Al^{3+}$ into the apoplast. Genes related to CW loosening or structure such as *XTH*, *beta-1,3-glucanase* (*BEG*), and *hydroxyproline-rich glycoprotein* (*HRGP*) appeared to play crucial roles (Fig. 6) in the PEG-induced modification of CW structure.

## 4.3 Blocking of Cell-to-Cell Trafficking by Callose

Cell wall deposited callose is a β-1,3-glucan with some β-1,6-branches and is produced by callose synthases and degraded by β-1,3-glucanases at the plasma membrane. Increased cytosolic $Ca^{2+}$ ($[Ca^{2+}]_{cyt}$) and modification of the plasma membrane (PM) are crucial factors for the induction of callose synthesis by activating 1,3-β-glucan synthase (Kauss et al. 1990; Kauss 1996). Callose plays important roles during a variety of processes in plant development and/or in response to biotic and abiotic stresses including Al stress (Stass and Horst 2009; Chen and Kim 2009). Al-induced higher callose formation in the outer cortex cells of the DTZ compared with the cells of the MZ and EZ might be related to a higher PM depolarization in maize (Sivaguru et al. 1999) and higher $[Ca^{2+}]_{cyt}$ in the DTZ indicated by fluorescence resonance energy transfer-sensitized emission of the yellow cameleon 3.60 reporter in Arabidposis (Rincón-Zachary et al. 2010). Al-induced callose deposition in the root tips is positively correlated with Al-induced inhibition of root growth and Al accumulation in the root tips (Larsen et al. 1996; Horst et al. 1997; Yang et al. 2012) and proved to be a sensitive indicator of Al injury in roots (Stass and Horst 2009). Al-induced callose formation

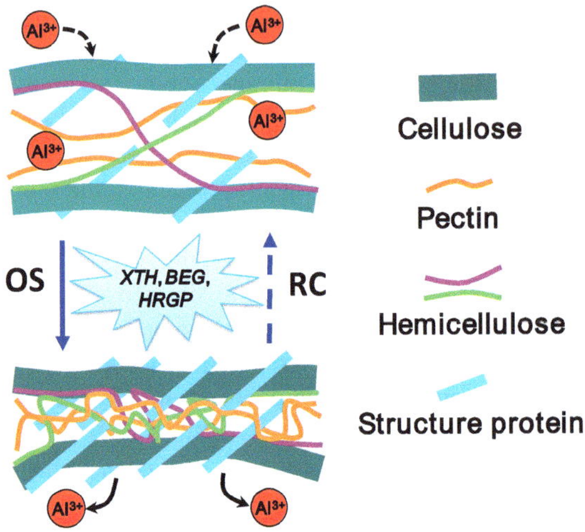


**Fig. 6** A model representing the effect of osmotic stress (OS) on cell wall (CW) structure and Al binding, and the possible role of CW modification-related genes or structure proteins in the OS-induced change in CW porosity and thus Al binding to the CW in common bean plants. Under polyethylene glycol (PEG)-6000-induced OS, loss of water from the CW matrix leads to reduced CW porosity and excludes $Al^{3+}$ from the apoplast, while the recovery of the CW from OS restores the pore size and allows $Al^{3+}$ entry into the apoplast and its binding to the pectic matrix and/or hemicellulose. CW modification genes *xyloglucan endotransglucosylase/hydrolase* (*XTH*), *β-1,3-glucanase* (*BEG*) and the structural protein *HRGP* (*hydroxyproline-rich glycoprotein*) are supposed to be involved in the modification of CW porosity. Based on Yang et al. (2013)

has been successfully used as a reliable parameter for the classification of genotypes of different plant species for Al sensitivity (Wissemeier et al. 1992; Horst et al. 1997) and for the screening of maize cultivars for adaptation to acid Al-toxic soils (Collet and Horst 2001; Eticha et al. 2005; Narro and Arcos 2010).

Callose formation is not only an indicator of Al stress, but has also been implicated in Al toxicity. A recent study by Zhang et al. (2014) found that heterologous expression of the sweet sorghum (*Sorghum bicolor*) gene *SbGlu1*, which encodes a β-1,3-glucanase, reduced callose deposition and Al accumulation and enhanced the Al resistance in Arabidopsis. Callose can be deposited at plasmodesmata (PD) to regulate the cell-to-cell movement of molecules by controlling the size exclusion limit (SEL) of PD (Chen and Kim 2009). Microinjection of the dye lucifer yellow carbohydrazide into peripheral root cells of an Al-sensitive wheat cultivar (*Triticum aestivum*, cv Scout 66) before or after Al treatment revealed that the Al-induced inhibition of root growth resulted from the Al-induced blockage of cell-to-cell trafficking via the PD (Sivaguru et al. 2000). Further immunofluorescence combined with immunoelectron microscopic techniques using monoclonal antibodies against callose demonstrated that the Al-induced callose deposition at the PD is responsible for the blockage of the symplastic transport, which was further verified using a callose synthesis inhibitor. In addition, the expression of PD-associated proteins such as calreticulin and unconventional myosin VIII was induced by Al and both proteins were co-localized with callose deposits. These results suggest that the extracellular Al-induced callose deposition at PD can effectively block symplastic transport and cell-to-cell signaling in higher plants.

## 5 Solute Flow

The rapid binding of Al in the root apoplast may reduce CW porosity and thus the mobility of particularly higher molecular solutes. This assumption has been confirmed by Schmohl and Horst (2000) who demonstrated a greatly Al-reduced release of acid phosphatase by maize suspension cells. However, these results can also be explained by a lower permeability of the plasma membrane for macromolecules. The study by Sivaguru et al. (2006) provided a more convincing evidence of the Al-induced inhibition of the apoplastic solute bypass flow in maize root apices by using the fluorescent probes HPTS (8-hydroxypyrene-1,3,6-trisulfonic acid, trisodium salt, molecular weight 524) and dextran-Texas Red conjugates (molecular weight 3000, 10,000, and 40,000) at the outer cortical cells, especially in the DTZ, and inhibition of transfer of these solutes to the xylem and finally the shoot. A contribution of Al-induced callose deposition to the inhibition of the apoplastic bypass flow could not be ruled out, since the inhibition of callose synthesis by pretreatment of the roots with 2-deoxy-D-glucose (DDG) prior to Al treatment partially alleviated the Al-induced inhibition of solute bypass flow.

It has been hypothesized that Al may not only affect the apoplastic flow of high-molecular solutes but also affect the root hydraulic conductivity (Kruger and Sucoff 1989; Maison and Bertsch 1997). Using artificial pectin membranes, Blamey et al. (1993) demonstrated that the binding of Al to pectin strongly reduced water permeability of the membranes in vitro. Gunsé et al. (1997) showed that Al decreased the hydraulic conductivity accompanied with reduced CW extensibility in an Al-sensitive maize cultivar. But Sivaguru et al. (2006) could not confirm an effect of Al on water flow from the roots to the shoot in maize.

However, Al may reduce the permeability of the plasma membrane for water and impede symplastic water transport, since Al strongly interacts also with membrane components affecting membrane structural properties such as fluidity and permeability (Vierstra and Haug 1978; Wagatsuma et al. 2005; Khan et al. 2009). In the root cortical cells of Northern red oak (*Quercus rubra* L.), Al decreased membrane permeability to water (Zhao et al. 1987; Chen et al. 1991) possibly by blocking aquaporins as suggested by gene expression analysis which suggested that Al suppressed the expression of genes coding for tonoplast aquaporins in rye (*Secale cereal* L.) (Milla et al. 2002). The role of Al on water transport is not yet well understood and urgently needs further experimental clarification given the importance of the Al/drought interaction for plant production on acid soils (Yang et al. 2013).

# 6 Conclusions

The apoplast of the most Al-sensitive apical root zone plays an important role in Al toxicity and resistance in plants. There is increasing evidence that inhibition of root growth is induced by Al directly and indirectly through interaction with CW structure and assembly mediated by phytohormones. An in-depth molecular characterization of hormone signaling regulating root growth plasticity via modification of cell wall properties in response to Al stress is urgently required and may represent a prerequisite for an improved understanding of general mechanisms of plant adaptation to a changing environment.

# References


Baluška F, Mancuso S (2013) Root apex transition zone as oscillatory zone. Front Plant Sci 2:354

Baluška F, Parker JS, Barlow PW (1992) Specific patterns of cortical and endoplasmic microtubules associated with cell growth and tissue differentiation in roots of maize (*Zea mays* L.). J Cell Sci 103:191–200

Baluška F, Volkmann D, Barlow PW (2001) A polarity crossroad in the transition growth zone of maize root apices: cytoskeletal and developmental implications. J Plant Growth Regul 20:170–181

Baluška F, Mancuso S, Volkmann D, Barlow PW (2010) Root apex transition zone: a signalling-response nexus in the root. Trends Plant Sci 15:402–408

Baron-Epel O, Gharyal PK, Schindler M (1988) Pectins as mediators of wall porosity in soybean cells. Planta 175:389–395

Bauchot AD, Hallett IC, Redgwell RJ, Lallu N (1999) Cell wall properties of kiwifruit affected by low temperature breakdown. Postharvest Biol Technol 16:245–255

Blamey FPC, Edmeades DC, Wheeler DM (1990) Role of root cation-exchange capacity in differential aluminium tolerance of *Lotus* species. J Plant Nutr 13:729–744

Blamey FPC, Asher CJ, Edwards DG, Kerven GL (1993) In vitro evidence of aluminium effects on solution movement through root cell walls. J Plant Nutr 16:555–562

Bordenave M (1996) Analysis of pectin methyl esterases. In: Linskens H, Jackson J (eds) Plant cell wall analysis. Springer, Berlin, pp 165–180

Boyer JS (2009) Cell wall biosynthesis and the molecular mechanism of plant enlargement. Funct Plant Biol 36:383–394

Bray EA (2004) Genes commonly regulated by water-deficit stress in *Arabidopsis thaliana*. J Exp Bot 55:2331–2341

Brett C, Waldron K (1996) Physiology and biochemistry of the plant cell wall. Chapman and Hall, London

Carpita NC, Gibeaut DM (1993) Structural models of primary cell walls in flowering plants: consistency of molecular structure with the physical properties of the walls during growth. Plant J 3:1–30

Carpita N, Sabularse D, Montezinos D, Delmer DP (1979) Determination of the pore size of cell walls of living plant cells. Science 205:1144–1147

Chang YC, Yamamoto Y, Matsumoto H (1999) Accumulation of aluminium in the cell wall pectin in cultured tobacco (*Nicotiana tabacum* L.) cells treated with a combination of aluminium and iron. Plant Cell Environ 22:1009–1017

Chen XY, Kim JY (2009) Callose synthesis in higher plants. Plant Signal Behav 4:489–492

Chen J, Sucoff EI, Stadelmann EJ (1991) Aluminum and temperature alteration of cell membrane permeability of *Quercus rubra*. Plant Physiol 96:644–649

Chen Q, Guo CL, Wang P, Chen XQ, Wu KH, Li KZ, Yu YX, Chen LM (2013) Up-regulation and interaction of the plasma membrane $H^+$-ATPase and the 14-3-3 protein are involved in the regulation of citrate exudation from the broad bean (*Vicia faba* L.) under Al stress. Plant Physiol Biochem 70:504–511

Chesson A, Gardner PT, Wood TJ (1997) Cell wall porosity and available surface area of wheat straw and wheat grain fractions. J Sci Food Agr 75:289–295

Cleland RE (1995) Auxin and cell elongation. In: Davies PJ (ed) Plant hormones: physiology, biochemistry and molecular biology. Kluwer, Dordrecht, pp 214–227

Collet L, Horst WJ (2001) Characterisation of maize cultivars in their adaptation to acid soils on the single plant level. In: Horst WJ, Schenk MK, Bürckert A et al (eds) Plant nutrition: food security and sustainability of agro-ecosystems through basic and applied research. Kluwer, Dordrecht, pp 86–87

Cosgrove DJ (1989) Characterization of long-term extension of isolated cell walls from growing cucumber hypocotyls. Planta 177:121–130

Cosgrove DJ (1997) Assembly and enlargement of the primary cell wall in plants. Annu Rev Cell Dev Biol 13:171–201

Cosgrove DJ (2005) Growth of the plant cell wall. Nat Rev Mol Cell Biol 6:850–861

Delhaize E, Ryan PR (1995) Aluminum toxicity and tolerance in plants. Plant Physiol 107:315–321

Delhaize E, Ma JF, Ryan PR (2012) Transcriptional regulation of aluminium tolerance genes. Trends Plant Sci 17:341–348

Doncheva S, Amenós M, Poschenrieder C, Barceló J (2005) Root cell patterning: a primary target for aluminium toxicity in maize. J Exp Bot 56:1213–1220

Eticha D, Staß A, Horst WJ (2005) Cell-wall pectin and its degree of methylation in the maize root-apex: significance for genotypic differences in aluminium resistance. Plant Cell Environ 28:1410–1420

Fleischer A, ÓNeill MA, Ehwald R (1999) The pore size of non-graminaceous plant cell walls is rapidly decreased by borate ester cross-linking of the pectic polysaccharide rhamnogalacturonan II. Plant Physiol 121:829–838

Gerendás J (2007) Significance of polyamines for pectin methylesterase activity and the ion dynamics in the apoplast. In: Sattelmacher B, Horst WJ (eds) The apoplast of higher plants: compartment of storage, transport, and reactions. Kluwer, Dordrecht, pp 67–83

Grabski S, Arnoys E, Busch B, Schindler M (1998) Regulation of actin tension in plant cells by kinases and phosphatases. Plant Physiol 116:279–290

Gunsé B, Poschenrieder C, Barceló J (1997) Water transport properties of roots and root cortical cells in proton- and Al-stressed maize varieties. Plant Physiol 113:595–602

Hager A (2003) Role of the plasma membrane $H^+$-ATPase in auxin-induced elongation growth: historical and new aspects. J Plant Res 116:483–505

Hartwell BL, Pember FR (1918) The presence of aluminium as a reason for the difference in the effect of so-called acid soil on barley and rye. Soil Sci 6:259–281

Horst WJ (1995) The role of the apoplast in aluminium toxicity and resistance of higher plants: a review. Zeitschrift für Pflanzenernährung und Bodenkunde 158:419–428

Horst WJ, Asher CJ, Cakmak J, Szulkiewicz P, Wissemeier AH (1992) Short-term responses of soybean roots to aluminium. J Plant Physiol 140:174–178

Horst WJ, Püschel A-K, Schmohl N (1997) Induction of callose formation is a sensitive marker for genotypic aluminium sensitivity in maize. Plant Soil 192:23–30

Horst WJ, Schmohl N, Kollmeier M, Baluska F, Sivaguru M (1999) Does aluminium affect root growth of maize through interaction with the cell wall–plasma membrane–cytoskeleton continuum? Plant Soil 215:163–174

Horst WJ, Kollmeier M, Schmohl N et al (2007) Significance of the root apoplast for aluminium toxicity and resistance of maize. In: Sattelmacher B, Horst W (eds) The apoplast of higher plants: compartment of storage, transport, and reactions. Kluwer, Dordrecht, pp 49–66

Horst WJ, Wang Y, Eticha D (2010) The role of the root apoplast in aluminium-induced inhibition of root elongation and in aluminium resistance of plants: a review. Ann Bot 106:185–197

Illéš P, Schlicht M, Pavlovkin J, Lichtscheidl I, Baluška F, Ovečka M (2006) Aluminium toxicity in plants: internalization of aluminium into cells of the transition zone in Arabidopsis root apices related to changes in plasma membrane potential, endosomal behavior, and nitric oxide production. J Exp Bot 57:4201–4213

Kauss H (1996) Callose synthesis. In: Smallwood M, Knox JP, Bowles DJ (eds) Membranes: specialized functions in plants. BIOS Scientific Publishers, Oxford, pp 77–92

Kauss H, Waldmann T, Quader H (1990) $Ca^{2+}$ as a signal in the induction of callose synthesis. In: Ranjeva R, Boudet AM (eds) Signal perception and transduction in higher plants, vol 47. Springer, Berlin, pp 117–131

Khan MS, Tawaraya K, Sekimoto H, Koyama H, Kobayashi Y, Murayama T, Chuba M, Kambayashi M, Shiono Y, Uemura M, Ishikawa S, Wagatsuma T (2009) Relative abundance of delta5-sterols in plasma membrane lipids of root-tip cells correlates with aluminum tolerance of rice. Physiol Plant 135:73–83

Kinraide TB, Ryan PR, Kochian LV (1992) Interactive effects of $Al^{3+}$, $H^+$, and other cations on root elongation considered in terms of cell-surface electrical potential. Plant Physiol 99:1461–1468

Klimashevskii EL, Dedov VM (1975) Localization of growth inhibiting action of aluminum ions in elongating cell walls. Fiziol Rast 22:1183–1190

Kobayashi Y, Kobayashi Y, Sugimoto M, Lakshmanan V, Iuchi S, Kobayashi M, Bais HP, Koyama H (2013) Characterization of the complex regulation of *AtALMT1* expression in response to phytohormones and other inducers. Plant Physiol 162:732–740

Kochian LV (1995) Cellular mechanisms of aluminum toxicity and resistance in plants. Annu Rev Plant Physiol Plant Mol Biol 46:237–260

Kochian LV, Hoekenga OA, Piñeros MA (2004) How do crop plants tolerate acid soils? Mechanisms of aluminum tolerance and phosphorous efficiency. Annu Rev Plant Biol 55:459–493

Kochian LV, Miguel A, Piñeros MA, Liu J, Magalhaes JV (2015) Plant adaptation to acid soils: the molecular basis for crop aluminum resistance. Annu Rev Plant Biol. doi:10.1146/annurev-arplant-043014-114822

Kohorn BD, Kohorn SL (2012) The cell wall-associated kinases, WAKs, as pectin receptors. Front Plant Sci 3:88

Kollmeier M, Felle HH, Horst WJ (2000) Genotypical differences in aluminum resistance of maize are expressed in the distal part of the transition zone. Is reduced basipetal auxin flow involved in inhibition of root elongation by aluminum? Plant Physiol 122:945–956

Kruger E, Sucoff E (1989) Growth and nutrient status of *Quercus rubra* L. in response to Al and Ca. J Exp Bot 40:653–658

Larsen PB, Tai C-Y, Kochian LV, Howell SH (1996) Arabidopsis mutants with increased sensitivity to aluminum. Plant Physiol 110:743–751

Li JY, Liu J, Dong D, Jia X, McCouch SR, Kochian LV (2014) Natural variation underlies alterations in Nramp aluminum transporter (NRAT1) expression and function that play a key role in rice aluminum tolerance. Proc Natl Acad Sci U S A 111:6503–6508

Liu J, Piñeros MA, Kochian LV (2014) The role of aluminum sensing and signaling in plant aluminum resistance. J Integr Plant Biol 56:221–230

Ljung K, Hull AK, Celenza J, Yamada M, Estelle M, Normanly J, Sandberg G (2005) Sites and regulation of auxin biosynthesis in Arabidopsis roots. Plant Cell 17:1090–1104

Ma JF, Ryan PR, Delhaize E (2001) Aluminium tolerance in plants and the complexing role of organic acids. Trends Plant Sci 6:273–278

Ma JF, Shen RF, Nagao S, Tanimoto E (2004) Aluminum targets elongating cells by reducing cell wall extensibility in wheat roots. Plant Cell Physiol 45:583–589

Maison A, Bertsch PM (1997) Aluminium speciation in the presence of wheat root cell walls: a wet chemical study. Plant Cell Environ 20:504–512
Maron LG, Kirst M, Mao C, Milner MJ, Menossi M, Kochian LV (2008) Transcriptional profiling of aluminum toxicity and tolerance responses in maize roots. New Phytol 179:116–128
Mattiello L, Kirst M, da Silva FR, Jorge RA, Menossi M (2010) Transcriptional profile of maize roots under acid soil growth. BMC Plant Biol 10:196
Milla MA, Butler E, Huete AR, Wilson CF, Anderson O, Gustafson JP (2002) Expressed sequence tag-based gene expression analysis under aluminum stress in rye. Plant Physiol 130:1706–1716
Narro LA, Arcos AL (2010) Genetics of aluminum-induced callose formation in maize roots, a selection trait for aluminum resistance. Crop Sci 50:1848–1853
Nishitani K (1997) The role of endoxyloglucan transferase in the organization of plant cell walls. Int Rev Cyt 173:157–206
Overvoorde P, Fukaki H, Beeckman T (2010) Auxin control of root development. Cold Spring Harb Perspect Biol 2:a001537
Panda S, Matsumoto H (2007) Molecular physiology of aluminum toxicity and tolerance in plants. Bot Rev 73:326–347
Panda SK, Baluska F, Matsumoto H (2009) Aluminum stress signaling in plants. Plant Signal Behav 4:592–597
Perrot-Rechenmann C (2010) Cellular responses to auxin: division versus expansion. Cold Spring Harb Perspect Biol 2:a001446
Petersson SV, Johansson AI, Kowalczyk M, Makoveychuk A, Wang JY, Moritz T, Grebe M, Benfey PN, Sandberg G, Ljung K (2009) An auxin gradient and maximum in the Arabidopsis root apex shown by high-resolution cell-specific analysis of IAA distribution and synthesis. Plant Cell 21:1659–1668
Petrášek J, Friml J (2009) Auxin transport routes in plant development. Development 136:2675–2688
Pitaksaringkarn W, Matsuoka K, Asahina M, Miura K, Sage-Ono K, Ono M, Yokoyama R, Nishitani K, Ishii T, Iwai H, Satoh S (2014) *XTH20* and *XTH19* regulated by *ANAC071* under auxin flow are involved in cell proliferation in incised Arabidopsis inflorescence stems. Plant J 80:604–614
Rajashekar CB, Lafta A (1996) Cell-wall changes and cell tension in response to cold acclimation and exogenous abscisic acid in leaves and cell cultures. Plant Physiol 111:605–612
Rangel AF, Rao IM, Horst WJ (2007) Spatial aluminium sensitivity of root apices of two common bean (*Phaseolus vulgaris* L.) genotypes with contrasting aluminium resistance. J Exp Bot 58:3895–3904
Rangel AF, Rao IM, Horst WJ (2009) Intracellular distribution and binding state of aluminum in root apices of two common bean (*Phaseolus vulgaris*) genotypes in relation to Al toxicity. Physiol Plant 135:162–173
Rayle DL, Cleland RE (1992) The acid growth theory of auxin-induced cell elongation is alive and well. Plant Physiol 99:1271–1274
Rincón-Zachary M, Teaster ND, Sparks JA, Valster AH, Motes CM, Blancaflor EB (2010) Fluorescence resonance energy transfer-sensitized emission of yellow cameleon 3.60 reveals root zone-specific calcium signatures in Arabidopsis in response to aluminum and other trivalent cations. Plant Physiol 152:1442–1458
Rose JK, Braam J, Fry SC, Nishitani K (2002) The XTH family of enzymes involved in xyloglucan endotransglucosylation and endohydrolysis: current perspectives and a new unifying nomenclature. Plant Cell Physiol 43:1421–1435
Ryan PR, DiTomaso JM, Kochian LV (1993) Aluminium toxicity in roots: an investigation of spatial sensitivity and the role of the root cap. J Exp Bot 44:437–446
Ryan PR, Tyerman SD, Sasaki T, Furuichi T, Yamamoto Y, Zhang WH, Delhaize E (2011) The identification of aluminium-resistance genes provides opportunities for enhancing crop production on acid soils. J Exp Bot 62:9–20

Sattelmacher B (2001) The apoplast and its significance for plant mineral nutrition. New Phytol 149:167–192
Schmohl N, Horst WJ (2000) Cell wall pectin content modulates aluminium sensitivity of Zea mays (L.) cell grown in suspension culture. Plant Cell Environ 23:735–742
Schmohl N, Pilling J, Fisahn J, Horst WJ (2000) Pectin methylesterase modulates aluminium sensitivity in *Zea mays* and *Solanum tuberosum*. Physiol Plant 109:419–427
Shen H, Ligaba A, Yamaguchi M, Osawa H, Shibata K, Yan X, Matsumoto H (2004) Effect of K-252a and abscisic acid on the efflux of citrate from soybean roots. J Exp Bot 55:663–671
Shen H, Hou NY, Schlicht M, Wan YL, Mancuso S, Baluska F (2008) Aluminium toxicity targets PIN2 in Arabidopsis root apices: Effects on PIN2 endocytosis, vesicular recycling, and polar auxin transport. Chin Sci Bull 53:2480–2487
Sivaguru M, Horst WJ (1998) The distal part of the transition zone is the most aluminum-sensitive apical root zone of maize. Plant Physiol 116:155–163
Sivaguru M, Baluska F, Volkmann D, Felle HH, Horst WJ (1999) Impacts of aluminum on the cytoskeleton of the maize root apex. Short-term effects on the distal part of the transition zone. Plant Physiol 119:1073–1082
Sivaguru M, Fujiwara T, Samaj J, Baluska F, Yang Z, Osawa H, Maeda T, Mori T, Volkmann D, Matsumoto H (2000) Aluminum-induced 1→3-beta-D-glucan inhibits cell-to-cell trafficking of molecules through plasmodesmata. A new mechanism of aluminum toxicity in plants. Plant Physiol 124:991–1005
Sivaguru M, Ezaki B, He ZH, Tong H, Osawa H, Baluška F, Volkmann D, Matsumoto H (2003) Aluminum-induced gene expression and protein localization of a cell wall-associated receptor kinase in Arabidopsis. Plant Physiol 132:2256–2266
Sivaguru M, Horst WJ, Eticha D, Matsumoto H (2006) Aluminum inhibits apoplastic flow of high-molecular weight solutes in root apices of *Zea mays* L. J Plant Nutr Soil Sci 169:679–690
Sivaguru M, Liu J, Kochian LV (2013) Targeted expression of *SbMATE* in the root distal transition zone is responsible for sorghum aluminum resistance. Plant J 76:297–307
Stass A, Horst WJ (2009) Callose in abiotic stress. In: Bacic A, Fincher GB, Stone BA (eds) Chemistry, biochemistry, and biology of (1→3)-β-glucans and related polysaccharides. Academic Press, Burlington, MA, pp 499–524
Stass A, Kotur Z, Horst WJ (2007) Effect of boron on the expression of aluminium toxicity in *Phaseolus vulgaris*. Physiol Plant 131:283–290
Tabuchi A, Matsumoto H (2001) Changes in cell-wall properties of wheat (*Triticum aestivum*) roots during aluminum-induced growth inhibition. Physiol Plant 112:353–358
Takahashi K, Hayashi K, Kinoshita T (2012) Auxin activates the plasma membrane $H^+$-ATPase by phosphorylation during hypocotyl elongation in Arabidopsis. Plant Physiol 159:632–641
Taylor GJ, McDonald-Stephens JL, Hunter DB, Bertsch PM, Elmore D, Rengel Z, Reid RJ (2000) Direct measurement of aluminum uptake and distribution in single cells of *Chara corallina*. Plant Physiol 123:987–996
Teale WD, Paponov IA, Ditengou F, Palme K (2005) Auxin and the developing root of *Arabidopsis thaliana*. Physiol Plant 123:130–138
Van HL, Kuraishi S, Sakurai N (1994) Aluminum-induced rapid root inhibition and changes in cell-wall components of squash seedlings. Plant Physiol 106:971–976
Verbelen JP, De Cnodder T, Le J, Vissenberg K, Baluška F (2006) The root apex of *Arabidopsis thaliana* consists of four distinct zones of growth activities: meristematic zone, transition zone, fast elongation zone and growth terminating zone. Plant Signal Behav 1:296–304
Vierstra R, Haug A (1978) The effect of $Al^{3+}$ on the physical properties of membrane lipids in *Thermoplasma acidophilum*. Biochem Bioph Res Co 84:138–143
von Uexküll HR, Mutert E (1995) Global extent, development and economic impact of acid soils. Plant Soil 171:1–15
Wagatsuma T, Ishikawa S, Uemura M, Mitsuhashi W, Kawamura T, Khan MSH, Tawaraya K (2005) Plasma membrane lipids are the powerful components for early stage aluminum tolerance in triticale. Soil Sci Plant Nutr 51:701–704

Wang Y, Staß A, Horst WJ (2004) Apoplastic binding of aluminum is involved in silicon-induced amelioration of aluminum toxicity in maize. Plant Physiol 136:3762–3770
Wehr JB, Menzies NW, Blamey FPC (2004) Inhibition of cell-wall autolysis and pectin degradation by cations. Plant Physiol Biochem 42:485–492
Wissemeier AH, Horst WJ (1995) Effect of calcium supply on aluminium-induced callose formation, its distribution and persistence in roots of soybean (*Glycine max* (L.) Merr.). J Plant Physiol 145:470–476
Wissemeier AH, Klotz F, Horst WJ (1987) Aluminium induced callose synthesis in roots of soybean (*Glycine max* L.). J Plant Physiol 129:487–492
Wissemeier AH, Diening A, Hergenröder A, Horst WJ, Mix-Wagner G (1992) Callose formation as parameter for assessing genotypical plant tolerance of aluminium and manganese. Plant Soil 146:67–75
Wu D, Shen H, Yokawa K, Baluška F (2014) Alleviation of aluminium-induced cell rigidity by overexpression of *OsPIN2* in rice roots. J Exp Bot 65:5305–5315
Xia J, Yamaji N, Kasai T, Ma JF (2010) Plasma membrane-localized transporter for aluminum in rice. Proc Natl Acad Sci U S A 107:18381–18385
Yang JL, You JF, Li YY, Wu P, Zheng SJ (2007) Magnesium enhances aluminum-induced citrate secretion in rice bean roots (*Vigna umbellata*) by restoring plasma membrane $H^+$-ATPase activity. Plant Cell Physiol 48:66–73
Yang JL, Li YY, Zhang YJ, Zhang SS, Wu YR, Wu P, Zheng SJ (2008) Cell wall polysaccharides are specifically involved in the exclusion of aluminum from the rice root apex. Plant Physiol 146:602–611
Yang ZB, Eticha D, Rao IM, Horst WJ (2010) Alteration of cell-wall porosity is involved in osmotic stress-induced enhancement of aluminium resistance in common bean (*Phaseolus vulgaris* L.). J Exp Bot 61:3245–3258
Yang JL, Zhu XF, Peng YX, Zheng C, Li GX, Liu Y, Shi YZ, Zheng SJ (2011) Cell wall hemicellulose contributes significantly to aluminum adsorption and root growth in Arabidopsis. Plant Physiol 155:1885–1892
Yang ZB, Eticha D, Albacete A, Rao IM, Roitsch T, Horst WJ (2012) Physiological and molecular analysis of the interaction between aluminium toxicity and drought stress in common bean (*Phaseolus vulgaris*). J Exp Bot 63:3109–3125
Yang ZB, Ro IM, Horst WJ (2013) Interaction of aluminium and drought stress on root growth and crop yield on acid soils. Marschner Review. Plant Soil 372:3–25
Yang ZB, Geng X, He C, Zhang F, Wang R, Horst WJ, Ding Z (2014) TAA1-regulated local auxin biosynthesis in the root-apex transition zone mediates the aluminum-induced inhibition of root growth in Arabidopsis. Plant Cell 26:2889–2904
Zakir Hossain AKM, Koyama H, Hara T (2006) Growth and cell wall properties of two wheat cultivars differing in their sensitivity to aluminum stress. J Plant Physiol 163:39–47
Zhang G, Taylor GJ (1989) Kinetics of aluminum uptake by excised roots of aluminum-tolerant and aluminum-sensitive cultivars of *Triticum aestivum* L. Plant Physiol 91:1094–1099
Zhang G, Taylor GJ (1990) Kinetics of aluminum uptake in Triticum aestivum L. Identity of the linear phase of Al uptake by excised roots of aluminum-tolerant and aluminum-sensitive cultivars. Plant Physiol 94:577–584
Zhang H, Shi WL, You JF, Bian MD, Qin XM, Yu H, Liu Q, Ryan PR, Yang ZM (2014) Transgenic *Arabidopsis thaliana* plants expressing a β-1,3-glucanase from sweet sorghum (*Sorghum bicolor* L.) show reduced callose deposition and increased tolerance to aluminium toxicity. Plant Cell Environ. doi:10.1111/pce.12472
Zhao Y (2012) Auxin biosynthesis: a simple two-step pathway converts tryptophan to indole-3-acetic acid in plants. Mol Plant 5:334–338
Zhao X-J, Sucoff E, Stadelmann EJ (1987) $Al^{3+}$ and $Ca^{2+}$ alteration of membrane permeability of *Quercus rubra* root cortex cells. Plant Physiol 83:159–162
Zhu XF, Shi YZ, Lei GJ, Fry SC, Zhang BC, Zhou YH, Braam J, Jiang T, Xu XY, Mao CZ, Pan YJ, Yang JL, Wu P, Zheng SJ (2012) XTH31, encoding an in vitro XEH/XET-active enzyme,

regulates aluminum sensitivity by modulating in vivo XET action, cell wall xyloglucan content, and aluminum binding capacity in Arabidopsis. Plant Cell 24:4731–4747

Zhu XF, Lei GJ, Wang ZW, Shi YZ, Braam J, Li GX, Zheng SJ (2013) Coordination between apoplastic and symplastic detoxification confers plant aluminum resistance. Plant Physiol 162:1947–1955

Zhu XF, Wan JX, Sun Y, Shi YZ, Braam J, Li GX, Zheng SJ (2014) Xyloglucan endotransglucosylase-hydrolase17 interacts with xyloglucan endotransglucosylase-hydrolase31 to confer xyloglucan endotransglucosylase action and affect aluminum sensitivity in Arabidopsis. Plant Physiol 165:1566–1574

MIX
Papier aus verantwortungsvollen Quellen
Paper from responsible sources
FSC® C105338

If you have any concerns about our products,
you can contact us on
**ProductSafety@springernature.com**

In case Publisher is established outside the EU,
the EU authorized representative is:
**Springer Nature Customer Service Center GmbH**
**Europaplatz 3, 69115 Heidelberg, Germany**

Printed by Libri Plureos GmbH
in Hamburg, Germany